1987

An Introduction to
Nonstandard Real Analysis

An Introduction to Nonstandard Real Analysis

ALBERT E. HURD

Department of Mathematics
University of Victoria
Victoria, British Columbia
Canada

PETER A. LOEB

Department of Mathematics
University of Illinois
Urbana, Illinois

1985

ACADEMIC PRESS, INC.

(Harcourt Brace Jovanovich, Publishers)

Orlando   San Diego   New York   London
Toronto   Montreal   Sydney   Tokyo

ACADEMIC PRESS, INC.
Orlando, Florida 32887

United Kingdom Edition published by
ACADEMIC PRESS INC. (LONDON) LTD.
24/28 Oval Road, London NW1 7DX

Library of Congress Cataloging in Publication Data

Main entry under title:

An introduction to nonstandard real analysis.

Includes bibliographical references and index.
1. Mathematical analysis, Nonstandard. I. Hurd, A. E.
(Albert Emerson), DATE . II. Loeb, P. A.
QA299.82.I58 1985 515 84-24563
ISBN 0-12-362440-1 (alk. paper)

PRINTED IN THE UNITED STATES OF AMERICA

85 86 87 88   9 8 7 6 5 4 3 2 1

AN INTRODUCTION TO NONSTANDARD REAL ANALYSIS

A. E. Hurd and P. A. Loeb

ERRATA

Page	Location	Change
19	l 16	For R, read R^n
23	l 19	For A_j, read $\underline{A}_j$
28	$l-2$	For $b \neq 0$, read $b \not\equiv 0$
31	l 4	For $\langle \psi(x) \rangle$, read $\langle {}^*\psi(x) \rangle$
33	l 9	Should read: . . . ${}^*s_n \simeq L$, ${}^*t_n \simeq M$, and hence ${}^*s_n + {}^*t_n \simeq L + M$
35	Examples 1–4	Read ${}^*s$ and ${}^*t$ for all infinitely indexed s and t
46	$l-1$	For ${}^*f(x_{n, {}^*\psi(n)}) \geq {}^*f(x_{n,k})$, read ${}^*f({}^*x_{n, {}^*\psi(n)}) \geq {}^*f({}^*x_{n,k})$
47	ll 1–3	Replace all x by ${}^*x$
50	$l-19$	Should read: . . . on a closed set A, then . . .
52	$ll-10, -9$	For $f(x+h)$, read ${}^*f(x+h)$
54	$l-3$	Should read: . . . and corresponding $\varepsilon \simeq 0$ and $\delta \simeq 0$.
58	l 8	Should read: . . . for any positive infinitesimal Δx.
61	l 5	For $\|f_n(x) - f(x)\|$, read $\|f_n(x) - \underline{f}(x)\|$
66	l 6	For ${}^*\sin \theta$, read ${}^*\sin \theta_2$
74	ll 8, 9	Should read: . . . let $\mu(I)$ be the area of the square $I \times I$. For which . . .
89	$l-14$	Should read: . . . $B \in {}^*\mathscr{P}_{\mathrm{F}}(V_n(X))$, $n \geq 1$. Now . . .
93	l 15	For $n \in N$, read $n \in {}^*N$
93	$l-9$	Should read: . . . if the intersection of any finite number of elements of A is nonempty. Show that . . .
96	$l-4$	For $A \cup B$, read $A \cap B$
97	ll 2, 3	For $A \cup B$, read $A \cap B$
103	$l-2$	For $x \in X$, read $x \in A$
107	l 2	Should read: . . . set of internal entities in . . .
111	l 15	Should read: . . . is a subbase at x, . . .
123	l 15	Should read: . . . form a base at x.
123	$l-14$	Should read: . . . is a metric; it is called . . .
129	l 6	For if, read iff
131	$l-9$	Should read: . . . is complete but Exercise 13 is not applicable.

131	$l-1$	For $x \in {}^*X$, read $x \in X$
142	$l-3$	For $x' \in {}^*X$, read $x' \in X'$
143	$l-11$	Should read: . . . of X' is compact in the weak* topology.
144	$l\ 6$	For e^{nk}, read e^{n_k}
145	$l\ 7$	Should read: (b) $m(a)x \subseteq m(a)\mu(x) = \mu(ax)$
147	$l\ 13$	For $\{\|x+y\| - \|x-y\|\}$, read $\{\|x+y\|^2 - \|x-y\|^2\}$
152	$l\ 15$	For Bernstein and Robinson [4], read Bernstein and Robinson [7]
155	$ll\ 4, 5$	Should read: . . . complete [even if (X, d) is not] . . .
157	$l\ 2$	Should read: . . . for each closed set $A \subset X$ and . . .
179	$l-14$	For $°I\psi_1 < \infty$, read $°I|\psi_1| < \infty$
181	$l-12$	For $\mu(A)$, read $\hat{\mu}(A)$
182	$l\ 9$	For $\{x_0\}$, read $\{\{x_0\}, \varnothing\}$
186	$l-4$	Should read: . . . treatment of "S-integrability", we . . .
189	$l-2$	For $\varkappa > \operatorname{card} \mathscr{T}$, read $\kappa > \operatorname{card} \mathscr{T}$
191	$l\ 4$	For $f(z)$, read $\tilde{f}(z)$
195	$l\ 9$	For $J_X f = °If$, read $J_X f = °I({}^*f)$
205	$l-12$	For Perkins [33], read Perkins [38]
208	$l\ 15$	For $X_k(\omega) = 0$, read $X_k(\omega) = 1$
210	$l\ 9$	For $X(\omega)$, read $X_n(\omega)$
211	$l\ 4$	For $[(\eta - 1)/\eta, \eta)$, read $[\eta - (1/\eta), \eta)$
211	$l-2$	For $(\gamma s)^k$, read $(\lambda s)^k$
212	$l\ 2$	For s_n, read s_{n-1}
212	$l\ 8$	For $(\gamma s_1)^{k_1}$, read $(\lambda s_1)^{k_1}$; for $(\lambda s_1)^{k_2}$, read $(\lambda s_2)^{k_2}$
212	$l\ 9$	For $(\gamma s_1)^{k_1}$, read $(\lambda s_1)^{k_1}$
212	$l-4$	The right-hand side of the equation should read: $$\frac{\gamma'!}{(\gamma'-k)!k!}\left(\frac{\lambda t}{\gamma'+j-t_0\lambda}\right)^k\left(1 - \frac{\lambda t}{\gamma'+j-t_0\lambda}\right)^{\gamma'-k}$$
213	$l\ 13$	For $\{\tilde{N}_t : t \in R\}$, read $\{\tilde{N}_t : t \in R^+\}$
215	$l-8$	For $\{X_n : n \in N\}$, read $\{X_n : n \in {}^*N\}$

Dedicated to the memory of
ABRAHAM ROBINSON

Contents

Chapter III
Nonstandard Theory of Topological Spaces

Chapter IV
Nonstandard Integration Theory

Appendix

Preface

The notion of an infinitesimal has appeared off and on in mathematics since the time of Archimedes. In his formulation of the calculus in the 1670s, the German mathematician Wilhelm Gottfried Leibniz treated infinitesimals as ideal numbers, rather like imaginary numbers, which were smaller in absolute value than any ordinary real number but which nevertheless obeyed all of the usual laws of arithmetic. Leibniz regarded infinitesimals as a useful fiction which facilitated mathematical computation and invention. Although it gained rapid acceptance on the continent of Europe, Leibniz's method was not without its detractors. In commenting on the foundations of calculus as developed both by Leibniz and Newton, Bishop George Berkeley wrote, "And what are these same evanescent increments? They are neither finite quantities, nor quantities infinitely small, nor yet nothing. May we not call them the ghosts of departed quantities?" The question was, How can there be a positive number which is smaller than any real number without being zero? Despite this unanswered question, the infinitesimal calculus was developed by Euler and others during the eighteenth and nineteenth centuries into an impressive body of work. It was not until the late nineteenth century that an adequate definition of limit replaced the calculus of infinitesimals and provided a rigorous foundation for analysis. Following this development, the use of infinitesimals gradually faded, persisting only as an intuitive aid to conceptualization.

There the matter stood until 1960 when Abraham Robinson gave a rigorous foundation for the use of infinitesimals in analysis. More specifically, Robinson showed that the set of real numbers can be regarded as a subset of a larger set of "numbers" (called hyperreal numbers) which contains infinitesimals and also, with appropriately defined artithmetic operations, satisfies all of the arithmetic rules obeyed by the ordinary real numbers. Even more, he demonstrated that the relational structure over the reals (sets, relations, etc.) can be extended to a similar structure over the hyperreals in such a way that all statements true in the real structure remain true, with a suitable interpretation, in the hyperreal structure. This latter property, known as the transfer principle, is the pivotal result of Robinson's discovery.

Robinson's invention, called nonstandard analysis, is more than a justification of the method of infinitesimals. It is a powerful new tool for mathematical research. Rather quickly it became apparent that every mathematical structure has a nonstandard model from which knowledge of the original structure can be gained by applications of the appropriate transfer principle. In the twenty-five years since Robinson's discovery, the use of nonstandard models has led to many new insights into traditional mathematics, and to solutions of unsolved problems in areas as diverse as functional analysis, probability theory, complex function theory, potential theory, number theory, mathematical physics, and mathematical economics.

Robinson's first proof of the existence of hyperreal structures was based on a result in mathematical logic (the compactness theorem). It was perhaps this aspect of his work, more than any other, which made it difficult to understand for those not adept at mathematical logic. At present, the most common demonstration of the existence of nonstandard models uses an "ultrapower" construction. But the use of ultrapowers is not restricted to nonstandard analysis. Indeed, the construction of ultrapower extensions of the real numbers dates back to the 1940s with the work of Edwin Hewitt [17] and others, and the use of ultrapowers to study Banach spaces [10,16] has become an important tool in modern functional analysis. Nonstandard analysis is a far-reaching generalization of these applications of ultrapowers. One essential difference between the method of ultrapowers and the method of nonstandard analysis is the consistent use of the transfer principle in the latter. To present this principle one needs a certain amount of mathematical logic, but the logic is used in an essential way only in stating and proving the transfer principle, and not in applying nonstandard analysis. We hope to demonstrate that the amount of logic needed is minimal, and that the advantages gained in the use of the transfer principle are substantial.

The aim of this book is to make Robinson's discovery, and some of the subsequent research, available to students with a background in undergraduate mathematics. In its various forms, the manuscript was used by the second author in several graduate courses at the University of Illinois at Urbana–Champaign. The first chapter and parts of the rest of the book can be used in an advanced undergraduate course. Research mathematicians who want a quick introduction to nonstandard analysis will also find it useful. The main addition of this book to the contributions of previous textbooks on nonstandard analysis [12, 37, 42, 46] is the first chapter, which eases the reader into the subject with an elementary model suitable for the calculus, and the fourth chapter on measure theory in nonstandard models.

A more complete discussion of this book's four chapters must begin by noting H. Jerome Keisler's major contribution to nonstandard analysis in the form of his 1976 textbook, "Elementary Calculus" [23] together with the instructor's volume, "Foundations of Infinitesimal Calculus" [24]. Keisler's book is an excellent calculus text (see the second author's review [30]) which makes that part of nonstandard analysis needed for the calculus available to freshman students. Keisler's approach uses equalities and inequalities to transfer properties from the real number system to the hyperreal numbers. In our first chapter, we have modified that approach to an equivalent one by formulating a simple transfer principle based on a restricted language.

The first chapter begins by using ultrafilters on the set of natural numbers to construct a simple ultrapower model of the hyperreal numbers. A formal language is then developed in which only two kinds of sentences are used to transfer properties from the real number system to the larger, hyperreal number system. The rest of the chapter is devoted to extensive applications of this simple transfer principle to the calculus and to more-advanced real analysis including differential equations. By working through these applications, the reader should acquire a good feeling for the basics of nonstandard analysis by the end of the chapter. Anyone who begins this book with no background in mathematical logic should have no problem with the logic in the first chapter and hence should easily pick up the background needed to proceed. Indeed, it is our hope that such a reader will grow quite impatient with the restrictions on the language we impose in the first chapter, and thus be more than ready for the general language introduced in Chapter II and used in the rest of the book. We will not comment on what might be in the mind of a logician at that point.

Chapter II extends the context of Chapter I to "higher-order" models appropriate to the discussion of sets of sets, sets of functions, etc., and covers the notions of internal and external sets and saturation. These topics, together with a general language and transfer principle, are held in abeyance until the second chapter so that the beginner can master the subject in reasonably easy steps. They are, however, essential to the applications of nonstandard analysis in modern mathematics. External constructions, such as the nonstandard hulls discussed in Chapter III and the standard measure spaces on nonstandard models described in Chapter IV, have been the principal tools through which new results in standard mathematics have been obtained using nonstandard analysis.

The general theory of Chapter II is applied in Chapter III to topological spaces. These are sets with an additional structure giving the notion of nearness. The presentation assumes no familiarity with topology but is rather brisk, so that acquaintance with elementary topological ideas would be useful. The chapter includes discussions of compactness and of metric, normed, and Hilbert spaces. We present a brief discussion of nonstandard hulls of metric spaces, which are important in nonstandard technique. Some of the more advanced topics in Kelley's "General Topology," such as function spaces and compactifications, are also included.

Finally, in Chapter IV, we introduce the reader to nonstandard measure theory, certainly one of the most active and fruitful areas of present-day research in nonstandard analysis. With measure theory one extends the notion of the Riemann integral. We shall take a "functional" approach to the integral on nonstandard spaces. This approach will produce both classical results in standard integration theory and some new results which have already proved quite useful in probability theory, mathematical physics, and mathematical economics. The development in this chapter does not assume familiarity with measure theory beyond the Riemann integral. Most of the results in [27, 29, 32, 33] are presented without further reference. We note here that the measures and measure spaces constructed on nonstandard models in Chapter IV are often referred to in the literature as Loeb measures and Loeb spaces.

With one exception (Section I.15), every section of the book has exercises. In designing the text, we have assumed the active participation of the reader, so some of the exercises are details of proofs in the text. At the back of the book there is a list of the notation used, together with the page where the notation is introduced. Of course, we freely use the symbols $\in$, $\cup$, and $\cap$ for set membership, union, and intersection. We have starred sections that can be skipped at the first reading. Every item in the book has three numbers, the number of the chapter (I, II, III or IV), the number of the section, and the number of the item in the section. Thus, Theorem IV.2.3 is the third item in the second section of the fourth chapter. In referring to an item, we shall omit the chapter number for items in the same chapter as the reference, and the section number for items in the same section as the reference.

CHAPTER I

Infinitesimals and
The Calculus

Our aim in this chapter is to introduce the reader to nonstandard analysis in the familiar context of the calculus. It was in this context that the concept of an infinitesimal was used by Leibniz and his followers to define the derivative, thus launching the infinitesimal calculus on its spectacular development. The notion of an infinitesimal is a cornerstone in all applications of nonstandard methods to analysis, and so an understanding of this chapter is basic to the rest of the book. Moreover, such an understanding will make the technical elaborations of the later chapters easier to appreciate.

In spite of the many technical advantages attending the use of infinitesimals as developed by Leibniz, the notion of infinitesimal was always controversial. The main question was whether infinitesimals actually existed. Since an infinitesimal real number was supposed to be smaller in absolute value than any ordinary positive one, it was clear that all infinitesimals other than zero were not ordinary real numbers. Leibniz regarded them as "numbers" in some ideal world. Further, he implicitly made the important but somewhat vague hypothesis that *the infinitesimals satisfied the same rules as the ordinary real numbers.* Consider how this hypothesis would work in the calculation of the derivative of the function e^x. Leibniz would write

$$\frac{d}{dx} e^x = \frac{e^{x+dx} - e^x}{dx} = e^x \left(\frac{e^{dx} - 1}{dx} \right),$$

where dx is an infinitesimal. A separate calculation (Example 11.3.2) would show that $(e^{dx} - 1)/dx = 1$. We will learn in this chapter that the foregoing calculation is correct as long as the equality signs are replaced by $\simeq$, where $a \simeq b$ means that a and b are infinitesimally close. Two facts should be noted:

(a) We need to be able to add infinitesimals to ordinary real numbers. This implies that both infinitesimals and ordinary reals are contained in a larger set of "numbers" for which the operations of arithmetic are defined.

1

(b) The function e^x needs to be extended to this larger set of numbers in such a way that the law of exponents is satisfied.

The example of the previous paragraph shows that to make Leibniz's approach to the calculus rigorous we must

A. construct a set $*R$ of "numbers" and define operations of addition, multiplication, and linear ordering on $*R$ so that (i) the field R of real numbers (or an isomorphic copy of R) is embedded as a subfield of $*R$ and (ii) the laws of ordinary arithmetic are valid in $*R$,

B. show how functions and relations on R are extended to functions and relations on $*R$, thus extending the "relational" structure on R to one on $*R$,

C. ensure that statements true in the relational structure on R are "extended" to statements true in the relational structure on $*R$.

A set $*R$ having the properties mentioned in A is developed in §I.1 using ultrafilters. We show in the Appendix that the existence of ultrafilters follows from Zorn's lemma, a form of the axiom of choice. In §I.2 we show how relations and functions on R are extended to relations and functions on $*R$.

To deal with C we must develop a very modest amount of mathematical logic (§§I.3 and I.4) in order to make precise what is meant by the words "statement" and "true." The sense in which true statements for R "extend" to true statements for $*R$ is made precise in the transfer principle, which is stated in §I.5. This principle is at the heart of nonstandard methods as developed by Abraham Robinson. Its proof is deferred to §I.15 since it is not necessary to know the proof in order to apply the transfer principle. In the intervening sections we show how to use the transfer principle to prove results in the calculus. The proofs are usually similar to those developed in the early days of the calculus except for the role played by mathematical logic.

As noted in the Preface, we have used a very simple formal language in this chapter in order to facilitate the initiation of readers not familiar with formal languages. Consideration of a more elaborate language and nonstandard model is deferred until Chapter II.

I.1 The Hyperreal Number System
as an Ultrapower

We assume that anyone reading this book is familiar with the real number system as a complete linearly ordered field $\mathscr{R} = (R, +, \cdot, <)$, where R denotes the set of real numbers and $+$, $\cdot$, and $<$ denote the usual algebraic operations and relations of addition, multiplication, and linear ordering on R. Our object in this section is to construct another linearly ordered field $\mathscr{R} = (\mathbf{R}, +, \cdot, <)$ which contains an isomorphic copy of $\mathscr{R}$ but is strictly larger than $\mathscr{R}$. $\mathscr{R}$ will be called a *nonstandard* or *hyperreal* number system.

The construction of $\mathscr{R}$ is reminiscent of the construction of the reals from the rationals by means of equivalence classes of Cauchy sequences. To begin the construction, let N denote the natural numbers and $\hat{R}$ denote the set of all sequences of real numbers (indexed by N); i.e., each element in $\hat{R}$ is of the form $r = \langle r_1, r_2, r_3, \ldots \rangle$. For convenience we denote $\langle r_1, r_2, r_3, \ldots \rangle$ by $\langle r_i : i \in N \rangle$ or simply $\langle r_i \rangle$. Operations of addition, $\oplus$, and multiplication, $\odot$, can be defined on $\hat{R}$ in the following way: If $r = \langle r_i \rangle$ and $s = \langle s_i \rangle$ are elements of $\hat{R}$, we define

$$r \oplus s = \langle r_i + s_i \rangle$$

and

$$r \odot s = \langle r_i \cdot s_i \rangle.$$

It is easy to check (Exercise 1) that $(\hat{R}, \oplus, \odot)$ is a commutative ring with an identity $\langle 1, 1, \ldots \rangle$ and a zero $\langle 0, 0, \ldots \rangle$ (where 1 and 0 are the unit and zero in R). However, the ring is *not* a field; for example,

$$\langle 1, 0, 1, 0, 1, \ldots \rangle \odot \langle 0, 1, 0, 1, 0, \ldots \rangle = \langle 0, 0, 0, \ldots \rangle,$$

so the product of nonzero elements can be zero. We remedy the situation by introducing an equivalence relation on $\hat{R}$ and defining operations and relations $+$, $\cdot$, and $<$ on the set $\mathbf{R}$ of equivalence classes which make $(\mathbf{R}, +, \cdot, <)$ into a linearly ordered field.

To introduce the equivalence relation we need the notion of an ultrafilter (for more on ultrafilters see the Appendix).

1.1 Definition Let I be a nonempty set. A *filter* on I is a nonempty collection $\mathscr{U}$ of subsets of I having the following properties:

(i) The empty set $\varnothing \notin \mathscr{U}$.
(ii) If $A, B \in \mathscr{U}$, then $A \cap B \in \mathscr{U}$.
(iii) If $A \in \mathscr{U}$ and $I \supseteq B \supseteq A$, then $B \in \mathscr{U}$.

A filter $\mathscr{U}$ is an *ultrafilter* if

(iv) for any subset A of I either $A \in \mathscr{U}$ or its complement $A' = I - A \in \mathscr{U}$ [but not both by (i) and (ii)].

For each $x \in I$ there is a *fixed* ultrafilter $\mathscr{U}_x = \{B \subseteq I : x \in B\}$. If I is an infinite set the collection $\mathscr{F}_I = \{A \subseteq I : I - A \text{ is finite}\}$ is a filter called the *cofinite* or *Fréchet* filter on I. An ultrafilter $\mathscr{U}$ on I is *free* if it contains $\mathscr{F}_I$.

A free ultrafilter $\mathcal{U}$ cannot contain any finite set F, since otherwise F' is cofinite and hence in $\mathcal{U}$ and so $F \cap F' = \varnothing$, contradicting 1.1(i). Intuitively an ultrafilter is a very large collection of subsets, but not too large, since, for example, it cannot contain two disjoint sets by 1.1(i) and (ii). Note that if $\mathcal{U}$ is an ultrafilter on I, then $I \in \mathcal{U}$. Note also that if $A_1, A_2, \ldots, A_n$ are a finite number of sets in I with $A_i \cap A_j = \varnothing$ for $i \neq j$ and $\bigcup A_i(1 \leq i \leq n) = I$, then one and only one of the sets A_i is in $\mathcal{U}$ (Exercise 2).

It is not at all obvious that free ultrafilters exist. These, however, are the important ultrafilters for our construction. Therefore we take as a basic assumption the following axiom.

1.2 Ultrafilter Axiom If $\mathcal{F}$ is a filter on I, then there is an ultrafilter $\mathcal{U}$ on I which contains $\mathcal{F}$.

We show in the Appendix that the ultrafilter axiom follows from Zorn's lemma (that is, the axiom of choice).

Now assume that we have chosen a free ultrafilter $\mathcal{U}$ on N. We define a relation $\equiv$ on $\hat{R}$ as follows ($\equiv$ will depend on $\mathcal{U}$, but this dependence will not be indicated explicitly).

1.3 Definition If $r = \langle r_i \rangle$ and $s = \langle s_i \rangle$ are in $\hat{R}$, then $r \equiv s$ if and only if $\{i \in N : r_i = s_i\} \in \mathcal{U}$. We then say that $\langle r_i \rangle = \langle s_i \rangle$ almost everywhere (a.e.).

1.4 Lemma The relation $\equiv$ is an equivalence relation on $\hat{R}$.

Proof: The relation $\equiv$ is reflexive ($r \equiv r$) because $N \in \mathcal{U}$, symmetric ($r \equiv s$ implies $s \equiv r$) because $=$ is a symmetric relation on R, and transitive ($r \equiv s$ and $s \equiv t$ imply $r \equiv t$) because of conditions 1.1(ii) and (iii) for a filter. The details are left to the reader (Exercise 3). □

Note that two sequences can have the same limit as $n \to \infty$ and not be equivalent. For example, $\langle 1, \frac{1}{2}, \frac{1}{3}, \ldots \rangle \not\equiv \langle 0, 0, 0, \ldots \rangle$ since $\varnothing \notin \mathcal{U}$; sequences like $\langle 1, \frac{1}{2}, \frac{1}{3}, \ldots \rangle$ will later be used to define "infinitesimal numbers" different from zero. We will see shortly that the equivalence relation also eliminates the problem that the product of nonzero elements can be zero. For example, consider again the sequences $\langle 1, 0, 1, 0, 1, \ldots \rangle$ and $\langle 0, 1, 0, 1, 0, \ldots \rangle$. By 1.1(iv), one of these two sequences is equivalent to $\langle 0, 0, \ldots \rangle$; which one depends on the particular ultrafilter $\mathcal{U}$ used to define $\equiv$ (there are many such).

The set $\hat{R}$ is divided into disjoint subsets called equivalence classes by $\equiv$. Each equivalence class consists of all sequences equivalent to any given sequence in the class. Thus r and s are in the same equivalence class iff $r \equiv s$. Of course, two sequences which differ at only a finite number of places are equivalent under $\equiv$.

1.5 Definition Let $\mathbf{R}$ denote the set of all the equivalence classes of $\hat{R}$ induced by $\equiv$. The equivalence class containing a particular sequence $s = \langle s_i \rangle$ is denoted by $[s]$ or $\mathbf{s}$. Thus if $r \equiv s$ in $\hat{R}$ then $\mathbf{r} = [r] = [s] = \mathbf{s}$. Elements of $\mathbf{R}$ are called *nonstandard* or *hyperreal* numbers.

$\mathbf{R}$ is technically known as an *ultrapower* (the general concept will be presented in Chapter II). Notice that if $\mathbf{r} = [\langle r_i \rangle]$ and $\mathbf{s} = [\langle s_i \rangle]$ are two elements of $\mathbf{R}$ then $\mathbf{r} = \mathbf{s}$ if and only if $\langle r_i \rangle = \langle s_i \rangle$ a.e. (Exercise 4). We use the same idea to define operations and relations which make $\mathbf{R}$ into an ordered field.

1.6 Definition Let $\mathbf{r} = [\langle r_i \rangle]$ and $\mathbf{s} = [\langle s_i \rangle]$. Then

(i) $\mathbf{r} + \mathbf{s} = [\langle r_i + s_i \rangle]$, i.e., $[r] + [s] = [r \oplus s]$,

(ii) $\mathbf{r} \cdot \mathbf{s} = [\langle r_i \cdot s_i \rangle]$, i.e., $[r] \cdot [s] = [r \odot s]$,

(iii) $\mathbf{r} < \mathbf{s}$ ($\mathbf{s} > \mathbf{r}$) if and only if $\{i \in N : r_i < s_i\} \in \mathscr{U}$, and $\mathbf{r} \leq \mathbf{s}$ ($\mathbf{s} \geq \mathbf{r}$) if and only if $\mathbf{r} < \mathbf{s}$ or $\mathbf{r} = \mathbf{s}$. The structure $(\mathbf{R}, +, \cdot, <)$ is denoted by $\mathscr{R}$.

We must check that the definitions are independent of the particular representatives chosen from the equivalence classes. That is, if $r \equiv \bar{r}$ and $s \equiv \bar{s}$ we must show that $[r \oplus s] = [\bar{r} \oplus \bar{s}]$, $[r \odot s] = [\bar{r} \odot \bar{s}]$, and $[\bar{r}] < [\bar{s}]$ if and only if $[r] < [s]$. We check the first equality and leave the rest to the reader (Exercise 5). Let $r = \langle r_i \rangle$, $\bar{r} = \langle \bar{r}_i \rangle$, $s = \langle s_i \rangle$, and $\bar{s} = \langle \bar{s}_i \rangle$. Then $\{i \in N : r_i = \bar{r}_i\}$ and $\{i \in N : s_i = \bar{s}_i\}$ are in $\mathscr{U}$. Obviously

(1.1) $\{i \in N : r_i + s_i = \bar{r}_i + \bar{s}_i\} \supseteq \{i \in N : r_i = \bar{r}_i\} \cap \{i \in N : s_i = \bar{s}_i\}$.

The right-hand side of (1.1) is in $\mathscr{U}$ by 1.1(ii), and so the left-hand side is in $\mathscr{U}$ by 1.1(iii).

The remarkable fact is the following.

1.7 Theorem The structure $\mathscr{R}$ is a linearly ordered field.

Proof: That $\mathscr{R}$ is a commutative ring with zero $\mathbf{0} = [\langle 0, 0, \ldots \rangle]$ and unit $\mathbf{1} = [\langle 1, 1, \ldots \rangle]$ is easy to check. For example, the distributive law $\mathbf{r} \cdot (\mathbf{s} + \mathbf{t}) = \mathbf{r} \cdot \mathbf{s} + \mathbf{r} \cdot \mathbf{t}$ is proved as follows: Let $\mathbf{r} = [r]$, $\mathbf{s} = [s]$, and $\mathbf{t} = [t]$.

Then

$$
\begin{aligned}
\mathbf{r} \cdot (\mathbf{s} + \mathbf{t}) &= [r] \cdot ([s] + [t]) \\
&= [r] \cdot [s \oplus t] \\
&= [r \odot (s \oplus t)] \\
&= [(r \odot s) \oplus (r \odot t)] \\
&= [r \odot s] + [r \odot t] \\
&= \mathbf{r} \cdot \mathbf{s} + \mathbf{r} \cdot \mathbf{t}.
\end{aligned}
$$

The other commutative ring axioms are left to the reader (Exercise 6).

To show that $\mathscr{R}$ is a field we need to prove in addition that every nonzero element in $\mathbf{R}$ has an inverse (i.e., if $\mathbf{r} \neq \mathbf{0}$ then there is an element $\mathbf{r}^{-1}$ in $\mathbf{R}$ so that $\mathbf{r} \cdot \mathbf{r}^{-1} = \mathbf{1}$).

Suppose that $\mathbf{r} = [\langle r_i \rangle] \neq [\langle 0, 0, \ldots \rangle]$. Then $\{i \in N : r_i = 0\} \notin \mathscr{U}$ and so $\{i \in N : r_i \neq 0\} \in \mathscr{U}$ by 1.1(iv). Define $\mathbf{r}^{-1} = [\langle \bar{r}_i \rangle]$, where $\bar{r}_i = r_i^{-1}$ if $r_i \neq 0$, and $\bar{r}_i = 0$ if $r_i = 0$. Then it is easy to check that $\mathbf{r} \cdot \mathbf{r}^{-1} = \mathbf{1}$. Here we have used the fact that $\mathscr{U}$ is an ultrafilter.

Finally we must show that $\mathscr{R}$ is a linearly ordered field with the ordering given by $<$. We say that an element $\mathbf{r}$ of $\mathbf{R}$ is positive if $\mathbf{r} > \mathbf{0}$. We must show that

(α) the sum of two positive elements is positive,
(β) the product of two positive elements is positive,
(γ) (Law of Trichotomy)   for a given element $\mathbf{r}$ either $\mathbf{r}$ is positive, $\mathbf{r} = \mathbf{0}$, or $-\mathbf{r}$ is positive (where $-\mathbf{r}$ is the additive inverse of $\mathbf{r}$).

(α) and (β) are left to the reader (Exercise 7). To demonstrate (γ), let $\mathbf{r} = [\langle r_i \rangle]$ and define $A = \{i \in N : r_i > 0\}$, $B = \{i \in N : r_i = 0\}$, and $C = \{i \in N : r_i < 0\}$. We want to show that one and only one of A, B, and C is in $\mathscr{U}$. This follows from Exercise 2 or the following equivalent proof: From the law of trichotomy in R, we see that $A \cup B \cup C = N \in \mathscr{U}$. Now one of A, B, and C must be in $\mathscr{U}$, for otherwise A', B', and C' are in $\mathscr{U}$ by 1.1(iv), and so $(A' \cap B' \cap C') = (A \cup B \cup C)' = \varnothing \in \mathscr{U}$, and this is a contradiction. Suppose finally that at least two (say, A and B) of the sets A, B, and C are in $\mathscr{U}$, and so $A \cap B \in \mathscr{U}$. By the law of trichotomy in R, $A \cap B = \varnothing$, contradicting 1.1(i). $\square$

The definition of absolute value in $\mathscr{R}$ is now clear.

1.8 Definition   If $\mathbf{r} \in \mathbf{R}$, then

$$
|\mathbf{r}| = \begin{cases} \mathbf{r} & \text{if } \mathbf{r} > \mathbf{0} \\ \mathbf{0} & \text{if } \mathbf{r} = \mathbf{0} \\ -\mathbf{r} & \text{if } \mathbf{r} < \mathbf{0}. \end{cases}
$$

This absolute value has all of the properties of the familiar absolute value in R. In Exercise 8 the reader is asked to shown that if $\mathbf{r} = [\langle r_i \rangle]$ then $|\mathbf{r}| = [\langle |r_i| \rangle]$.

Next we want to show that $\mathcal{R}$ can be embedded isomorphically as a linearly ordered subfield of $\mathbf{R}$. To be precise, we define a mapping $*: R \to \mathbf{R}$ as follows.

1.9 Definition If $r \in R$, we define $*(r) = {}^*r$, where ${}^*r = [\langle r, r, \ldots \rangle] \in \mathbf{R}$.

1.10 Theorem The mapping $*$ is an order-preserving isomorphism of R into $\mathbf{R}$.

Proof: The mapping $*$ is 1–1, for if ${}^*r = {}^*s$ then $[\langle r, r, \ldots \rangle] = [\langle s, s, \ldots \rangle]$ and so $r = s$. It is a trivial matter to show that $*$ preserves the field and order properties. For example, the equation $[\langle r, r, \ldots \rangle] + [\langle s, s, \ldots \rangle] = [\langle r + s, r + s, \ldots \rangle]$ establishes ${}^*(r + s) = {}^*r + {}^*s$. The details are left to the reader (Exercise 9). $\square$

Of particular interest are the standard numbers in $\mathbf{R}$; these are the images of elements of R under $*$.

1.11 Definition If $A \subseteq R$ then $(A)_*$ is the set of all elements ${}^*a$, where $a \in A$; $(R)_*$ is the set of *standard* numbers in $\mathbf{R}$.

Finally we want to show that $\mathbf{R}$ contains numbers other than standard numbers. In order to do so we use for the first time the assumption that $\mathcal{U}$ is a *free* ultrafilter. Consider the number $\omega = [\langle 1, 2, 3, \ldots \rangle]$. This number cannot equal any standard number ${}^*r = [\langle r, r, r, \ldots \rangle]$, for the set $\{i \in N : r = i\}$ consists of at most one natural number. Thus $\mathbf{R}$ is a strictly larger set than $(R)_*$. In §1.6, ω will be called an *infinite* number. Similarly the number $\omega^{-1} = (\langle 1, \frac{1}{2}, \frac{1}{3}, \ldots \rangle)$ is not in $(R)_*$ and is called an *infinitesimal*. We will see that there are many other distinct infinite and infinitesimal numbers in $\mathbf{R}$.

To sum up, we have shown that the structure $\mathcal{R}$ is at least an ordered field. The proof of this fact has involved simple but tedious manipulations involving the ultrafilter $\mathcal{U}$. One might ask whether other properties of $\mathcal{R}$ are likewise true of $\mathbf{R}$. For example, $\mathcal{R}$ has the property that if $r < s$ then there is a number t so that $r < t < s$. It turns out that $\mathbf{R}$ also has this property (Exercise 10). After checking this and a few more properties, one begins to suspect that all reasonable statements that are true in $\mathcal{R}$ are also true in $\mathcal{R}$ if the statements are suitably interpreted. This is the content of the *transfer principle*, which will be stated in a simple form in §1.5 and proved at the end of the chapter. With the transfer principle the proofs of Theorem 1.7 and similar results become trivial.

Exercises I.1

1. Show that $(\hat{R}, \oplus, \odot)$ is a ring with identity $\langle 1, 1, 1, \ldots \rangle$ and zero $\langle 0, 0, 0, \ldots \rangle$.
2. Fix an ultrafilter $\mathscr{U}$ in a set I and show that if $A_1, A_2, \ldots, A_n$ are a finite number of subsets of I with $A_i \cap A_j = \varnothing$ for $i \neq j$ and $\bigcup A_i (1 \leq i \leq n) = I$, then one and only one of the sets A_i is in $\mathscr{U}$.
3. Complete the proof of Lemma 1.4.
4. Show that if $\mathbf{r} = [\langle r_i \rangle]$ and $\mathbf{s} = [\langle s_i \rangle]$, then $\mathbf{r} = \mathbf{s}$ (equality of equivalence classes as sets) if and only if $\langle r_i \rangle = \langle s_i \rangle$ a.e.
5. Show that parts (ii) and (iii) of Definition 1.6 are independent of the representatives chosen for the equivalence classes. Also show that $\mathbf{r} \leq \mathbf{s}$ if and only if $\{ i \in N : r_i \leq s_i \} \in \mathscr{U}$.
6. Prove that $\mathscr{R}$ is a ring.
7. Establish the properties (α) and (β) of the ordering $<$ which are stated in the proof of Theorem 1.7.
8. Show that if $\mathbf{r} = [\langle r_i \rangle]$ then $|\mathbf{r}| = [\langle |r_i| \rangle]$.
9. Complete the proof of Theorem 1.10.
10. For any $\mathbf{r}, \mathbf{s} \in \mathbf{R}$ with $\mathbf{r} < \mathbf{s}$, show that there exists a $\mathbf{t} \in \mathbf{R}$ with $\mathbf{r} < \mathbf{t} < \mathbf{s}$.
11. Show directly (without using Theorem 1.7) that if $\mathbf{r} < \mathbf{s}$ and $\mathbf{s} < \mathbf{t}$, then $\mathbf{r} < \mathbf{t}$.
12. Show that $|\mathbf{r} + \mathbf{s}| \leq |\mathbf{r}| + |\mathbf{s}|$ and $|\mathbf{rs}| = |\mathbf{r}||\mathbf{s}|$ for all $\mathbf{r}, \mathbf{s} \in \mathbf{R}$.
13. Show that there are infinitely many distinct elements of $\mathbf{R}$ greater than $\omega = [\langle 1, 2, 3, \ldots \rangle]$.
14. Show that an ultrafilter $\mathscr{U}$ is free iff it is not fixed at any $x \in I$.
15. Show that if one lets $\mathscr{U}$ be an ultrafilter fixed at some $n \in N$ in the construction of $\mathscr{R}$, then $\mathscr{R}$ is isomorphic to $\mathscr{R}$.

I.2   *-Transforms of Relations

In order to do calculus we must introduce sets and functions into discussions involving $\mathscr{R}$ and $\mathscr{R}$. Of course, sets and functions are just special types of relations. We will show how to extend relations from $\mathscr{R}$ to $\mathscr{R}$. The procedure generalizes what we have done for the relation $<$.

2.1 Definition For any set S, the set $S^n = S \times S \times \cdots \times S$ (n factors) is the set of ordered n-tuples $\langle a^1, a^2, \ldots, a^n \rangle$, $a^i \in S$. An n-ary relation P on S is a subset of S^n. If $\langle a^1, \ldots, a^n \rangle$ is an element of P we write either $\langle a^1, \ldots, a^n \rangle \in P$ or $P\langle a^1, \ldots, a^n \rangle$. The *complement* of a relation P is the relation $P' = (S \times S \times \cdots \times S) - P$. In particular, a *subset* A of S is a unary relation on

S, and we write $c \in A$ or $A\langle c \rangle$ if c is in A. The *domain* of P is the subset of S^{n-1} consisting of those $(n-1)$-tuples $\langle a^1, \ldots, a^{n-1} \rangle$ in S^{n-1} for which there exists an $a \in S$ so that $P\langle a^1, \ldots, a^{n-1}, a \rangle$. The set of all such elements a is called the *range* of P. We write dom P and range P for the domain and range of P.

An S-valued *function* f of n variables on S is an $(n+1)$-ary relation with the special property that if $f\langle a^1, a^2, \ldots, a^n, a \rangle$ and $f\langle a^1, a^2, \ldots, a^n, b \rangle$ then $a = b$. Here a is called the *image* of $\langle a^1, \ldots, a^n \rangle$ under f. We also frequently write $f(a^1, \ldots, a^n) = a$ if $f\langle a^1, \ldots, a^n, a \rangle$ (notice the different brackets). If $f(a^1, \ldots, a^n) = f(b^1, \ldots, b^n)$ implies $\langle a^1, \ldots, a^n \rangle = \langle b^1, \ldots, b^n \rangle$, then we say that f is *one-to-one* $(1-1)$ or *injective*.

As examples, we see that $=$ and $<$ are 2-ary (usually written "binary") relations on R; we will usually write $a = b$ and $a < b$ rather than $=\langle a, b \rangle$ and $<\langle a, b \rangle$. Similarly $+$ and $\cdot$ are functions of two variables; we will write $a + b = c$ and $a \cdot b = c$ rather than $+(a, b) = c$ (or $+\langle a, b, c \rangle$) and $\cdot(a, b) = c$. Following common practice, we will often leave out the dot denoting multiplication altogether.

Next, we generalize Definition 1.9 by introducing a mapping $*$ from the set of n-ary relations on R to the set of n-ary relations on $\mathbf{R}$.

2.2 Definition Let P be an n-ary relation on R. The *$*$-transform* $*P$ of P is the set of all n-tuples $\langle \mathbf{r}^1, \ldots, \mathbf{r}^n \rangle$ in $\mathbf{R}^n$ such that if $\mathbf{r}^k = [\langle r_1^k, r_2^k, \ldots \rangle]$, $1 \leq k \leq n$, then $P\langle r_i^1, r_i^2, \ldots, r_i^n \rangle$ holds a.e.; that is, $\{i \in N : P\langle r_i^1, \ldots, r_i^n \rangle\} \in \mathcal{U}$.

The set $*P$ is well defined, for if $[\langle r_i^k \rangle] = [\langle \bar{r}_i^k \rangle]$, $k = 1, 2, \ldots, n$, and $A = \{i \in N : P\langle r_i^1, \ldots, r_i^n \rangle\} \in \mathcal{U}$, then there is a set B in $\mathcal{U}$ such that $r_i^k = \bar{r}_i^k$, $1 \leq k \leq n$, for each $i \in B$ (Exercise 1), and so $P\langle \bar{r}_i^1, \ldots, \bar{r}_i^n \rangle$ for $i \in A \cap B \in \mathcal{U}$.

The definitions in 1.6 for $+$, $\cdot$, and $<$ are easily seen to be special cases of Definition 2.2, in the sense that $*+$ is $+$, etc. (Exercise 2). Notice that $*=$ is equality on $\mathbf{R}$, and not just a relation satisfying the formal properties of equality (see Exercise 4 in §1.1).

For any set $A \subseteq R$, it follows from Definition 2.2 that $*A = \{[\langle s_i \rangle] \in \mathbf{R} : \{i \in N : s_i \in A\} \in \mathcal{U}\}$ and therefore $*A \supseteq (A)_*$ (Exercise 3). We will show later that, for an infinite set A, $*A$ contains points not in $(A)_*$. As an example, we see that $*R = \mathbf{R}$ since $\{i \in N : r_i \in R\} = N \in \mathcal{U}$, and $\mathbf{R}$ properly contains $(R)_*$.

For a slightly less trivial example, let $A = [a, b] = \{x \in R : a \leq x \leq b\}$. Then $*A = \{[\langle s_i \rangle] \in \mathbf{R} : \{i \in n : a \leq s_i \leq b\} \in \mathcal{U}\}$, and it follows that $*[a, b] = \{\mathbf{x} \in \mathbf{R} : *a \leq \mathbf{x} \leq *b\}$. It will again be a consequence of the transfer principle that $*A$ has all of the properties (appropriately interpreted) of A.

To take another example, let Z denote the set of integers, and consider
$*Z = \{[\langle s_i \rangle] \in \mathbf{R} : \{i \in N : s_i \in Z\} \in \mathscr{U}\}$. If the transfer principle is true, then
$*Z$ must be a subring of $\mathscr{R}$ under the induced operations of addition and
multiplication. To establish this without the transfer principle would require
more tedious calculations with the ultrafilter $\mathscr{U}$ (Exercise 4).

Let $\mathbf{r}^k = [\langle r_i^k \rangle]$, $1 \leq k \leq n$, and $\mathbf{s} = [\langle s_i \rangle]$. If f is a function of n variables then the relation $*f\langle [\langle r_i^1 \rangle], \ldots, [\langle r_1^n \rangle], [\langle s_i \rangle] \rangle$ holds in $\mathbf{R}^{n+1}$ iff
$f\langle r_i^1, \ldots, r_i^n, s_i \rangle$ holds a.e., in which case $f(r_i^1, \ldots, r_i^n) = s_i$ almost every-
where. It follows that $*f$ is a function of n variables on $\mathbf{R}$. Moreover,
$*f(\mathbf{r}^1, \ldots, \mathbf{r}^n)$ is defined iff $f(r_i^1, \ldots, r_i^n)$ is defined a.e., and, if defined, then
$*f(\mathbf{r}^1, \ldots, \mathbf{r}^n) = [\langle s_i \rangle]$ where $f(r_i^1, \ldots, r_i^n) = s_i$ almost everywhere.

It should now be easy to prove the following result. The proof is left as
an exercise.

2.3 Theorem If P is an n-ary relation on R and $P\langle r^1, \ldots, r^n \rangle$ for $r^k \in R$, then
$*P\langle *r^1, \ldots, *r^n \rangle$. Thus if we identify each r^k with $*r^k$, the relation $*P$ extends
P (where R is regarded as embedded in $\mathbf{R}$).

In particular, if f is a function of n variables on R and $f(r^1, \ldots, r^n) = s$
with r^k and s in R, then $*f(*r^1, \ldots, *r^n) = *s$. Consequently $*f$ is an extension
of f if we think of R as embedded in $\mathbf{R}$ by the mapping $*$.

An n-ary relation P can be defined by its *characteristic function*

$$\chi_P\langle x_1, \ldots, x_n \rangle = \begin{cases} 1, & \langle x_1, \ldots, x_n \rangle \in P, \\ 0, & \langle x_1, \ldots, x_n \rangle \notin P. \end{cases}$$

The proof of the following proposition is also left to the reader.

2.4 Proposition $*\chi_P = \chi_{*P}$.

2.5 Notational Convention In order to eliminate the use of boldface symbols
and so conform to Robinson's notation in the rest of Chapter I, we adopt
the following conventions:

(a) $\mathbf{R}$ and $\mathscr{R}$ will be denoted by $*R$ and $*\mathscr{R}$, respectively.

(b) Numbers in R *and* $*R$ will be denoted by lowercase letters, e.g., r, s.
The context will make clear whether a number is in R or $*R$.

(c) We regard R as embedded in $*R$, and hence identify R and $(R)_*$. For
example, we will write 5 instead of $*5$ for $[\langle 5, 5, \ldots \rangle]$.

(d) We will use the usual notation $<$, $+$, and $\cdot$ for $\mathbf{<}$, $\mathbf{+}$, and $\mathbf{\cdot}$. Again
the context will settle any possible confusion.

(e) $\mathscr{R}$ will from now on denote the structure consisting of the set R together with all relations and in particular all functions on R. $*\mathscr{R}$ will denote

the structure consisting of *R and the extensions of n-ary relations and functions on R. Any relation or function in *$\mathscr{R}$ is the extension of a standard one.

We will deal with the structure $\mathscr{R}$ and *$\mathscr{R}$ as special cases of structures known as *relational systems*.

2.6 Definition A *relational system* is a structure $\mathscr{S} = (S, \{P_i : i \in I\}, \{f_j : j \in J\})$ consisting of a set S, a collection of relations P_i ($i \in I$) on S, and a collection of functions f_j ($j \in J$) on S.

It should be noted that a given relational system does not necessarily contain all relations and functions on S. The formal exhibition of functions in $\mathscr{S}$ is needed for the definition of "terms" in the next section. In the next two sections we present the elements of symbolic logic for relational systems to facilitate work with the transfer principle.

Exercises I.2

1. Complete the proof that *P, as defined in 2.2, is well defined.
2. Show that *$+$, *$\cdot$, and *$<$ are $+$, $\cdot$, and $<$, respectively.
3. If A is a subset of R, show that *$A \supseteq (A)_*$.
4. Show that if Z denotes the set of integers, then *Z is a subring of *$\mathscr{R}$ under the induced operations of addition and multiplication from $+$ and $\cdot$.
5. Prove Theorem 2.3.
6. Prove Proposition 2.4.
7. Show that if A is finite, then *$A = (A)_*$.
8. Let A and B be two subsets of R, and show that *$(A \cap B) = {}^*A \cap {}^*B$ and *$(A \cup B) = {}^*A \cup {}^*B$.
9. Show that *$(\text{dom } P) = \text{dom } {}^*P$ for any n-ary relation P.
10. Let f be a 1–1 function, and show that *f is 1–1.

I.3   Simple Languages for Relational Systems

From this point (with the exception of §I.15) we suppress the construction of *$\mathscr{R}$ and work instead with the properties obtained through use of the transfer principle. This facilitates our work in the same way that work with the set of real numbers is helped by the suppression of its construction via Dedekind cuts or Cauchy sequences.

It is not enough to extend functions and relations to *R; the extensions must follow the same rules as the original functions and relations. For example, we want *$\sin(x + 2\pi) = {}^*\sin(x)$ for all $x \in {}^*R$. To make this precise, we must make the notion of "rule" precise, and for that we need the formal language which we begin to develop here.

At the end of the last section we remarked that both $\mathscr{R}$ and $^*\mathscr{R}$ could be regarded as relational systems, that is, structures of the form $\mathscr{S} = (S, \{P_i : i \in I\}, \{f_j : j \in J\})$, where S is some basic set, the $P_i (i \in I)$ are relations, and the $f_j (j \in J)$ are functions on S. Given any such structure $\mathscr{S}$ (it need not be $\mathscr{R}$ or $^*\mathscr{R}$), we present in this section a symbolic language $L_{\mathscr{S}}$ in which we may make mathematical assertions about $\mathscr{S}$. In the next section we will show how to interpret formal sentences in $L_{\mathscr{S}}$ in the relational structure $\mathscr{S}$, and also indicate how to translate informal mathematical statements about $\mathscr{S}$ into sentences in $L_{\mathscr{S}}$. The transfer principle is stated (in §I.5) and proved (in §I.15) in the context of a formal language. Later, with a little experience in translating mathematical statements into formal sentences and vice versa, this formality will usually be unnecessary. Thus, in subsequent chapters of this book the transfer principle will be applied directly to informal mathematical statements; at this stage, however, it is important to develop confidence in the translation procedure.

The language consists of (a) a basic set of symbols and (b) combinations of those basic symbols into formal sentences of a particular type which we will call simple sentences. The basic symbols of $L_{\mathscr{S}}$ fall into two categories. The first category consists of *logical symbols* which are common to any simple language and do not vary if $\mathscr{S}$ is changed. These are

3.1 Logical Connectives The symbols $\wedge$ and $\rightarrow$, to be interpreted later as "and" and "implies."

3.2 Quantifier Symbol The symbol $\forall$, to be interpreted as "for all."

3.3 Parentheses The symbols $[\,,\,]$, $(\,,\,)$, and $\langle\,,\,\rangle$, to be used as usual in mathematics for bracketing.

3.4 Variable Symbols A countable collection of symbols like x, y, x_1, x_2, and m, n, to be used as "variables." Only a countable collection is necessary since each sentence will involve only finitely many variables, and we write down only finitely many sentences in any proof.

The symbols in the second category depend on $\mathscr{S}$ and will be called *parameters*. They consist of

3.5 Constant Symbols A symbol $\underline{s}$ called the name of s for each s in S.

3.6 Relation Symbols A relation symbol $\underline{P}$ for each relation P in S. We call $\underline{P}$ the name of P.

3.7 Function Symbols A function symbol $\underline{f}$ for each function f in S. We call $\underline{f}$ the name of f.

Whenever possible in this and the next section we *underline* to denote constant, relation, and function symbols. But, for convenience, familiar constants, relations, and functions like 1, 2, π, $<$, $+$, and sin will be named by the same symbol. Note that we do not rule out the possibility of more than one name for each entity.

We next describe how to make up meaningful combinations of the above symbols. The first step is to show how to build up familiar expressions like $\underline{f}(x)$, $\underline{g}(2, \underline{f}(x, y))$, and $\sin(x + \pi)$. Such expressions are like composite functions in usual mathematical notation, and use constant, variable, and function symbols. They are special cases of terms, which are defined as follows.

3.8 Definition *Terms* are defined inductively as follows:

(i) Each constant and variable symbol is a term.
(ii) If $\underline{f}$ is the name of a function of n variables and $\tau^1, \ldots, \tau^n$ are terms, then $\underline{f}(\tau^1, \ldots, \tau^n)$ is a term.

A term containing no variables is called a *constant term*.

In $L_{\mathscr{R}}$, for example, the expression $xy + xz$ is a term which could be written more formally as $\underline{S}(\underline{P}(x, y), \underline{P}(x, z))$, where S and P are the functions defined by $S(a, b) = a + b$ and $P(a, b) = ab$ for $a, b \in R$.

We can now be precise about the form of the mathematical sentences to be used in the rest of the chapter; they will be called simple sentences in $L_{\mathscr{S}}$.

3.9 Definition A *simple sentence* is a string of symbols in $L_{\mathscr{S}}$ which takes either of the following forms:

(A) *Atomic sentences.* Such sentences are of the form $\underline{P}\langle \tau^1, \ldots, \tau^n \rangle$, where $\underline{P}$ is the name of an n-ary relation and the τ^i ($i = 1, \ldots, n$) are constant terms.
(B) *Compound sentences.* Such sentences are of the form

$$(\forall x_1) \cdots (\forall x_n) \left[\bigwedge_{i=1}^{k} \underline{P}_i \langle \tilde{\tau}_i \rangle \to \bigwedge_{j=1}^{l} \underline{Q}_j \langle \tilde{\sigma}_j \rangle \right],$$

where $\bigwedge_{i=1}^{n} \underline{S}_i$ denotes $\underline{S}_1 \wedge \cdots \wedge \underline{S}_n$, $\tilde{\tau}_i$ and $\tilde{\sigma}_j$ are n_i-tuples and n_j-tuples of terms (e.g., $\tilde{\tau}_i = \langle \tau_i^1, \ldots, \tau_i^{n_i} \rangle$) involving no other variables than $x_1, \ldots, x_n$, n_i and n_j being the orders of the relation named by the symbols $\underline{P}_i$ and $\underline{Q}_j$.

For example, if $\underline{I}$ is the name of the inequality relation I (so that $I\langle a, b\rangle$ if $a < b$), then the expression $\underline{I}\langle 1, 2\rangle$ is an atomic sentence. The sentence $\underline{I}\langle 1, 2\rangle$ will usually be written in the more familiar form $1 < 2$. This and similar conventions will be used here and throughout the text. The expression

$$(\forall x)(\forall y)(\forall z)[\underline{R}\langle x\rangle \wedge \underline{R}\langle y\rangle \wedge \underline{R}\langle z\rangle \rightarrow x \cdot (y + z) = x \cdot y + x \cdot z],$$

where $\underline{R}$ is the name of the unary relation R defined by $R\langle a\rangle$ iff a is a real number, is a compound sentence which, as we shall see in the next section, expresses the fact that the distributive law holds in $\mathscr{R}$. On the other hand, the expression

$$(\forall x)[\underline{R}\langle x\rangle \wedge x < 1 \rightarrow x < y]$$

is not a simple sentence, since the variable y occurs within the square brackets without a corresponding $(\forall y)$ outside.

We will use the symbol $\leftrightarrow$ to abbreviate pairs of compound sentences which differ only in that $\bigwedge_{i=1}^{k} \underline{P}_i$ and $\bigwedge_{j=1}^{l} \underline{Q}_j$ are reversed. For example, the pair of sentences $(\forall x)[(x = x) \rightarrow \underline{R}\langle x\rangle]$ and $\forall(x)[\underline{R}\langle x\rangle \rightarrow (x = x)]$ will be abbreviated as $(\forall x)[(x = x) \leftrightarrow \underline{R}\langle x\rangle]$.

The use of simple sentences as the only type of sentence in the language and the corresponding use of "Skolem functions" (defined in §I.4) was suggested by a similar use of equalities, inequalities, and Skolem functions in H. J. Keisler's work [23, 24].

Exercises I.3

In the following we let $\mathscr{S} = \mathscr{R}$, where $\mathscr{R}$ consists of the basic set R, the collection $\{P_i : i \in I\}$ of all relations on R, and the collection $\{f_j : j \in J\}$ of all functions on R.

1. Let $\underline{f}, \underline{g}$, and $\underline{h}$ be names of functions of 1, 2, and 3 variables, respectively, and x, y, and z denote variables. Which of the following are terms in $L_{\mathscr{S}}$?

   (a) $\underline{h}(x, 2, \underline{f}(\underline{g}(x, z)))$
   (b) $\underline{f}(\underline{g}(y, 1)\underline{h}(1, 1, z))$
   (c) $\underline{h}(1, \underline{f}(2), \underline{g}(1, 3))$

2. In Exercise 1, which of the expressions are constant terms?
3. Let P and S denote the binary functions of product and sum in R (see remarks after Definition 3.8). Also let I and R denote the relations defined in the text. Which of the following expressions are atomic sentences and which are compound sentences?

   (a) $\underline{I}\langle \underline{S}(2, \underline{P}(2, 3)), 9\rangle$
   (b) $\underline{I}\langle 2, \underline{S}(x, 2)\rangle$

(c) $(\forall x)(\forall y)(\forall z)[\underline{R}\langle x\rangle \wedge \underline{R}\langle y\rangle \wedge \underline{R}\langle z\rangle \wedge \underline{I}\langle 0, x\rangle \wedge \underline{I}\langle y, z\rangle \rightarrow$
$\underline{I}\langle \underline{P}(x, y), \underline{P}(x, z)\rangle]$
(d) $(\forall x)[\underline{R}\langle x\rangle \rightarrow \underline{P}(x, 1)]$
(e) $(\forall x)[\underline{R}\langle x\rangle \rightarrow \underline{I}\langle x, 1\rangle]$

4. Show how the term $\sqrt{x^2 + 2xy}$ can be built up inductively from constant
   and variable symbols using the function symbols $\underline{S}, \underline{P}$ and the function
   symbol $\underline{Q}$ for the function $Q(x)$ defined by $Q(x) = \sqrt{x}$.

I.4 Interpretation of Simple Sentences

In preparation for the statement of the transfer principle, which will be
presented in the next section, we now show how a simple sentence in $L_{\mathscr{S}}$ is
interpreted in a relational system $\mathscr{S} = (S, \{P_i : i \in I\}, \{f_j : j \in J\})$; i.e., we show
how to determine when such a sentence is true or false.

The process of interpretation was already implicitly begun in §I.3 by
designating symbols like $\underline{s}$, $\underline{P}$, and $\underline{f}$ in $L_{\mathscr{S}}$ as names of the corresponding
entities s, P, and f in $\mathscr{S}$. It will be seen that the interpretation of simple
sentences is completely determined once this naming has been specified. The
interpretation is, in fact, the obvious one consistent with the interpretation
of $\forall$, $\wedge$, and $\rightarrow$ as "for all," "and," and "implies," respectively.

We first show how to interpret constant terms, i.e., those containing no
variable symbols. In defining "terms" we have taken no account of the domain
of definition of the functions whose names may occur in the terms, and this
must be accounted for in the interpretation. Suppose that the constant term
is just a constant symbol s naming $s \in S$. Then this term will be interpreted
as the corresponding s. If the term is of the form $\underline{f}(\underline{s}^1, \ldots, \underline{s}^n)$, where $\underline{f}$ is a
symbol for a function of n variables and names the function f, and the $\underline{s}^i$
name $s^i \in S$ $(i = 1, \ldots, n)$, then we obviously should interpret $\underline{f}(\underline{s}^1, \ldots, \underline{s}^n)$ as
the constant $f(s^1, \ldots, s^n)$ as long as $f(s^1, \ldots, s^n)$ is defined, i.e., $\langle s^1, \ldots, s^n\rangle$
is in the domain of f. We proceed inductively to more general constant terms
as follows.

4.1 Definition A constant term is *interpretable* in $\mathscr{S}$ if either

(i) it is a constant symbol $\underline{s}$ naming an element $s \in S$, in which case it is
interpreted as s, or
(ii) it is of the form $\underline{f}(\tau^1, \ldots, \tau^n)$, where the terms $\tau^1, \ldots, \tau^n$ are interpretable in $\mathscr{S}$ and hence can be interpreted as the elements $s^1, \ldots, s^n \in S$, and

the n-tuple $\langle s^1, \ldots, s^n \rangle$ is in the domain of the function f named by $\underline{f}$; in this case $\underline{f}(\tau^1, \ldots, \tau^n)$ is interpreted as $f(s^1, \ldots, s^n)$.

Note that terms σ which are interpretable contain no variable symbols and are interpreted as the unique elements in S which can be obtained by imitating in $\mathscr{S}$ the process of constructing σ. In $\mathscr{R}$, for example, the term $\sin\sqrt{1 + (\pi/2)^2}$ is interpretable since, inductively, 1 is defined by (i), $1 + (\pi/2)^2$ is interpretable by (ii), $\sqrt{1 + (\pi/2)^2}$ is interpretable by (ii), and finally $\sin\sqrt{1 + (\pi/2)^2}$ is interpretable by (ii). On the other hand, $1 + \tan(\pi/2)$ is not interpretable since $\pi/2$ is not in the domain of the function tan.

We can now state when a simple sentence in $\mathscr{S}$ is true. A sentence of the form "P implies Q" is regarded as true if whenever P is true then Q is true. More precisely, such a sentence is false only if P is true and Q is false. Thus the following definition is natural.

4.2 Definition In the notation of Definition 3.9,

(A) the atomic sentence $\underline{P}\langle \tau^1, \ldots, \tau^n \rangle$ is *true* (or *holds*) in $\mathscr{S}$ if each of the terms τ^i, $1 \le i \le n$, is interpretable in $\mathscr{S}$ as s^i, $1 \le i \le n$, and the n-tuple $\langle s^1, \ldots, s^n \rangle$ is in the relation P named by $\underline{P}$ (thus $\underline{P}\langle \tau^1, \ldots, \tau^n \rangle$ does not hold in $\mathscr{S}$ either if one of the τ^i is not interpretable in $\mathscr{S}$ or if all of the τ^i are interpretable in $\mathscr{S}$ but the corresponding n-tuple $\langle s^1, \ldots, s^n \rangle$ is not in the relation P),

(B) the sentence

$$(\forall x_1)(\forall x_2) \cdots (\forall x_n) \left[\bigwedge_{i=1}^{k} \underline{P}_i \langle \tilde{\tau}_i \rangle \to \bigwedge_{j=1}^{l} \underline{Q}_j \langle \tilde{\sigma}_j \rangle \right]$$

is *true* (or *holds*) in $\mathscr{S}$ if, for each replacement of the variable symbols $x_1, \ldots, x_n$ with constant symbols $\underline{s}^1, \ldots, \underline{s}^n$ such that, with this replacement, the atomic sentences $\underline{P}_i \langle \tau_i^1, \ldots, \tau_i^{n_i} \rangle$ are all true in $\mathscr{S}$ ($1 \le i \le k$), the atomic sentences $\underline{Q}_j \langle \sigma_j^1, \ldots, \sigma_j^{n_j} \rangle$, $1 \le j \le l$, are also true in $\mathscr{S}$ with the same replacement (we are assuming that each of the constant symbols $\underline{s}^i$, $1 \le i \le n$, names an element of $\mathscr{S}$).

This scheme for interpreting simple sentences in $\mathscr{S}$ is only the obvious one consistent with the interpretation of $\wedge$ and $\to$ as "and" and "implies," respectively. In most cases of interest it will be clear whether a given sentence is true in $\mathscr{S}$ or not. For example, the atomic sentence

(4.1) $1 + (\pi/2)^2 > 2$

is true in $\mathscr{R}$. Likewise, the sentence

(4.2) $(\forall x)(\forall y)[x > 0 \wedge y > 0 \to xy > 0]$

is true in $\mathcal{R}$, and expresses the fact that the product of positive real numbers is positive. Note that

(4.3) $$(\forall x)[(\sqrt{x} > -1) \to (\sqrt{x} \geq 0)]$$

is true in $\mathcal{R}$, since $\sqrt{a}$ is interpretable only for $a \geq 0$, and if $a \geq 0$ then $\sqrt{a} \geq -1$ and $\sqrt{a} \geq 0$. On the other hand,

(4.4) $$(\forall x)[\underline{R}\langle x \rangle \to \sqrt{x} \geq 0]$$

is not true in $\mathcal{R}$, since $\sqrt{a}$ is not defined for all real numbers a. Many examples will be presented in this and the succeeding sections which will provide practice in deciding on the truth in $\mathcal{S}$ of simple sentences.

It is even more important to be able to translate an informal mathematical statement about a relational system $\mathcal{S}$ (involving English phrases like "for all," "there exists," "and," and "or") into a sentence in $L_{\mathcal{S}}$ which has the same interpretation. The rest of this section is devoted to some examples and remarks concerning this problem.

A basic problem in translation is that the simple language of this chapter involves only formal analogues of the phrases "for all," "and," and "implies" and these must occur in certain formal combinations in a simple sentence, whereas informal mathematical statements often involve phrases like "there exists," "or," and "not." It is not always easy to decide whether there exists a corresponding sentence in our simple language having the same interpretation. Fortunately, the specific translations which are necessary to do calculus will not cause any difficulty. Some typical examples follow.

Sometimes the translation is direct. Take, for example, the statement "The distributive law holds in R," or, more precisely, "For all real numbers x, y, and z, $x(y + z) = xy + xz$." Some simple sentences in $L_{\mathcal{R}}$ which each correspond to this statement are

(4.5) $$(\forall x)(\forall y)(\forall z)[\underline{R}\langle x \rangle \wedge \underline{R}\langle y \rangle \wedge \underline{R}\langle z \rangle \to x(y + z) = xy + xz]$$

and

(4.6) $$(\forall x)(\forall y)(\forall z)[1 = 1 \to x(y + z) = xy + xz].$$

In the latter sentence we have used the fact that the only substitutions allowed for variables in $L_{\mathcal{R}}$ are names of elements in R. We often, however, write $\underline{R}\langle x \rangle$ for clarity.

Mathematical statements involving "not" attached to a given n-ary relation P on S can often be restated using the complement P' of P. For example, corresponding to the true statement "2 is not less than 1" is the atomic sentence $\underline{I}'\langle 2, 1 \rangle$, where I is the relation $\{\langle x, y \rangle \in R^2 : x < y\}$ and I' is the complement of I given by $\{\langle x, y \rangle \in R^2 : x \geq y\}$. Note, however, that if a term in $\underline{P}\langle\tau^1,\ldots,\tau^n\rangle$ is not interpretable in $\mathscr{S}$, then neither $\underline{P}\langle\tau^1,\ldots,\tau^n\rangle$ nor $P'\langle\tau^1,\ldots,\tau^n\rangle$ is true in $\mathscr{S}$.

Statements involving "or" are sometimes more difficult to deal with. Consider the law of trichotomy for the ordering on R, "For all real x, either $x > 0$ or $x = 0$ or $x < 0$." This can be translated by any one of the three simple sentences

$$(\forall x)[\underline{R}\langle x\rangle \wedge x \not< 0 \wedge x \neq 0 \rightarrow x > 0],$$
(4.7)
$$(\forall x)[\underline{R}\langle x\rangle \wedge x \not< 0 \wedge x \not> 0 \rightarrow x = 0],$$
$$(\forall x)[\underline{R}\langle x\rangle \wedge x \not> 0 \wedge x \neq 0 \rightarrow x < 0],$$

where we have written $x \not< 0$ and $x \not> 0$ for the expressions $\underline{I}'\langle x,0\rangle$ and $\underline{I}'\langle 0,x\rangle$.

Statements involving "there exists" are translated by the technical artifice of introducing so-called *Skolem functions*, which will occur frequently in later sections of this chapter. Consider, for example, the true statement "For each nonzero x in R there exists a y in R so that $xy = 1$." Notice that the statement asserts the existence of a special function ψ of one variable whose domain is the set of nonzero reals, and which satisfies $x\psi(x) = 1$ [so that $\psi(x) = x^{-1}$]. An equivalent mathematical statement using this special function ψ is thus "For all nonzero real x, $x\psi(x) = 1$," which translates to the simple sentence

(4.8) $(\forall x)[\underline{R}\langle x\rangle \wedge x \neq 0 \rightarrow x\underline{\psi}(x) = 1].$

The function ψ is an example of a Skolem function. Clearly the same trick can be used systematically in other situations to remove the expression "there exists." As another example of a Skolem function, let f be a function of n variables with domain A and range $B \subset R$. The fact that B is the range of f can be expressed using n Skolem functions ψ_i, $1 \leq i \leq n$, with the two simple sentences

(4.9)
$$(\forall x_1)(\forall x_2)\cdots(\forall x_n)[\underline{f}(x_1,\ldots,x_n) = \underline{f}(x_1,\ldots,x_n) \rightarrow \underline{B}\langle\underline{f}(x_1,\ldots,x_n)\rangle],$$
$$(\forall y)[\underline{B}\langle y\rangle \rightarrow \underline{f}(\underline{\psi}_1(y),\ldots,\underline{\psi}_n(y)) = y].$$

How could the ψ_i be defined?

The ideas just presented do not constitute a general translation scheme between statements and simple sentences, but will suffice for the problems presented in this chapter. In the next chapter we present a richer formal language for more general mathematical structures which will involve formal analogues of "there exists," "or," and "not," and so will avoid Skolem functions. We have restricted ourselves to simple sentences in this chapter because the transfer principle is easier to state and prove for these sentences and because this restriction allows a more gradual introduction to the general techniques of nonstandard analysis.

Exercises I.4

1. Show in detail that the sentence (4.1) is true when interpreted in $\mathscr{R}$.
2. Show that the sentences (4.9) express the fact that B is the range of the function f of n variables. In doing so, define the Skolem functions ψ_i, $1 \leq i \leq n$.
3. Let $[x]$ denote the greatest integer less than or equal to $x \in R$. Write one or more simple sentences whose interpretation in $\mathscr{R}$ characterizes this property of $[x]$.
4. Write a simple sentence whose interpretation in $\mathscr{R}$ asserts that for each real x there is a nonnegative integer $m \geq x$ (this is the Archimedean property of $\mathscr{R}$ as an ordered field).
5. Write sentences in $L_{\mathscr{R}}$ characterizing the fact that a given nonempty set $A \subseteq N$ has a first element.
6. Characterize the fact that $A \subseteq R$ is the domain of a given function f of n variables using one or more simple sentences.
7. Write a simple sentence in $L_{\mathscr{R}}$ which characterizes the fact that the function sin defined on R is

    (a) bounded by 1 in absolute value,
    (b) periodic with period 2π.

8. Write a simple sentence in $L_{\mathscr{R}}$ which expresses the fact that the function f on R is continuous, i.e., given $\varepsilon > 0$ there is a $\delta > 0$ so that $|x - a| < \delta$ implies $|f(x) - f(a)| < \varepsilon$.
9. Let A, B, and C denote unary relations defining subsets of R. Write simple sentences whose interpretation in $\mathscr{R}$ asserts that

    (a) $A \subseteq B$,
    (b) $A = B$,
    (c) $C = A \cap B$,
    (d) $C = A \cup B$.

I.5 The Transfer Principle for
Simple Sentences

We are now able to state accurately the transfer principle for simple sentences in $L_{\mathscr{R}}$. The proof will be deferred to the end of the chapter. In the intervening sections we will present many applications of the principle which should convince the reader that it is a very powerful tool. Moreover, it will be clear that one need not know the proof of the transfer principle to apply it successfully. A transfer principle for more general sentences and more general mathematical structures will be presented in Chapter II.

We first introduce the notion of the ∗-transform of a sentence in $L_{\mathscr{R}}$. Here, we adopt the following conventions.

5.1 Conventions

(a) If r is a name in $L_{\mathscr{R}}$ of $r \in R$ then $\underline{r}$ is also a name in $L_{*\mathscr{R}}$ of $*r \in *R$ (remember that we identify r and $*r$).

(b) If $\underline{P}$ is a name in $L_{\mathscr{R}}$ of the relation P on R then $*\underline{P}$ is a name in $L_{*\mathscr{R}}$ of the relation $*P$ on $*R$. In particular,

(c) If $\underline{f}$ is a name in $L_{\mathscr{R}}$ of the function f on R then $*\underline{f}$ is a name in $L_{*\mathscr{R}}$ of the function $*f$ on $*R$.

(d) The symbols $<$, $+$, and $\cdot$ will denote the corresponding relation and functions in $\mathscr{R}$ and $*\mathscr{R}$.

Stated briefly, the ∗-transform of a simple sentence Φ in $L_{\mathscr{R}}$ is the simple sentence $*\Phi$ in $L_{*\mathscr{R}}$ obtained by starring all function and relation symbols in the sentence Φ. This is made more precise in the following definitions.

5.2 Definition The ∗-transform of terms is defined by induction as follows:

(i) If τ is a constant or variable symbol then $*\tau = \tau$.
(ii) If $\tau = \underline{f}(\tau^1, \ldots, \tau^n)$ then $*\tau = *\underline{f}(*\tau^1, \ldots, *\tau^n)$.

For example, the ∗-transform of the term $\underline{f}(x, \underline{g}(y + \pi))$ is the term $*\underline{f}(x, *\underline{g}(y + \pi))$ [notice again that $+$ should also be starred, but we avoid the awkward notation $*+$ by Convention 5.1(d)].

5.3 Definition If Φ is a simple sentence in $L_{\mathscr{R}}$ we define the ∗-transform $*\Phi$ of Φ as follows:

(a) If Φ is the atomic sentence $\underline{P}\langle \tau^1, \ldots, \tau^n \rangle$ then $*\Phi$ is $*\underline{P}\langle *\tau^1, \ldots, *\tau^n \rangle$.
(b) If Φ is the sentence

$$(\forall x_1) \cdots (\forall x_n) \left[\bigwedge_{i=1}^{k} \underline{P}_i \langle \tilde{\tau}_i \rangle \to \bigwedge_{j=1}^{l} \underline{Q}_j \langle \tilde{\sigma}_j \rangle \right]$$

then $*\Phi$ is the sentence

$$(\forall x_1) \cdots (\forall x_n) \left[\bigwedge_{i=1}^{k} *\underline{P}_i \langle *\tilde{\tau}_i \rangle \to \bigwedge_{j=1}^{l} *\underline{Q}_j \langle *\tilde{\sigma}_j \rangle \right],$$

where $*\tilde{\tau} = \langle *\tau^1, \ldots, *\tau^n \rangle$ if $\tilde{\tau} = \langle \tau^1, \ldots, \tau^n \rangle$.

Thus, for example, the ∗-transform of the simple sentence

(5.1) $(\forall x)(\forall y)(\forall z)[\underline{R}\langle x\rangle \wedge \underline{R}\langle y\rangle \wedge \underline{R}\langle z\rangle \wedge x < y \rightarrow x + z < y + z]$

is the sentence

(5.2) $(\forall x)(\forall y)(\forall z)[*\underline{R}\langle x\rangle \wedge *\underline{R}\langle y\rangle \wedge *\underline{R}\langle z\rangle \wedge x < y \rightarrow x + z < y + z],$

which expresses the fact (true in $*\mathscr{R}$) that if $x, y, z \in *R$ and $x < y$ then $x + z < y + z$.

We now come to the main result of this section.

5.4 Theorem   (Transfer Principle)   If Φ is a simple sentence in $L_{\mathscr{R}}$ which is true in $\mathscr{R}$, then $*\Phi$ is true in $*\mathscr{R}$.

As examples of the application of Theorem 5.4 we will establish some elementary results which will be useful later. Some of these have already been presented in §§I.1 and I.2. Throughout the rest of this book we will often say that (the truth of) a certain mathematical statement $*\Phi$ about a nonstandard structure $*\mathscr{S}$ follows *by transfer* of a sentence Φ in $L_{\mathscr{S}}$. This means that the truth of the statement follows by the transfer principle from the fact that Φ is true in $\mathscr{S}$.

5.5 Proposition Let P be an n-ary relation on R and χ_P its characteristic function on R^n. Then $*P$ is an extension of P. Also $*\chi_P = \chi_{*P}$, and $*(P') = (*P)'$.

Proof: Suppose $\langle c^1, \ldots, c^n\rangle \in P$. Then the atomic sentence $P\langle c^1, \ldots, c^n\rangle$ is true in $\mathscr{R}$. By transfer, $*P\langle c^1, \ldots, c^n\rangle$; i.e., $*P$ is an extension of P.

The rest follows by transfer of the four sentences

(5.3) $(\forall x_1) \cdots (\forall x_n)[\underline{R}\langle x_1\rangle \wedge \ldots \wedge \underline{R}\langle x_n\rangle \rightarrow \chi_P(x_1, \ldots, x_n)$
$$= \chi_P(x_1, \ldots, x_n)],$$

(5.4) $(\forall x_1) \cdots (\forall x_n)[\chi_P(x_1, \ldots, x_n) \neq 1 \rightarrow \chi_P(x_1, \ldots, x_n) = 0],$

(5.5) $(\forall x_1) \cdots (\forall x_n)[\underline{P}\langle x_1, \ldots, x_n\rangle \leftrightarrow \chi_P(x_1, \ldots, x_n) = 1],$

(5.6) $(\forall x_1) \cdots (\forall x_n)[\underline{P'}\langle x_1, \ldots, x_n\rangle \leftrightarrow \chi_P(x_1, \ldots, x_n) = 0],$

which are true in $\mathscr{R}$.   □

5.6 Proposition If f is a function of n variables on R, then $*f$ is a function of n variables and is an extension of f with $*(\mathrm{dom}\, f) = \mathrm{dom}\, *f$ and $*(\mathrm{range}\, f) = \mathrm{range}\, *f$.

Proof: That *f is a function follows from the definition of *f. It also follows by transfer of the sentence

(5.7) $(\forall x_1) \cdots (\forall x_n)(\forall y)(\forall z)[\underline{f}\langle x_1, \ldots, x_n, y\rangle \wedge \underline{f}\langle x_1, \ldots, x_n, z\rangle \to y = z]$.

That *f is an extension of f follows from Proposition 5.5. Transfer of the sentence

(5.8) $(\forall x_1) \cdots (\forall x_n)[\underline{\mathrm{dom}\, f}\langle x_1, \ldots, x_n\rangle \leftrightarrow \underline{R}\langle f(x_1, \ldots, x_n)\rangle]$

yields *$(\mathrm{dom}\, f) = \mathrm{dom}\, {}^*f$.

To show that *$(\mathrm{range}\, f) = \mathrm{range}\, {}^*f$ is a little tricky, and so we consider the case $n = 1$ first. The trick is in noticing that there is a Skolem function ψ defined on range f so that $f(\psi(b)) = b$ for each $b \in$ range f. Thus the sentences

(5.9) $(\forall x)(\forall y)[\underline{f}(x) = y \to \underline{\mathrm{range}\, f}\langle y\rangle]$

and

(5.10) $(\forall y)[\underline{\mathrm{range}\, f}\langle y\rangle \to \underline{f}(\underline{\psi}(y)) = y]$

are true in $\mathscr{R}$. By transfer, the first says that $y \in$ range *f implies that $y \in$ *$(\mathrm{range}\, f)$, and the second yields the inverse implication. The general case is now clear, and is left to the reader [see sentences (4.9)].   □

5.7 Proposition If f and g are functions of n variables on R, then for $x = \langle x_1, \ldots, x_n\rangle$ we have

(i) *$(f + g)(x) = {}^*f(x) + {}^*g(x)$ and *$(f \cdot g)(x)$
                    $= {}^*f(x) \cdot {}^*g(x)$ when $x \in \mathrm{dom}\, {}^*f \cap \mathrm{dom}\, {}^*g$,

(ii) *$|f(x)| = |{}^*f(x)|$ when $x \in \mathrm{dom}\, {}^*f$.

Proof: (i) follows from Exercise 6(b) and by transfer of

(5.11) $(\forall x_1) \cdots (\forall x_n)[\underline{\mathrm{dom}\, f} \cap \underline{\mathrm{dom}\, g}\langle x_1, \ldots, x_n\rangle \to (\underline{f + g})(x_1, \ldots, x_n)$
                            $= \underline{f}(x_1, \ldots, x_n) + \underline{g}(x_1, \ldots, x_n)]$;

a similar sentence holds for the product.

(ii) follows by transfer of the sentences

(5.12) $(\forall x)[x \geq 0 \to |x| = x]$

and

(5.13) $(\forall x)[x < 0 \to |x| = -x]$.   □

5.8 Proposition

(i) $*\varnothing = \varnothing$.

(ii) If A and B are two sets in R^n then $*(A \cup B) = *A \cup *B$, $*(A \cap B) = *A \cap *B$, and $*(A') = (*A)'$.

(iii) Let $A_i \, (i \in I)$ be a family of sets in R^n. Then $\bigcup *A_i \, (i \in I) \subseteq *[\bigcup A_i (i \in I)]$ and $\bigcap *A_i \, (i \in I) \supseteq *[\bigcap A_i (i \in I)]$.

Proof: (i) Since $\chi_\varnothing$ is identically zero, it follows from Proposition 5.5 that $*\chi_\varnothing = \chi_{*\varnothing}$ is identically zero, so $*\varnothing$ is empty.

(ii) Using Propositions 5.5 and 5.7, we have

$$
\begin{aligned}
\chi_{*(A \cup B)} &= *\chi_{A \cup B} \\
&= *(\chi_A + \chi_B - \chi_A \chi_B) \\
&= *\chi_A + *\chi_B - *\chi_A *\chi_B \\
&= \chi_{*A} + \chi_{*B} - \chi_{*A}\chi_{*B} \\
&= \chi_{*A \cup *B},
\end{aligned}
$$

with a similar proof for the intersection (Exercise 1). That $*(A') = (*A)'$ follows directly from Proposition 5.5.

(iii) Let $j \in I$ be fixed. Then $[\bigcup A_i \, (i \in I)]$ is the name of a set in R^n and the sentence

$$(\forall x)[x \in A_j \to x \in [\bigcup A_i \, (i \in I)]$$

is true in $\mathscr{R}$. By transfer we see that $*A_j \subseteq *[\bigcup A_i \, (i \in I)]$. This is true for each $j \in I$, and the result for unions follows by elementary set theory. The result for intersections is similarly established (Exercise 1). $\square$

We conclude these examples by indicating how to prove Theorem 1.7 using the transfer principle. The field properties are easy. For example, the distributive law follows by transfer of the sentence

(5.14) $(\forall x)(\forall y)(\forall z)[\underline{R}\langle x\rangle \wedge \underline{R}\langle y\rangle \wedge \underline{R}\langle z\rangle \to x \cdot (y + z) = x \cdot y + x \cdot z]$,

and the existence of multiplicative inverses follows by transfer of

(5.15) $(\forall x)[x \neq 0 \to x\psi(x) = 1]$,

where $\psi(x)$ is the Skolem function defined on the nonzero elements of R by $\psi(x) = x^{-1}$. To prove the law of trichotomy one can use parts (i) and (ii) of Proposition 5.8. The details are left to the reader (Exercise 2).

Exercises I.5

1. Finish the proofs for 5.8(ii) and (iii).
2. Finish the proof of Theorem 1.7 using the transfer principle.
3. Show, using the transfer principle, that if $B \neq \emptyset$ is an upper-bounded set in R and C is the set of upper bounds of B with least element c, then $*C$ is the set of upper bounds of $*B$ and c is the least upper bound of $*B$.
4. Use the transfer principle to show that for each $x \in *R$ there is an $m \in *N$ so that $m \geq x$ (i.e., the $*$-Archimedean property holds in $*\mathscr{R}$).
5. Let f be a function on R which is continuous at each point $x \in R$. Show that for each $x \in *R$ and each $\varepsilon > 0$ in $*R$ there is a $\delta > 0$ in $*R$ so that if $y \in *R$ and $|x - y| < \delta$ then $|*f(x) - *f(y)| < \varepsilon$. In this case we say that $*f$ is $*$-continuous at each $x \in *R$.
6. Let A and B be subsets of R. Use the transfer principle directly to show that

   (a) if $A \subseteq B$ then $*A \subseteq *B$,
   (b) $*(A \cap B) = *A \cap *B$,
   (c) $*(A \cup B) = *A \cup *B$,
   (d) $*(A') = (*A)'$.

7. Use the transfer principle to establish Exercises 9 and 10 in § I.2.
8. Use the transfer principle to show that if $f(x) = \sin x$ then $*f$ is periodic with period 2π.
9. Let S be a real-valued function defined on the natural numbers N (a sequence). Suppose that for each $m \in N$ there is an $n \in N$ so that $S(n) \geq m$. Using the transfer principle, show that for each $m \in *N$ there is an $n \in *N$ so that $*S(n) \geq m$.
10. Use the transfer principle to show that if $|f(x)| \leq M$ for all $x \in \text{dom } f$, then $|*f(x)| \leq M$ for all $x \in \text{dom } *f$.
11. Use the transfer principle to show that if f is a real-valued function on R and $c \in R$, then $*\{x \in R : f(x) > c\} = \{x \in *R : *f(x) > c\}$.
12. Show that if P and Q are relations on R and $P \neq Q$ then $*P \neq *Q$.

I.6 Infinite Numbers, Infinitesimals, and the Standard Part Map

We have already noted in constructing $*R$ that it contains numbers not in R [we are regarding R as embedded as a subset of $*R$; i.e., R and $(R)_*$ are identified]. For example, the numbers $\omega = [\langle 1, 2, 3, \ldots \rangle]$ and $1/\omega$ are not in R. They are called non-standard numbers.

6.1 Definition A hyperreal number is called *standard* if the number is in R and *non-standard* otherwise. An n-tuple $\langle a^1, \ldots, a^n \rangle$ is standard if each $a^i \in R$ and non-standard otherwise.

Since $*R$ is called the set of nonstandard real numbers, we use a hyphen to indicate elements of $*R - R$. We further classify the numbers in $*R$ by the following definition.

6.2 Definition

(i) A number $s \in *R$ is *infinite* if $n < |s|$ for all standard natural numbers n.

(ii) A number $s \in *R$ is *finite* if $|s| < n$ for some standard natural number n.

(iii) A number $s \in *R$ is *infinitesimal* if $|s| < 1/n$ for all standard natural numbers n.

In the construction of §I.1, we see that the number $\omega = [\langle 1, 2, 3, \ldots \rangle]$ is infinite since, for any $r \in R$, $\{i \in N : i > r\}$ is cofinite and thus in $\mathcal{U}$, showing that $\omega > r$ for any standard $r > 0$. There are many more infinite numbers. For example, $\omega + r$ is infinite for *any* positive $r \in *R$ since $\omega + r > \omega$ (we are using the properties of linear ordering in $*R$). Similarly $1/\omega = [\langle 1, \frac{1}{2}, \frac{1}{3}, \ldots \rangle]$ is an infinitesimal number, but again there are many more since the reciprocal of any infinite number is infinitesimal (Exercise 1). The number 0 is the only standard infinitesimal number. Clearly, every infinitesimal number is finite, but the sets of finite and infinite numbers are disjoint.

6.3 Theorem

(i) The finite and infinitesimal numbers in $*R$ each form subrings of $*R$; i.e., sums, differences, and products of finite (infinitesimal) numbers are finite (infinitesimal).

(ii) The infinitesimals are an ideal in the finite numbers; i.e., the product of an infinitesimal and a finite number is infinitesimal.

Proof: (i) Let ε and δ be infinitesimal, and let $r > 0$ be standard. Then $|\varepsilon| < r/2$ and $|\delta| < r/2$ and so $|\varepsilon + \delta| < r$ and $|\varepsilon - \delta| < r$. Also $|\varepsilon|$ and $|\delta|$ are both $< \sqrt{r}$, so that $|\varepsilon\delta| < r$. Here we have used the familiar properties of the absolute value for numbers in $*R$ which are valid by transfer (see Exercise 3). This shows that the infinitesimals in $*R$ form a subring. A similar argument works for the finite numbers (Exercise 4).

(ii) Let ε be infinitesimal and b be finite. Then $|b| < s$ for some standard $s > 0$. Also, $|\varepsilon| < r/s$ for any standard $r > 0$. Therefore $|\varepsilon b| < r$, and so εb is infinitesimal. □

We next introduce two important equivalence relations and the associated notions of monad and galaxy. Monads are central to the nonstandard treatment of convergence and continuity.

6.4 Definition Let x and y be numbers in $*R$.

(i) x and y are *near* or *infinitesimally close* if $x - y$ is infinitesimal. We write $x \simeq y$ if x and y are near and $x \not\simeq y$ otherwise. The *monad* of x is the set $m(x) = \{y \in *R : x \simeq y\}$.

(ii) x and y are *finitely close* if $x - y$ is finite. We write $x \sim y$ if x and y are finitely close and $x \nsim y$ otherwise. The *galaxy* of x is the set $G(x) = \{y \in *R : x \sim y\}$.

The monadic and galactic structure of $*R$ is easily visualized. To aid in the visualization, we present the following facts. Clearly $m(0)$ is the set of infinitesimals and $G(0)$ is the set of finite numbers. It follows easily from 6.3 that any two monads $m(x)$ and $m(y)$ are either equal (if $x \simeq y$) or disjoint (if $x \not\simeq y$) and the relation $\simeq$ is an equivalence relation on $*R$. Likewise any two galaxies $G(x)$ and $G(y)$ are either equal (if $x - y$ is finite) or disjoint. It is equally easy to prove the somewhat disconcerting fact that between any two disjoint monads or galaxies is a third, disjoint from the first two. If $x \not\simeq 0$ we see easily that $m(x)$ is a translate of $m(0)$; i.e., for any x,

$$m(x) = \{y \in *R : y = x + z, z \in m(0)\}.$$

Similarly

$$G(x) = \{y \in *R : y = x + z, z \in G(0)\}.$$

We leave the proofs of these facts as exercises. These remarks show that the structure of $*R$ with respect to infinite, finite, and infinitesimal numbers is somewhat complicated but easily visualized. Some authors say that x is infinitely close to y if $x - y$ is infinitesimal.

We continue with the following basic fact about the structure of $*R$.

6.5 Theorem If $\rho \in *R$ is finite, there is a unique standard real number $r \in R$ with $\rho \simeq r$; i.e., every finite number is near a unique standard number.

Proof: Let $A = \{x \in R : \rho \leq x\}$ and $B = \{x \in R : x < \rho\}$. Since ρ is finite, there exists a standard number s such that $-s < \rho < s$. It follows that B is nonempty and has an upper bound. Let r be the least upper bound of B (the existence of r is assured by the completeness of R). For each $\varepsilon > 0$ in R, $(r + \varepsilon) \in A$ and $(r - \varepsilon) \in B$, so $r - \varepsilon < \rho \le r + \varepsilon$, and hence $|r - \rho| \le \varepsilon$. It follows that $r \simeq \rho$. If $r_1 \simeq \rho$ then $|r_1 - r| \le |r_1 - \rho| + |\rho - r| < 2\varepsilon$ for each standard $\varepsilon > 0$, whence $r = r_1$. □

6.6 Definition If $\rho \in {}^*R$ is finite, the unique standard number $r \in R$ such that $\rho \simeq r$ is called the *standard part* of ρ and is denoted by $\text{st}(\rho)$ or ${}^\circ\rho$. This defines a map $\text{st}: G(0) \to R$ called the *standard part map*.

Clearly st maps $G(0)$ onto R since $\text{st}(r) = r$ when $r \in R$. That the map also preserves algebraic structure is shown by the following theorem.

6.7 Theorem The map st is an order-preserving homomorphism of $G(0)$ onto R, i.e.,

  (i) $\text{st}(x \pm y) = \text{st}(x) \pm \text{st}(y)$,
  (ii) $\text{st}(xy) = \text{st}(x)\,\text{st}(y)$,
  (iii) $\text{st}(x/y) = \text{st}(x)/\text{st}(y)$ if $\text{st}(y) \ne 0$,
  (iv) $\text{st}(x) \le \text{st}(y)$ if $x \le y$.

Proof: Let $x = {}^\circ x + \varepsilon$, $y = {}^\circ y + \delta$ with ε and δ infinitesimal. Then $x \pm y = ({}^\circ x \pm {}^\circ y) + (\varepsilon \pm \delta)$, which establishes (i) using 6.3. Parts (ii) and (iii) are left to the reader (Exercise 6). To prove (iv), we have ${}^\circ x + \varepsilon \le {}^\circ y + \delta$, so that ${}^\circ x \le {}^\circ y + (\delta - \varepsilon) < {}^\circ y + r$ for any positive $r \in R$; from this we conclude that ${}^\circ x \le {}^\circ y$. □

6.8 Corollary The quotient field $G(0)/m(0)$ is isomorphic to the standard field $\mathscr{R}$.

Proof: $m(0)$ is the kernel of the linear (over R) map st, i.e., $m(0) = \{x \in G(0) : \text{st}(x) = 0\}$. □

6.9 Corollary If x, x', y, and y' are finite and $x \simeq x'$, $y \simeq y'$ then

  (i) $x \pm y \simeq x' \pm y'$,
  (ii) $xy \simeq x'y'$,
  (iii) $x/y \simeq x'/y'$ if $y \not\simeq 0$ (and hence $y' \not\simeq 0$).

From Definition 6.2 we see that the set of infinite hyperreal numbers is the complement of the set $G(0)$ of finite numbers. Since various subsets of the set of infinite numbers (especially the set of infinite integers) will occur frequently in the sequel, we adopt the following definition.

6.10 Definition Given a set $A \subset R$, the set of infinite numbers in $*A$ is the set $*A_\infty = *A \cap (*R - G(0))$.

6.11 Theorem If $A \subseteq N$ and A is infinite, then $*A$ contains infinite natural numbers, i.e., $*A \cap *N_\infty \neq \emptyset$.

Proof: For each $n \in N$, there is an element $k \in A$ with $k \geq n$, and so we may define a Skolem function $\psi: N \to A$ with $\psi(n) \geq n$. Thus the sentence $(\forall n)[\underline{N}\langle n \rangle \to \underline{A}\langle \psi(n) \rangle \wedge n \leq \psi(n)]$ is true in R. By transfer, $*\psi(n) \in *A$ and $n \leq *\psi(n)$ for all $n \in *N$ including $n = \omega \in *N_\infty$. Thus, $*\psi(\omega) \in *A \cap *N_\infty$.   □

Note that the proof of Theorem 6.11 shows that $*A$ contains arbitrarily large infinite natural numbers.

Exercises I.6

1. Show that the reciprocal of an infinite number is infinitesimal and the reciprocal of a nonzero infinitesimal number is infinite.
2. Show that if r is an infinitesimal standard number, then $r = 0$.
3. Write simple sentences for $L_{\mathscr{R}}$ which yield the properties of the absolute value function on $*R$ used in the proof of Theorem 6.3(i) for infinitesimal numbers.
4. Prove Theorem 6.3(i) for finite numbers.
5. Fill in the details in the remarks following Definition 6.4.
6. Prove Theorem 6.7, parts (ii) and (iii).
7. Show that it does *not* follow from $°x \leq °y$ that $x \leq y$ in $G(0)$. What can be said if $°x < °y$?
8. Prove Corollary 6.9.
9. Show that Corollary 6.9(iii) need not be true if $y \simeq 0$.
10. Start with the fact that every finite element of $*R$ is near a standard $r \in R$ and show that R is complete.
11. Show that if $x \in *R$ then $m(x) = \bigcup \{y \in *R : |x - y| < \varepsilon, \varepsilon > 0 \text{ infinitesimal in } *R\} = \bigcap \{y \in *R : |x - y| < \varepsilon, \varepsilon > 0 \text{ in } R\}$.
12. Show that if $x_i \in *R$, $1 \leq i \leq n$, then $\sqrt{x_1^2 + \cdots + x_n^2} \simeq 0$ iff $x_i \simeq 0$ for all i, $1 \leq i \leq n$.
13. Show that if a and b are finite numbers in $*R$ with $b \neq 0$, and n is infinite in $*N$, then $a + nb$ is infinite.

I.7 The Hyperintegers

The set of integers, which we denote by Z, and the set N of natural numbers play central roles in analysis. We therefore pay particular attention to the structure of the $*$-transforms $*Z$ and $*N$ of these sets; we will call elements of $*Z$ and $*N$ *hyperintegers* and *hypernatural numbers*, respectively. In the literature, $*Z$ and $*N$ are often called the nonstandard integers and nonstandard natural numbers, respectively.

The first obvious fact is the following.

7.1 Proposition $*Z$ is a linearly ordered subring of $*R$.

Proof: To show that $*Z$ is a subring of $*R$, we need only check that it is closed under addition and multiplication. This fact follows from the interpretation in $*R$ of the $*$-transform of the simple sentence

$$(7.1) \qquad (\forall x)(\forall y)[\underline{Z}\langle x\rangle \wedge \underline{Z}\langle y\rangle \to \underline{Z}\langle x+y\rangle \wedge \underline{Z}\langle x\cdot y\rangle],$$

which is true in $\mathscr{R}$. Finally, notice that $*Z$ inherits the linear ordering on $*R$. □

In $\mathscr{R}$ there is a greatest integer function $[\,\cdot\,]:R \to Z$ which satisfies

$$(7.2) \qquad\qquad [x] \le x < [x] + 1$$

for all $x \in R$. Therefore the extended function $*[\,\cdot\,]:*R \to *Z$ satisfies $*[x] \le x < *[x] + 1$ for all $x \in *R$ by the transfer principle. Thus we have

7.2 Proposition For each $x \in *R$ there is an element $k \in *Z$ so that $k \le x < k + 1$.

7.3 Corollary There are positive and negative infinite integers.

Proof: If x is positive infinite then the hyperinteger $k + 1$ of Proposition 7.2 is positive infinite and the hyperinteger $-(k + 1)$ is negative infinite. □

7.4 Corollary If $x \in *R$, there is an $n \in *N$ so that $|x| < n$.

The following result shows that the hyperintegers in $*Z$ are a unit distance apart.

7.5 Proposition For each $n \in *Z$, $n + 1$ is the smallest hyperinteger greater than n.

Proof: The simple sentence

(7.3) $(\forall x)(\forall y)[\underline{Z}\langle x\rangle \wedge \underline{Z}\langle y\rangle \wedge (x \leq y \leq x + 1) \wedge (y \neq x) \to (y = x + 1)]$

is true in $\mathscr{R}$. The interpretation of its *-transform in *$\mathscr{R}$ yields the desired conclusion. $\square$

7.6 Corollary $*Z \cap G(0) = Z$ (i.e., any finite hyperinteger is an ordinary integer).

Proof: Let k be a finite hyperinteger. Then st(k) is a real number and so $n \leq \text{st}(k) < n + 1$ for some $n \in Z$. It is easy to see (Exercise 1) that $0 \leq |n - k| < 1$. But $|n - k| \in *Z$ and so $n = k$ by Proposition 7.5. $\square$

7.7 Corollary If $x \in *Z$, then $*Z \cap m(x) = \{x\}$.

Proof: Exercise 2. $\square$

If we let Q denote the standard set of rational numbers, then the set $*Q$ will be called the set of *hyperrationals* or nonstandard rationals. In contrast to Corollary 7.7 we see that if $x \in *Q$ then $*Q \cap m(x)$ contains many other hyperrationals distinct from x; for example, if $\omega \in *N_\infty$ then $1/\omega \in *Q \cap m(0)$ (proof?). An interesting exercise, which we leave to the reader, shows that, in analogy with Corollary 6.8, the real number system is isomorphic to $[*Q \cap G(0)]/m(0)$. Notice that only the rational numbers are used in defining $[*Q \cap G(0)]/m(0)$. Although this would not be a recommended way of defining the real numbers from the rationals, the result is a prototype of many results in nonstandard analysis which construct standard mathematical structures from nonstandard structures.

We end this section with some remarks which will clarify the nature of the *-mapping and the transfer principle. Similar considerations will be crucial in correctly applying the more powerful transfer principle of Chapter II (see Remark 3.5 of §II.3 and §II.6 in Chapter II).

7.8 Remarks on Sets Which Are Nonstandard Extensions of Standard Sets
We first show that there are subsets of $*R$ which are not the *-transforms (i.e., nonstandard extensions) of sets in R. A typical example is the set R itself, regarded as embedded in $*R$. For suppose that $*A = R$ for some subset $A \subset R$. Two cases are possible:

(i) A is bounded above by a number $a \in R$. But in this case the sentence $(\forall x)[\underline{A}\langle x\rangle \to x \leq a]$ is true in $L_\mathscr{R}$. By transfer, $(\forall x)[*\underline{A}\langle x\rangle \to x \leq a]$, i.e., every element of $*A$ is $\leq a$, and $*A$ cannot equal R;

(ii) *A* is not bounded above. Then for all $x \in R$ there is a $y \in A$ with $y \geq x$. Thus there is a Skolem function $\psi : R \to A$ so that the sentence $(\forall x)[\underline{R}\langle x \rangle \to \psi(x) \geq x \wedge \underline{A}\langle \psi(x)\rangle]$ is true in $L_{\mathscr{R}}$. By transfer, $(\forall x)[{}^*\underline{R}\langle x \rangle \to {}^*\psi(x) \geq x \wedge {}^*\underline{A}\langle \psi(x)\rangle]$. In particular, if x is an infinite natural number then there is an element $y = {}^*\psi(x) \geq x$. Since $y \in {}^*A$ we see that ${}^*A$ contains infinite numbers and cannot equal R. Thus there is no $A \subset R$ so that ${}^*A = R$. A similar argument shows that there is no $A \subset R$ so that ${}^*A = N$. Thus the $*$-mapping of Definition 2.2 does not map the collection of all subsets of R onto the collection of all subsets of ${}^*R$ but only onto a subcollection of them.

It is obvious from Definition 5.3 that the $*$-transform ${}^*\Phi$ of a sentence Φ in $L_{\mathscr{R}}$ can contain only the names of the $*$-transforms of sets and n-ary relations on R. A lack of attention to this fact can lead to an incorrect understanding of the transfer principle. As an example, recall that R is Archimedean; this means that given any $x \in R$ there is an $n \in N$ so that $|x| \leq n$. We might naively expect that ${}^*R$ is Archimedean by transfer; i.e., for all $x \in {}^*R$ there is an $n \in N$ so that $|x| \leq n$. But this statement is obviously false, as we see by taking x to be an infinite integer. The mistake is in transferring the sentence but forgetting to replace N by ${}^*N$, thus leaving the name of the set N in the transferred statement. The correct transfer of the Archimedean property is in Corollary 7.4.

Even though only sets which are $*$-transforms of standard sets arise in the application of the transfer principle, other subsets of ${}^*R$ occur regularly in the nonstandard characterization of standard concepts. For example, we will show (Proposition 8.1) that a sequence $\langle s_n \rangle$ converges to the limit L if and only if ${}^*s_n$ is infinitesimally close to L for all $n \in {}^*N_\infty$. Neither ${}^*N_\infty$ nor the monad of zero is the $*$-transform of a standard set.

Exercises I.7

1. Show that $0 \leq |n - k| < 1$ in the proof of Corollary 7.6.
2. Prove Corollary 7.7.
3. Show that the real number system is isomorphic to $[{}^*Q \cap G(0)]/m(0)$.
4. Show that there is no $A \subset R$ so that ${}^*A = {}^*N_\infty$.
5. Show that there is no function f on R so that ${}^*f = \chi_R$.
6. Show that there is no $A \subset R$ so that ${}^*A = m(0)$.
7. Show that there is no $A \subset R$ so that ${}^*A = G(0)$.
8. Show that if A is a finite set $\{a_1, a_2, \ldots, a_n\}$ in R then ${}^*A = A$.
9. Show that if A is an unbounded set in R then ${}^*A \neq A$.
10. Show that if $x < y$ in R and $t \in {}^*R$ with $x \leq \mathrm{st}(t) < y$ then $|x - t| < y - x$ and $^\circ|x - t| < y - x$.

I.8 Sequences and Series

The first task in applying nonstandard analysis to a given theory is to find nonstandard equivalents for the basic definitions in the theory. The non-standard equivalents can then be applied to produce (often shorter) proofs of standard results. In this section we will illustrate these remarks by considering the basic theory of limits for real sequences and series. In this and the next few sections, nonstandard equivalents of the standard definitions will be presented as propositions. These results are due to Robinson [40, 42]. Familiarity with the standard definitions is assumed.

Let $S: N \to R$ be a standard sequence. As usual we write $s(n) = s_n$ and denote the sequence by $\langle s_n : n \in N \rangle$ or simply $\langle s_n \rangle$. The sequence $s: N \to R$ has a $*$-transform $*s: *N \to *R$ and we let $*s(n) = *s_n$ for $n \in *N$. From 2.3 or 5.5 we see that $*s_n = s_n$ for $n \in N$. Applying the remark preceding Theorem 2.3, we see, for example, that if $\omega = [\langle 2, 4, \dots, 2^n, \dots \rangle]$ then $*s_\omega = [\langle s_2, s_4, \dots, s_{2^n}, \dots \rangle]$.

8.1 Proposition The sequence $\langle s_n \rangle$ converges to L iff $*s_n \simeq L$ for all infinite n.

Proof: Recall the condition for convergence of $\langle s_n \rangle$ to L:

(8.1)   Given $\varepsilon > 0$ in R there is a $k \in N$ (depending in general
               on ε) so that $|s_n - L| < \varepsilon$ for all $n > k$.

Suppose $\langle s_n \rangle$ converges to L, let $\varepsilon > 0$ be given, and find the corresponding k from (8.1). Then the sentence

(8.2) $(\forall n)[\underline{N}\langle n \rangle \wedge n > k \to |\underline{s}_n - L| < \varepsilon]$

is true in $\mathcal{R}$. By transfer,

(8.3) $(\forall n)[*\underline{N}\langle n \rangle \wedge n > k \to |*\underline{s}_n - L| < \varepsilon]$,

and so if $n \in *N$ and $n > k$, then $|*s_n - L| < \varepsilon$. But *all* infinite $n \in *N$ are larger than k, so $|*s_n - L| < \varepsilon$ for all infinite n in $*N$. The core of our argument is that the latter conclusion could have been derived for *any* standard $\varepsilon > 0$. Thus $|*s_n - L| \simeq 0$ for all infinite $n \in *N$.

Conversely, suppose that $\langle s_n \rangle$ does not converge to L. Then there is a standard $\varepsilon > 0$ and a Skolem function $\psi: N \to N$ satisfying $\psi(k) \geq k$ and $|s_{\psi(k)} - L| \geq \varepsilon$ for all $k \in N$. Thus the sentence

(8.4) $(\forall k)[\underline{N}\langle k \rangle \to \underline{\psi}(k) \geq k \wedge |\underline{s}_{\psi(k)} - L| \geq \varepsilon]$

is true in $\mathcal{R}$. By transfer,

(8.5) $(\forall k)[*\underline{N}\langle k \rangle \to *\underline{\psi}(k) \geq k \wedge |*\underline{s}_{*\psi(k)} - L| \geq \varepsilon]$, and so $\left|{}^*s_{*\psi(k)} - L\right| \geq \varepsilon$ for all $k \in {}^*N$ and, in particular, for an infinite $k = \omega$. Since $n = {}^*\psi(\omega) \geq \omega$ is infinite and $\left|{}^*s_n - L\right| \geq \varepsilon$, ${}^*s_m \simeq L$ is not true for all infinite m. $\square$

If $\langle s_n \rangle$ converges to L we write $\lim_{n \to \infty} s_n = L$. We then get

8.2 Theorem If $\lim_{n \to \infty} s_n = L$ and $\lim_{n \to \infty} t_n = M$ then

   (i) $\lim_{n \to \infty} (s_n + t_n) = L + M$,
   (ii) $\lim_{n \to \infty} (s_n t_n) = LM$,
   (iii) $\lim_{n \to \infty} (s_n/t_n) = L/M$ if $M \neq 0$.

Proof: To prove (i), we have $s_n \simeq L$, $t_n \simeq M$, and hence $s_n + t_n \simeq L + M$ for any infinite n by Corollary 6.9. The proofs of (ii) and (iii) are left as exercises. $\square$

By variants of the arguments in the proof of Proposition 8.1 we can establish the following results.

8.3 Proposition The sequence $\langle s_n \rangle$ is a Cauchy sequence iff ${}^*s_n \simeq {}^*s_m$ for all infinite n and m.

Proof: Recall the standard condition for $\langle s_n \rangle$ to be a Cauchy sequence:

(8.6)   Given $\varepsilon > 0$ in R there is a $k \in N$ so that $|s_n - s_m| < \varepsilon$
             for all $n, m > k$.

Then proceed as in the proof of Proposition 8.1 (Exercise 2). $\square$

8.4 Proposition The sequence $\langle s_n \rangle$ is bounded iff ${}^*s_n$ is finite for all infinite n.

Proof: If $\langle s_n \rangle$ is bounded then there is a $k \in N$ so that the sentence

(8.7) $(\forall n)[\underline{N}\langle n \rangle \to |s_n| \leq k]$

is true in $\mathcal{R}$. By transfer, $\left|{}^*s_n\right| \leq k$ for all $n \in {}^*N$, and hence $\left|{}^*s_n\right|$ is finite for all infinite n.

Conversely if $\langle s_n \rangle$ is not bounded then there is a Skolem function $\psi: N \to N$ satisfying $\psi(k) > k$, $k \in N$, so that $\left|s_{\psi(k)}\right| > k$ for all $k \in N$. By transfer of the appropriate sentence (which the reader is invited to write down), $\left|{}^*s_{*\psi(k)}\right| > k$ for all $k \in {}^*N$ and, in particular, $\left|{}^*s_n\right|$ is infinite if $n = {}^*\psi(k)$ and k is infinite.
$\square$

Using Proposition 8.3, one can prove the standard result that any Cauchy sequence $\langle s_n \rangle$ is bounded (Exercise 3).

8.5 Theorem The sequence s_n converges iff it is a Cauchy sequence.

Proof: If $\langle s_n \rangle$ converges to L then $*s_n \simeq L \simeq *s_m$ for all infinite n, m by 8.1, so $\langle s_n \rangle$ is a Cauchy sequence by 8.3.

Conversely if $\langle s_n \rangle$ is a Cauchy sequence then $\langle s_n \rangle$ is bounded and so $*s_n$ is finite for all infinite n. Define $L = \mathrm{st}(*s_\omega)$, where ω is a specific infinite natural number. Then $*s_n \simeq *s_\omega \simeq L$ for all infinite n by 8.3, and so $\langle s_n \rangle$ converges to L by 8.1. □

8.6 Corollary A monotonic bounded sequence $\langle s_n \rangle$ converges.

Proof: We may assume that the sequence is increasing, and we need only show that $\langle s_n \rangle$ is a Cauchy sequence. If not then there exists an $\varepsilon > 0$ in R and a Skolem function $\psi : N \to N$ so that $\psi(1) = 1$, $\psi(n+1) > \psi(n)$, $s_{\psi(k)} + \varepsilon < s_{\psi(k+1)}$ for $k \in N$, and hence $s_{\psi(k)} > s_1 + k\varepsilon$ for all $k \in N$. By transfer (of what sentence?), $*s_{*\psi(k)} > s_1 + k\varepsilon$ for $k \in *N$, and so $*s_{*\psi(k)}$ is infinite if k is infinite. By 8.4, $\langle s_n \rangle$ is not bounded (contradiction). □

The notion of a limit point of a sequence $\langle s_n \rangle$ can be treated in much the same way as we have treated limits.

8.7 Proposition L is a limit point of the sequence $\langle s_n \rangle$ if and only if $s_n \simeq L$ for some infinite n.

Proof: Suppose that L is a limit point of $\langle s_n \rangle$. The standard definition of a limit point states that for a given $\varepsilon > 0$ in R and $k \in N$ there is an $n \in N$ with $n \geq k$ so that $|s_n - L| < \varepsilon$. Thus there is a Skolem function $\psi : R^+ \times N \to N$ (R^+ the positive reals) with $\psi(\varepsilon, k) \geq k$ so that the sentence

(8.8) $(\forall \varepsilon)(\forall k)[R^+ \langle \varepsilon \rangle \wedge \underline{N}\langle k \rangle \to \underline{\psi}(\varepsilon, k) \geq k \wedge |\underline{s}_{\psi(\varepsilon, k)} - L| < \varepsilon]$

is true in $\mathscr{R}$. By transfer, for all $\varepsilon \in *R^+$ and $k \in *N$ (and, in particular, $k \in *N_\infty$), $|*s_n - L| < \varepsilon$ if $n = *\psi(\varepsilon, k) \geq k$; thus $*s_n \simeq L$ if ε is infinitesimal.

Conversely, if L is not a limit point of $\langle s_n \rangle$ then there is a standard $\varepsilon > 0$ and $k \in N$ so that $|s_n - L| > \varepsilon$ for all $n \geq k$. By transfer of the appropriate sentence (Exercise 4), $|*s_n - L| \geq \varepsilon$ for all $n \geq k$ and, in particular, for all infinite n; i.e., $s_n \not\simeq L$ for any infinite n. □

8.8 Theorem (Bolzano–Weierstrass) Every bounded sequence $\langle s_n \rangle$ has a limit point L.

Proof: If $\langle s_n \rangle$ is bounded then $^*s_n$ is finite for all infinite n by Proposition 8.4. If $L = \mathrm{st}(^*s_\omega)$ for some infinite $\omega \in {}^*N$, then $^*s_\omega \simeq L$ and L is a limit point by Proposition 8.7. $\square$

8.9 Examples

1. If $s_n = (2n^2 + 3n)/(5n^2 + 1)$ for $n \in N$, then by the transfer principle, $^*s_n = (2n^2 + 3n)/(5n^2 + 1)$ for $n \in {}^*N$. Thus, for all infinite $n \in {}^*N$, $s_n = (2 + 3/n)/(5 + 1/n^2) \simeq \frac{2}{5}$, so $\langle s_n \rangle$ converges to $\frac{2}{5}$.

2. If $\lim_{n \to \infty} s_n = L \neq 1$ and $t_n = (1 + s_n)/(1 - s_n)$, then $t_n \simeq (1 + L)/(1 - L)$ for all infinite n, and so $\lim_{n \to \infty} t_n = (1 + L)/(1 - L)$.

3. Let s_n be defined recursively by $s_{n+1} = \frac{1}{2}(s_n + a/s_n)$, $n \geq 1$, $a > 0$, $s_1 \geq \sqrt{a}$. It is easy to see that $\langle s_n \rangle$ is a decreasing and positive sequence, i.e., $0 \leq s_{n+1} \leq s_n$, and $s_n \geq \sqrt{a}$ for all n (check); hence s_n converges to a limit L by Corollary 8.6. Thus $L \simeq s_{n+1} \simeq \frac{1}{2}(s_n + a/s_n) \simeq \frac{1}{2}(L + a/L)$ if n is infinite, so $2L = L + a/L$, and hence $L = \sqrt{a}$.

4. Let $s_n = (-1)^n(1 - 1/n)$. Then $s_n = 1 - 1/n \simeq 1$ for all even infinite n and $s_n = 1/n - 1 \simeq -1$ for all odd infinite n. Thus the numbers 1 and -1 are the only limit points of $\langle s_n \rangle$ by Proposition 8.7.

The methods just developed can be used effectively in the study of double sequences. A double sequence is a mapping $s: N \times N \to R$; we write $s(n, m) = s_{nm}$ and denote the sequence by $\langle s_{nm} \rangle$. The sequence $\langle s_{nm} \rangle$ converges to L, and we write $\lim_{n,m \to \infty} s_{nm} = L$, if, given $\varepsilon > 0$ in R, there is a $k \in N$ so that $|s_{nm} - L| < \varepsilon$ if $n, m > k$. The *-transform of s yields the numbers $^*s(n, m) = {}^*s_{nm}$ $(n, m \in {}^*N)$. The proofs of the following results are analogous to those of 8.1 and 8.5 and are left as exercises.

8.10 Proposition $\lim_{n,m \to \infty} s_{nm} = L$ iff $^*s_{nm} \simeq L$ for all infinite n, m.

8.11 Proposition $\langle s_{nm} \rangle$ converges iff $^*s_{nm} \simeq {}^*s_{n'm'}$ for all infinite n, m, n', m' and then converges to $L = \mathrm{st}(^*s_{nm})$ for n, m infinite.

We may want to compute the limit of $\langle s_{nm} \rangle$ by first computing $\lim_{n \to \infty} s_{nm} = s_m$ and then computing $\lim_{m \to \infty} s_m$. It may happen that the above limits exist but $\langle s_{nm} \rangle$ is not convergent. For example, if $s_{nm} = n/(n + m)$ then $\lim_{n \to \infty} s_{nm} = 1$, $m \in N$, so $\lim_{m \to \infty} (\lim_{n \to \infty} s_{nm}) = 1$. If $\omega \in {}^*N$ is infinite, however, then $\omega/(\omega + \omega) \not\simeq \omega/(\omega + 2\omega)$ and so $\langle s_{nm} \rangle$ is not convergent by 8.11.

Notice that if $\lim_{n \to \infty} s_{nm} = s_m$ then

(8.9) $^*s_{nm} \simeq s_m$ for $n \in {}^*N_\infty$, $m \in N$.

If, moreover, $\lim_{m \to \infty} s_m = L$ then

(8.10) $^*s_m \simeq L$ for $m \in {}^*N_\infty$.

If in place of (8.9) we could establish

(8.11) $^*s_{nm} \simeq {}^*s_m$ for $n \in {}^*N_\infty$, $m \in {}^*N$

then (8.10) and (8.11) would yield $^*s_{nm} \simeq {}^*s_m \simeq L$ for all infinite n, m, and hence the convergence of $\langle s_{nm} \rangle$ to L. Condition (8.11) is equivalent to the uniformity in m of the convergence of the sequences $\langle s_{nm} : n \in N \rangle$, as shown by the following result.

8.12 Proposition $\langle s_{nm} : n \in N \rangle$ converges to s_m uniformly in m iff $^*s_{nm} \simeq {}^*s_m$ for all $n \in {}^*N_\infty$, $m \in {}^*N$.

   Proof: Recall that $\lim_{n \to \infty} s_{nm} = s_m$ uniformly in m iff, given $\varepsilon > 0$, there exists a $k \in N$, possibly depending on ε but not on m, so that $|s_{nm} - s_m| < \varepsilon$ for all $n \geq k$. Let $\varepsilon > 0$ be specified and find the corresponding k. Then the sentence

(8.12) $(\forall n)(\forall m)[\underline{N}\langle n \rangle \wedge \underline{N}\langle m \rangle \wedge n \geq k \to |\underline{s}_{nm} - \underline{s}_m| < \varepsilon]$

is true in $\mathscr{R}$. By transfer, $|^*s_{nm} - {}^*s_m| < \varepsilon$ if $n, m \in {}^*N$ with $n > k$, and in particular for all infinite n. The latter conclusion is valid for any standard $\varepsilon > 0$, and so $^*s_{nm} \simeq {}^*s_m$ for $n \in {}^*N_\infty$, $m \in {}^*N$.
   The converse is left to the reader.   □

   The preceding discussion yields the first part of the following result. The second is left as an exercise.

8.13 Theorem If $\lim_{n \to \infty} s_{nm} = s_m$, uniformly in m, and $\lim_{m \to \infty} s_m = L$, then $\lim_{n,m \to \infty} s_{nm} = L$. If, moreover, $\lim_{m \to \infty} s_{nm} = s_n$ exists for each $n \in N$, then $\lim_{n \to \infty} s_n$ exists and equals L.

   Note in passing that $\lim_{n,m \to \infty} s_{nm}$ may exist even though $\lim_{n \to \infty} s_{nm}$ does not exist; For example, let $s_{nm} = [(-1)^n + (-1)^m]/m$.
   We continue with a consideration of infinite series. Recall that the infinite series $\sum_{i=1}^\infty a_i$ converges (to L) if the sequence $s_n = \sum a_i (1 \leq i \leq n)$ converges (to L). Both sequences $\langle a_n \rangle$ and $\langle s_n \rangle$ have $*$-transforms $\langle {}^*a_n : n \in {}^*N \rangle$ and $\langle {}^*s_n : n \in {}^*N \rangle$. We will write $^*s_n = {}^*\sum {}^*a_i (1 \leq i \leq n)$, thus defining the non-standard "summation" operation on the right-hand side. This operation has

all of the familiar properties of ordinary summation, as we easily check by transfer from the properties of $\langle s_n \rangle$. For example, $*\sum *a_i(1 \leq i \leq m) - *\sum *a_i(1 \leq i \leq n) = *\sum *a_i(n + 1 \leq i \leq m)$ for $m > n$ in $*N$. From the previous results in this section we immediately obtain

8.14 Proposition

  (1) $\sum_{i=1}^{\infty} a_i$ converges to L iff $*\sum *a_i(1 \leq i \leq n) \simeq L$ for all $n \in *N_\infty$;

  (2) $\sum_{i=1}^{\infty} a_i$ converges iff $*\sum *a_i(n \leq i \leq m) \simeq 0$ for both m and n in $*N_\infty$;

  (3) If $a_n \geq 0$ then $\sum a_i(i \in N)$ converges iff $*\sum *a_i(i \in N)$ is finite for some $n \in *N_\infty$, in which case $*\sum *a_i(1 \leq i \leq n)$ is finite for all $n \in *N_\infty$.

The series $\sum a_i(i \in N)$ converges *absolutely* if $\sum |a_i|(i \in N)$ converges. The comparison test and its consequences, the ratio and root tests, are important in the standard theory of absolute convergence. The following result is a nonstandard version of the limit comparison test.

8.15 Theorem If $\sum b_i(i \in N)$ converges absolutely and $|*a_n| \leq |*b_n|$ for all $n \in *N_\infty$, then $\sum a_i(i \in N)$ converges absolutely.

Proof: There is a $k \in N$ so that if $n \geq k$ then $|a_n| \leq |b_n|$ (why?). By transfer of the appropriate sentence, $*\sum |*a_i|(n \leq i \leq m) \leq *\sum |*b_i|(n \leq i \leq m) \simeq 0$ if $m > n$ are in $*N_\infty$. $\square$

The notion of the limit of a sequence $\langle s_n \rangle$ can be extended to the notions of lim sup s_n and lim inf s_n, called the limit superior and limit inferior, respectively. Here, for a change, we define lim sup and lim inf using nonstandard notions, and show that these definitions coincide with (one of) the standard definitions in Proposition 8.17.

8.16 Definition Let $\langle s_n \rangle$ be a sequence in R. For lim sup s_n we consider three cases:

  (i) If $*s_n$ is positive infinite for some $n \in *N_\infty$, then lim sup $s_n = +\infty$.

  (ii) If $*s_n$ is negative infinite for all $n \in *N_\infty$, then lim sup $s_n = -\infty$.

  (iii) If neither case (i) nor case (ii) holds, then lim sup $s_n = \sup\{\text{st}(*s_n) : n \in *N_\infty, *s_n \text{ finite}\}$.

We define lim inf s_n in a similar way, or equivalently we set lim inf $s_n = $ lim sup$(-s_n)$.

If neither case (i) nor case (ii) of Definition 8.16 holds, then, as in Proposition 8.4, the sequence $\langle s_n \rangle$ is bounded above. If $r = \sup\{\text{st}(*s_n): n \in {}^*N_\infty, *s_n \text{ finite}\}$, then r is a limit point of $\langle s_n \rangle$ (why?), so by Proposition 8.7, $r = \text{st}(*s_n)$ for some $n \in {}^*N_\infty$. Thus $r = \max\{\text{st}(*s_n): n \in {}^*N_\infty, *s_n \text{ finite}\} = \limsup s_n$. Clearly, r is the largest limit point of $\langle s_n \rangle$.

8.17 Proposition Let $\langle s_n \rangle$ be a sequence in R and let $u_n = \sup\{s_k: k \geq n$ in $N\}$ and $v_n = \inf\{s_k: k \geq n$ in $N\}$ for each $n \in N$. Then $\langle u_n \rangle$ is a nonincreasing sequence and $\langle v_n \rangle$ is a nondecreasing sequence. Moreover, $\limsup s_n = \inf\{u_m: m \in N\}$ and $\liminf s_n = \sup\{v_m: m \in N\}$.

Proof: If case (i) of Definition 8.16 holds, then u_n is $+\infty$ for all n (why?) and so $\inf u_n = +\infty$. If case (ii) holds, then for any $n_0 \in N$ there is an $m \in N$ so that $s_n \leq -n_0$ for all $n \geq m$ (why?). In this case, $u_m \leq -n_0$. Since this is true for each $n_0 \in N$, $\inf u_m = -\infty$. If case (iii) holds and r is the largest limit point of $\langle s_n \rangle$, then, for any $\varepsilon > 0$ in R, $u_m \geq r - \varepsilon$ for each $m \in N$ and $u_m \leq r + \varepsilon$ for some $m \in N$, so $r = \inf u_m$. The proof for $\liminf s_n$ is left to the reader. $\square$

Let $\langle s_n \rangle$ and $\langle t_n \rangle$ be bounded sequences. The reader should verify that

$$\liminf s_n + \liminf t_n \leq \liminf(s_n + t_n)$$

$$\leq \limsup(s_n + t_n) \leq \limsup s_n + \limsup t_n.$$

Moreover, $\langle s_n \rangle$ has limit $L \in R$ if and only if $\limsup s_n = \liminf s_n = L$.

8.18 Theorem (Ratio Test) A series $\sum_{i=1}^\infty a_i$ converges absolutely if $\limsup(|a_{i+1}|/|a_i|) < 1$. A series $\sum_{i=1}^\infty a_i$ diverges if $\liminf(|a_{i+1}|/|a_i|) > 1$.

Proof: Left to reader. $\square$

Exercises I.8

1. Prove Theorem 8.2, parts (ii) and (iii).
2. Prove Proposition 8.3.
3. Using Proposition 8.3, prove that a Cauchy sequence is bounded.
4. What sentence must be transferred for the proof of the second part of Proposition 8.7?
5. Use Exercises I.7.7 and I.7.8, Proposition 8.7, and Theorem 8.8 to show that if $A \subset R$ then $*A \subset A$ if and only if A is a finite set.
6. Prove Proposition 8.10.
7. Prove Proposition 8.11.
8. Finish the proof of Proposition 8.12.
9. Finish the proof of Theorem 8.13.
10. Fill in the details in the proof of Theorem 8.15.

11. Show that if $\langle s_n \rangle$ is bounded above and $r = \sup\{\text{st}(*s_n) : n \in *N_\infty, \ *s_n \text{ fi-}$ nite$\}$, then r is a limit point of (s_n).
12. Fill in the details and finish the proof of Theorem 8.17.
13. Fill in the details in the remark preceding Theorem 8.18.
14. Prove Theorem 8.18.
15. Use Exercise 13 in I.6 to show that if a, b are real and $b \neq 0$ then the sequence $\langle s_n \rangle$ given by $s_n = 1/(a + nb)$ converges to 0.
16. Suppose that $\langle s_n \rangle$ and $\langle t_n \rangle$ converge to L and M, respectively. Show that

    (a) $\langle s_n + t_n \rangle$ converges to $L + M$,
    (b) $\langle as_n \rangle$ converges to aL for $a \in R$,
    (c) $\langle s_n t_n \rangle$ converges to LM,
    (d) $\langle s_n / t_n \rangle$ converges to L/M if $M \neq 0$.

17. Show that if $\langle s_n \rangle$ and $\langle t_n \rangle$ converge to L and M, respectively, and $s_n \leq t_n$ for $n \in N$, then $L \leq M$. Prove as a consequence that the limit of a sequence is unique.
18. Show that if $r_n \leq s_n \leq t_n$ for all $n \in N$ and $\lim_{n \to \infty} r_n = \lim_{n \to \infty} t_n = s$, then $\langle s_n \rangle$ converges to s.
19. Show that if $\lim_{n \to \infty} (s_n - 1)/(s_n + 1) = 0$ then $\lim_{n \to \infty} s_n = 1$.
20. Investigate the limits $\lim_{n,m} s_{nm}$, $\lim_n s_{nm}$, $\lim_m s_{nm}$ and the iterated limits for the sequences

    (i) $s_{nm} = n/(n + m)$,
    (ii) $s_{nm} = (-1)^n n/(n + m)$,
    (iii) $s_{nm} = (-1)^{n+m}(1/n + 1/m)$.

21. Show that $\sum a_i(i \in N)$ converges iff $\sum *a_i(n \leq i \leq m) \simeq 0$ for all n and m $(n < m)$ in $*N_\infty$. Conclude that if $\sum a_i(i \in N)$ converges than $*a_i \simeq 0$ for all $n \in *N_\infty$.
22. Prove the formulas $\lim_{n \to \infty} (n!/b^n) = \infty$ $(b \geq 1)$, $\lim_{n \to \infty} (b^n/n^c) = \infty$ $(b > 1, c \geq 0)$, $\lim_{n \to \infty} (n^c/\ln n) = \infty$ $(c > 0)$ by using transfer of familiar properties of logs.

I.9 Topology on the Reals

In this section we present nonstandard characterizations of the basic topological notions of open, closed, and compact set, and use these characterizations to prove a few standard results. Familiarity with the standard definitions is assumed.

9.1 Proposition Let A be a subset of R. Then

(i) A is open iff $m(a) \subset {}^*A$ for each $a \in A$,

(ii) A is closed iff $m(a) \cap {}^*A$ is empty for each $a \in A'$.

Proof: (i) Suppose that A is open and let $a \in A$. By the definition of openness, there exists a standard $\varepsilon > 0$ so that

(9.1) $$(\forall x)[\underline{R}\langle x\rangle \wedge |x - a| < \varepsilon \to \underline{A}\langle x\rangle]$$

is true in $\mathscr{R}$. By transfer, if $x \in {}^*R$ and $|x - a| < \varepsilon$ then $x \in {}^*A$. In particular, if $|x - a| \simeq 0$ then $x \in {}^*A$ and so $m(a) \subset {}^*A$.

Conversely, suppose that $m(a) \subset {}^*A$ for each $a \in A$. If A is not open, there exists an $a \in A$ so that for each $n \in N$ we can find an $x_n \in A'$ with $|x_n - a| < 1/n$. Define a Skolem function $\psi : N \to R$ by $\psi(n) = x_n$, where x_n is a specifically chosen element of A' with $|x_n - a| < 1/n$. Then the sentence

(9.2) $$(\forall n)[\underline{N}\langle n\rangle \to \underline{A}'\langle \psi(n)\rangle \wedge |\psi(n) - a| < 1/n]$$

is true in $\mathscr{R}$. By transfer, for all $n \in {}^*N$, ${}^*\psi(n) \in {}^*A'$ and $|{}^*\psi(n) - a| < 1/n$. In particular, for $n = \omega$ where ω is infinite, the number $x_\omega = {}^*\psi(\omega)$ satisfies $x_\omega \in {}^*A'$ and $|x_\omega - a| < 1/\omega \simeq 0$, i.e., $x_\omega \in m(a)$ (contradiction).

(ii) This assertion can be proved by noting that, by definition, A is closed iff A' is open (exercise). $\square$

9.2 Theorem

(i) If $\{A_i : i \in I\}$ is a collection of open sets in R, then $\bigcup A_i (i \in I)$ is open.

(ii) If $A_1, \ldots, A_n$ are open in R, then $\bigcap A_i (1 \leq i \leq n)$ is open.

(iii) If $\{A_i : i \in I\}$ is a collection of closed sets in R, then $\bigcap A_i (i \in I)$ is closed.

(iv) If $A_1, \ldots, A_n$ are closed in R, then $\bigcup A_i (1 \leq i \leq n)$ is closed.

Proof: We prove (i) and (ii) and leave the proofs of (iii) and (iv) to the reader.

(i) Let $x \in \bigcup A_i (i \in I)$. Then $x \in A_j$ for some $j \in I$ and so $m(x) \subset {}^*A_j$ by 9.1(i). Thus $m(x) \subset \bigcup {}^*A_i (i \in I) \subseteq {}^*[\bigcup A_i (i \in I)]$, the last inclusion by Proposition 5.8(iii). This shows that $\bigcup A_i (i \in I)$ is open by 9.1(i).

(ii) Let $x \in \bigcap A_i (1 \leq i \leq n)$. Then $x \in A_i$ and so $m(x) \subset {}^*A_i$ for each i, $1 \leq i \leq n$, by 9.1(i). Thus $m(x) \subset {}^*A_1 \cap \cdots \cap {}^*A_n = {}^*[\bigcap A_i (1 \leq i \leq n)]$, the last equality by Proposition 5.8(ii). Thus $\bigcap A_i (1 \leq i \leq n)$ is open by 9.1(i). $\square$

Recall that a point $x \in R$ is an *accumulation point* of a set $A \subseteq R$ if, for every $n \in N$, there is a point y in A different from x with $|y - x| < 1/n$. The set of accumulation points of A is denoted by $\hat{A}$, and the *closure* of A is the set $\bar{A} = A \cup \hat{A}$.

9.3 Proposition A point $x \in R$ is an accumulation point of $A \subseteq R$ iff there is a $y \neq x$ in $*A$ with $y \simeq x$.

Proof: Suppose that x is an accumulation point of A. Then for each $n \in N$ we can find a $y \neq x$ in A with $|x - y| < 1/n$. Let $\psi: N \to A$ be a Skolem function obtained by associating a $y \in A$ with each $n \in N$ so that the sentence

(9.3) $(\forall n)[\underline{N}\langle n\rangle \to \underline{\psi}(n) \neq x \wedge \underline{A}\langle \underline{\psi}(n)\rangle \wedge |x - \underline{\psi}(n)| < 1/n]$

is true in $\mathcal{R}$. By transfer we see that, for each $n \in *N$, $*\psi(n) \neq x$, $*\psi(n) \in *A$, and $|x - *\psi(n)| < 1/n$. We need only choose $y = *\psi(\omega) \in *A$ for $\omega \in *N_\infty$.
    The converse is left to the reader.   □

9.4 Proposition The closure $\bar{A}$ of a set A in R consists of those $x \in R$ for which $m(x) \cap *A$ is not empty.

Proof: If $x \in \bar{A}$ then $x \in A$ or $x \in \hat{A}$. If $x \in A$ then $x \in *A$ and $x \in m(x)$. If $x \in \hat{A}$ then $m(x) \cap *A$ is not empty by Proposition 9.3.
    The converse is established by reversing the argument.   □

Proposition 9.4 can be expressed in a more graphic way. The standard part map st: $G(0) \to R$ defines a mapping, also denoted by st, from subsets of $G(0)$ to subsets of R by the obvious definition, For each $B \subset G(0)$, st$(B) = \{$st$(y): y \in B\} = \{x \in R:$ there exists a $y \in B$ with $y \simeq x\}$. Proposition 9.4 can be restated as asserting that st$(*A \cap G(0)) = \bar{A}$ for any subset A of R, and thus it shows how to construct the closure of any set A by constructing the *-transform of A and then collapsing back to R by a standard part operation. In this form, Proposition 9.4 is a prototype of similar results obtained in more complicated situations later in this book.

9.5 Theorem For any subsets A and B of R,

(a) $A \subseteq \bar{A}$,
(b) $\bar{\bar{A}} = \bar{A}$,
(c) $\overline{A \cup B} = \bar{A} \cup \bar{B}$,
(d) $\bar{A}$ is closed,

(e) if B is closed and $A \subseteq B$ then $\bar{A} \subseteq B$,

(f) if A is closed then $\bar{A} = A$.

Proof: (a) Immediate from the definition.

(b) $\bar{A} \subseteq \bar{\bar{A}}$ from (a). If $x \in \bar{\bar{A}}$ but $x \notin \bar{A}$ then $x \in \hat{\bar{A}}$. Thus, for any $n \in N$, there is a $y \in \bar{A}$ with $|x - y| < 1/n$; by Proposition 9.4 there is a $z \in {}^*A$ with $|x - z| < 1/n$. On the other hand, if $x \notin \bar{A}$ there is an $n \in N$ so that $|x - z| > 1/n$ for all $z \in A$. By transfer (check) this is true for all $z \in {}^*A$ (contradiction).

(d) If $b \notin \bar{A}$ then $m(b) \cap {}^*A = \varnothing$, for otherwise $b \in \bar{A}$ by 9.4, and then $b \in \bar{A}$ by part (b).

Parts (c), (e), and (f) are left as exercises.   □

Next we present an important characterization of compactness due to Robinson. Recall that, by definition, the collection $A_i (i \in I)$ of sets is a *covering* of the set $A \subseteq R$ if $A \subseteq \bigcup A_i (i \in I)$, and that A is compact if each covering $A_i (i \in I)$ by open sets contains a finite subcovering $A_i (i \in I')$ (i.e., $I' \subseteq I$ is finite). To obtain Robinson's characterization we need the following standard result.

9.6 Lemma Each covering of $A \subseteq R$ by open sets $A_i (i \in I)$ contains a finite subcovering if each covering of A by a collection of open intervals (a_n, b_n) with rational end points contains a finite subcovering.

Proof: Let $A_i (i \in I)$ be a covering of A by open sets. If $x \in A$ then $x \in A_j$ for some $j \in I$. Since the rationals are dense in R and A_j is open, we can find rationals a and b so that $x \in (a, b) \subset A_j$ (why?). The corresponding countable collection covers A. Select a finite subcovering from this latter covering. Each interval in the finite subcovering is contained in some A_j, and so we may find a finite collection of the $A_i (i \in I)$ which also covers A.   □

9.7 Robinson's Theorem The set $A \subset R$ is compact iff for each $y \in {}^*A$ there is an $x \in A$ with $x \simeq y$, i.e., every point in ${}^*A$ is near a point in A.

Proof: Suppose that A is compact but $y \in {}^*A$ is not near any $x \in A$. Then for each $x \in A$ there is a $\delta_x > 0$ in R such that $|x - y| \geq \delta_x$. Since A is compact we can extract a finite subcovering $A_i = \{z \in R : |x_i - z| < \delta_{x_i}\}$ ($i = 1, 2, \ldots, n$) from the covering of A by the sets $A_x = \{z \in R : |x - z| < \delta_x\}$ ($x \in A$). It follows that

(9.4) $(\forall y)[A\langle y \rangle \wedge |x_1 - y| \geq \delta_{x_1} \wedge \cdots \wedge |x_{n-1} - y| \geq \delta_{x_{n-1}} \rightarrow |x_n - y| < \delta_{x_n}]$

is true in $\mathscr{R}$. Transferring to $*\mathscr{R}$, we obtain a contradiction with the fact that $y \in *A$ and $|x_i - y| \geq \delta_{x_i}$ for $i = 1, 2, \ldots, n$.

Assume now that a covering A_i $(i \in I)$ contains no finite subcovering. By Lemma 9.6 there exists a covering of A by a countable collection $I_n = \{x \in R : a_n < x < b_n\}$, $n \in N$, of open intervals with rational end points which has no finite subcovering. Thus there is a Skolem function $\psi : N \to A$ so that

$$(9.5) \qquad (\forall n)(\forall k)[\underline{N}\langle n \rangle \wedge \underline{N}\langle k \rangle \wedge k \leq n \wedge a_k < \underline{\psi}(n) \to b_k \leq \underline{\psi}(n)]$$

is true in $\mathscr{R}$ (check). By transfer we see that if ω is infinite, then $*\psi(\omega) \notin *(a_k, b_k)$ for any $k \in N$. Thus, $*\psi(\omega) \in *A$ is not near a point x in A since $m(x) \subset *(a_k, b_k)$ for some $k \in N$. $\square$

In Chapter III we generalize this result to topological spaces, and the proof given there avoids an analogue of Lemma 9.6.

As an application of Robinson's theorem, we prove the following famous result.

9.8 Theorem (Heine–Borel) A set $A \subset R$ is compact iff it is closed and bounded.

Proof: If A is not closed then, by Proposition 9.1(ii), there is an $x \in A'$ and a $y \in *A$ with $y \simeq x$; since $\mathrm{st}(y) = x$ it follows by Theorem 9.7 that A is not compact. If A is not bounded there is a Skolem function $\psi : N \to A$ so that

$$(9.6) \qquad (\forall n)[\underline{N}\langle n \rangle \to n \leq |\underline{\psi}(n)| \wedge \underline{A}\langle \underline{\psi}(n) \rangle]$$

is true in $\mathscr{R}$. Transferring to $*\mathscr{R}$, and choosing ω infinite, we see that $\omega \leq |*\psi(\omega)|$, and so the point $y = *\psi(\omega) \in *A$ is not near any standard point. Hence A is not compact by Theorem 9.7.

If A is closed and bounded there is an $M \in R$ so that

$$(9.7) \qquad (\forall y)[\underline{A}\langle y \rangle \to |y| \leq M].$$

By transfer, if $y \in *A$ then $|y| \leq M$, and so $°y = x$ is in $\bar{A}$ (why?). Since A is closed, Theorem 9.5(f) shows that $x \in A$. Thus, A is compact by Theorem 9.7. $\square$

The nonstandard characterizations of topological notions on the real line developed in this section can easily be extended to n-dimensional space R^n. Observe that all characterizations are stated in terms of the notions of near points or monads. To extend our characterizations to subsets of R^n we make the following definition.

9.9 Definition If $x = (x_1, \ldots, x_n)$ and $y = (y_1, \ldots, y_n)$ are points in $*R^n$ then $x \simeq y$ iff $x_i \simeq y_i$, $1 \le i \le n$, and $m(x) = \{y \in *R^n : x \simeq y\}$.

With this definition the results of the section apply also to subsets of R^n. We return to these problems (in more generality) in Chapter III.

Exercises I.9

1. Finish the proof of Proposition 9.1(ii).
2. Finish the proof of Proposition 9.3 by showing that if x is not an accumulation point of A then for $y \in *A - \{x\}$ we have $y \not\simeq x$.
3. Prove parts (c), (e), and (f) of Theorem 9.5.
4. Show that a set $A \subseteq R$ is closed iff whenever $\langle x_n : n \in N \rangle$ is a sequence of points in A which converges to x, then $x \in A$.
5. Show that if $A_1, A_2, \ldots, A_n$ are open (closed) subsets of R then $A_1 \times A_2 \times \cdots \times A_n$ is open (closed) in R^n.
6. Use Robinson's theorem to show that if $K \subset R$ is compact and $A \subset R$ is closed then $K \cap A$ is compact.
7. Show that Robinson's theorem holds also for subsets A of R^n (with the obvious definition of compactness). Hence show that if $A_1, \ldots, A_n$ are compact subsets of R then $A_1 \times \cdots \times A_n$ is a compact subset of R^n.
8. Prove that R is connected. That is, show that R cannot be of the form $A \cup B$, where $\bar{A} \cap B$ and $A \cap \bar{B}$ are both empty. [Hint: Assume the contrary, choose $x \in A$, $y \in B$, and consider the points $x_n = x + (y - x)k/n$, $0 \le k \le n$. There is a largest k—say, k_0—, such that $x_k \in A$ for all $k \le k_0$ and $x_{k_0+1} \in B$.]
9. A set $A \subset R$ is bounded if there exists an $n \in N$ so that $A \subseteq [-n, n]$. Show that A is bounded iff every $x \in *A$ is finite.
10. Show that if A is compact in R and $x \notin A$, then there is a $y \in A$ such that for all $z \in A$, $|x - y| \le |x - z|$.
11. Let $F_1 \supseteq F_2 \supseteq \cdots$ be a decreasing sequence of non-empty compact sets in R. Show that $\bigcap F_i \, (i \in N) \ne \varnothing$ by choosing $x_n \in F_n$ for each $n \in N$ (so then $x_n \in F_m$ for $m \le n$.)
12. Use Theorem 9.7 to show that if A and B are compact subsets of R then $A + B = \{x + y; x \in A, y \in B\}$ is compact.

I.10 Limits and Continuity

It should now be clear that the notions of limit and continuity can be characterized nonstandardly in much the same way as were the notions of the previous sections; therefore we will be brief in the following discussion.

10.1 Proposition Let f be defined on $A \subseteq R$ and let $a \in \hat{A}$. Then

(a) $\lim_{x \to a} f(x) = L$ iff $*f(x) \simeq L$ for all $x \in *A$ with $x \simeq a$ but $x \neq a$,

(b) $\lim_{x \to a^+} f(x) = L$ $[\lim_{x \to a^-} f(x) = L]$ iff $*f(a + \varepsilon) \simeq L$ for all $\varepsilon > 0$ $[\varepsilon < 0]$ with $\varepsilon \simeq 0$, $a + \varepsilon \in *A$, and at least one such ε exists,

(c) $\lim_{x \to a} f(x) = \infty$ $(-\infty)$ iff $*f(x)$ is positive (negative) infinite for all $x \in *A$ with $x \simeq a$, $x \neq a$.

(d) $\lim_{x \to +\infty \ (-\infty)} f(x) = L$ iff $*f(x) \simeq L$ for all positive (negative) infinite $x \in *A$, and at least one such x exists.

Proof: We prove (a) and leave the remaining proofs to the reader as exercises. Recall that $\lim_{x \to a} f(x) = L$ if and only if, given $\varepsilon > 0$ in R, there exists a $\delta > 0$ in R so that $|f(x) - L| < \varepsilon$ if $0 < |x - a| < \delta$ and $x \in A$. Suppose that $\lim_{x \to a} f(x) = L$, and find the δ corresponding to some $\varepsilon > 0$ in R. Then

(10.1) $(\forall x)[\underline{A}\langle x \rangle \wedge 0 < |x - a| < \delta \to |f(x) - L| < \varepsilon].$

By transfer, if $x \in *A = \text{dom} *f$ and $0 < |x - a| < \delta$, then $|*f(x) - L| < \varepsilon$. In particular, $|*f(x) - L| < \varepsilon$ if $x \simeq a$ but $x \neq a$ for *any* $\varepsilon > 0$ in R and so $*f(x) \simeq L$.

Conversely, if $\lim_{x \to a} f(x)$ does not exist or $\lim_{x \to a} f(x)$ exists but is not equal to L, then there exists a standard $\varepsilon > 0$ and a Skolem function $\psi: N \to A - \{a\}$ so that $|\psi(n) - a| < 1/n$ and $|f(\psi(n)) - L| \geq \varepsilon$. Thus

(10.2) $(\forall n)[\underline{N}\langle n \rangle \to \underline{A}\langle \psi(n) \rangle \wedge 0 < |\psi(n) - a| < 1/n \wedge |f(\psi(n)) - L| \geq \varepsilon].$

By transfer, $|*f(*\psi(n)) - L| \geq \varepsilon$, $*\psi(n) \in *A$, and $0 < |*\psi(n) - a| < 1/n$ for all $n \in *N$. In particular, if $n \in *N_\infty$ then $x = *\psi(n)$ satisfies $x \in *A$, $x \simeq a$, $x \neq a$, and $|*f(x) - L| \geq \varepsilon$, i.e., $*f(x) \not\simeq L$.   □

10.2 Proposition Let f be defined on A and choose $a \in \hat{A}$. Then the limit $\lim_{x \to a} f(x)$ exists iff $*f(x) \simeq *f(y)$ for all $x, y \in *A$ with $x \simeq a$, $y \simeq a$ but $x \neq a$, $y \neq a$.

Proof: Exercise.   □

10.3 Theorem If $\lim_{x \to a} f(x) = L$, $\lim_{x \to a} g(x) = M$, then

(a) $\lim_{x \to a} (f + g)(x) = L + M$,

(b) $\lim_{x \to a} (fg)(x) = LM$,

(c) $\lim_{x \to a} (f/g)(x) = L/M$ if $M \neq 0$.

Proof: Exercise.   □

10.4 Proposition Let f be defined on $A \subseteq R$. Then f is continuous at $a \in A$ iff $*f(x) \simeq f(a)$ for all $x \in *A$ with $x \simeq a$, i.e., $*f(m(a) \cap *A) \subseteq m(f(a))$.

Proof: Immediate from 10.1 and the definition of continuity. □

Proposition 10.4 says that if f is continuous at $x \in A$, and $x + \Delta x \in *A$ where $\Delta x \simeq 0$, then $\Delta y = *f(x + \Delta x) - f(x) \simeq 0$. For example, if $f(x) = x^2$, then $\Delta y = (x + \Delta x)^2 - x^2 = 2x\,\Delta x + (\Delta x)^2 \simeq 0$.

10.5 Theorem If f and g are defined on A and continuous at $a \in A$, then so are $f + g$, fg, and $[$if $g(a) \neq 0] \ f/g$.

Proof: Immediate from 10.3 and 10.4. □

The preceding propositions can be used to prove the intermediate and extreme value theorems.

10.6 Intermediate Value Theorem If f is continuous on the closed and bounded interval $[a, b]$ and $f(a) < d < f(b)$ for some d, then there exists a $c \in (a, b)$ with $f(c) = d$.

Proof: Consider the points $x_k = a + k(b - a)/n$, $0 \le k \le n$. Considering the values of f at x_k, we see that there exists a Skolem function $\psi: N \to [a, b)$ satisfying $f(\psi(n)) < d$ and $f(\psi(n) + (b - a)/n) \ge d$ (check). Hence the sentence

(10.3) $(\forall n)[\underline{N\langle n \rangle} \to a \le \psi(n) < b \wedge \underline{f(\psi(n))} < d \wedge \underline{f}(\psi(n) + (b - a)/n) \ge d]$

is true in $\mathcal{R}$. Transferring to $*\mathcal{R}$, and letting $n \in *N_\infty$, we have

(10.4) $*f(*\psi(n)) < d$ and $*f(*\psi(n) + (b - a)/n) \ge d$.

Let $c = \mathrm{st}(*\psi(n)) = \mathrm{st}(*\psi(n) + (b - a)/n)$. By continuity we have $f(c) \le d$ and $f(c) \ge d$, and hence $f(c) = d$. Also c cannot equal either a or b, since otherwise $f(c) = f(a)$ or $f(b)$. □

10.7 Extreme Value Theorem If f is continuous on the closed and bounded interval $[a, b]$, then there exists a $c \in [a, b]$ so that $f(c) \ge f(x)$ for all $x \in [a, b]$.

Proof: For each $n \in N$ construct the points $x_{n,k} = a + k(b - a)/n, 0 \le k \le n$. There is a Skolem function $\psi: N \to N \cup \{0\}$ satisfying $\psi(n) \le n$ such that, for each $n \in N$, $f(x_{n,\psi(n)}) \ge f(x_{n,k})$, $0 \le k \le n$, since the finite set of numbers $f(x_{n,k})$, $0 \le k \le n$, has a maximum for some k satisfying $0 \le k \le n$. By transfer, $*f(x_{n,*\psi(n)}) \ge *f(x_{n,k})$, $0 \le k \le n$, for $k \in *N$ and n fixed and infinite. Then $c =$

$\text{st}(x_{n,*\psi(n)})$ satisfies the conditions of the theorem. To see this, fix $d \in [a, b]$. Then $d \simeq x_{n,k}$ for some $k \in *N$ with $0 \le k \le n$ (exercise), so, using continuity, $f(d) \simeq *f(x_{n,k}) \le *f(x_{n,*\psi(n)}) \simeq f(c)$. If $f(d) \simeq f(c)$ then $f(d) = f(c)$ since both numbers are real. Otherwise $f(d) < f(c)$. $\square$

Proposition 10.4 shows that f is continuous on A iff $*f(m(a) \cap *A) \subseteq m(f(a))$ for all $a \in A$. Uniform continuity on A results if an analogous condition holds for all $a \in *A$.

10.8 Proposition The function f is uniformly continuous on a set A iff $*f(m(a) \cap *A) \subseteq m(*f(a))$ for all $a \in *A$; i.e., a, $b \in *A$ and $a \simeq b$ implies $*f(a) \simeq *f(b)$.

Proof: Recall that f is uniformly continuous on A iff, given $\varepsilon > 0$ in R, there exists a $\delta > 0$ in R so that, for all $a \in A$, $|f(x) - f(a)| < \varepsilon$ if $|x - a| < \delta$ and $x \in A$.

Suppose that f is uniformly continuous on A, let $\varepsilon > 0$ in R be given, and find the corresponding $\delta > 0$ in R. Then the sentence

$$(10.5) \qquad (\forall a)(\forall b)[\underline{A}\langle a\rangle \wedge \underline{A}\langle b\rangle \wedge |a - b| < \delta \rightarrow |\underline{f}(a) - \underline{f}(b)| < \varepsilon]$$

is true in $\mathscr{R}$. By transfer, for all a and b in $*A$, $|a - b| < \delta$ implies $|*f(a) - *f(b)| < \varepsilon$. In particular, this is true for any $\varepsilon > 0$ in R if $a \simeq b$, and hence a, $b \in *A$ and $a \simeq b$ implies $*f(a) \simeq *f(b)$.

Conversely, suppose f is not uniformly continuous on A. Then there is an $\varepsilon > 0$ in R so that, for each $n \in N$, there are points $\psi_1(n) = a_n \in A$ and $\psi_2(n) = b_n \in A$ with $|a_n - b_n| < 1/n$ but $|f(a_n) - f(b_n)| \ge \varepsilon$. By transfer of the appropriate sentence (the reader is invited to write one down), for each $n \in *N$ there are points a_n and $b_n \in *A$ with $|a_n - b_n| < 1/n$ but $|*f(a_n) - *f(b_n)| \ge \varepsilon$. With $n \in *N_\infty$ we have $a_n \simeq b_n$ but $*f(a_n) \not\simeq *f(b_n)$. $\square$

10.9 Examples

1. $\lim_{x \to 3} x^2 = 9$ since if $h \simeq 0$, we have $(3 + h)^2 = 9 + 6h + h^2 \simeq 9$.
2. $\lim_{h \to 0} \{[(x + h)^2 - x^2]/h\} = 2x$ since if $h \simeq 0$, $h \ne 0$, $[(x + h)^2 - x^2]/h = 2x + h \simeq 2x$.
3. $\lim_{x \to \infty} (\sqrt{x + 1} - \sqrt{x}) = 0$ since for h positive infinite in $*R$

$$\frac{(\sqrt{h + 1} - \sqrt{h})(\sqrt{h + 1} + \sqrt{h})}{\sqrt{h + 1} + \sqrt{h}} = \frac{1}{\sqrt{h + 1} + \sqrt{h}} \simeq 0.$$

4. $f(x) = 1/x$ is continuous on $(0, 1)$ since if $a \in (0, 1)$ and $h \simeq 0$, $1/a - 1/(a + h) = h/a(a + h) \simeq 0$. However, f is not uniformly continuous on $(0, 1)$

since if $n \in \,^*N_\infty$, $1/n$ and $1/(n-1)$ are in $^*(0,1)$ and $1/n \simeq 1/(n-1)$ but $^*f(1/n) - \,^*f(1/(n-1)) = 1 \not\simeq 0$.

Proposition 10.8 can be used effectively to prove standard results.

10.10 Theorem If f is continuous on the compact set A, then f is uniformly continuous on A.

Proof: If $x, y \in \,^*A$ and $x \simeq y$, then both x and y are near a standard point $a \in A$ since A is compact (Theorem 9.7). Thus $^*f(x) \simeq f(a) \simeq \,^*f(y)$ by continuity (Proposition 10.4), so f is uniformly continuous by Proposition 10.8. □

10.11 Theorem If $A \subset R$ is compact and f is continuous on A, then $f(A)$ is compact.

Proof: If $y \in \,^*[f(A)] = \,^*f(^*A)$ (Proposition 5.6) then there is an $x \in \,^*A$ with $^*f(x) = y$. Since A is compact there is a point $a \in A$ with $x \simeq a$ (Theorem 9.7). Then $^*f(x) = y \simeq f(a)$ since f is continuous at a, and so $f(A)$ is compact by Theorem 9.7. □

10.12 Theorem Suppose that f is uniformly continuous on each bounded subset of its domain A. Then f has a unique extension g defined on $\bar{A}$ (i.e., f agrees with g on A) such that g is uniformly continuous on every bounded subset of $\bar{A}$.

Proof: Every standard point $y \in \bar{A}$ is near a finite point $x \in \,^*A$ and we define $g(y) = \mathrm{st}(^*f(x))$. This definition is independent of the x we choose since if $x' \simeq y$ then $x \simeq x'$, and both x and x' are in $^*B$, where $B = A \cap [-|y|-1, |y|+1]$ is bounded. Therefore, $^*f(x) \simeq \,^*f(x')$ by uniform continuity on B. We leave as an exercise the proof that $^*f(x)$ is finite.

If $C = A \cap [-2n, 2n]$, $n \in N$, then, given $\varepsilon > 0$, there exists a $\delta > 0$ so that $|f(x) - f(x')| < \varepsilon/2$ if $|x - x'| < \delta$ and $x, x' \in C$. By transfer, $|^*f(x) - \,^*f(x')| < \varepsilon/2$ for all $x, x' \in \,^*C$ satisfying $|x - x'| < \delta$. Now if $y, y' \in \bar{A} \cap [-n, n]$ are such that $|y - y'| < \delta/2$ and $y \simeq x$, $y' \simeq x$ for some $x, x' \in \,^*C$, then $|x - x'| < \delta$, and so $|g(y) - g(y')| \simeq |^*f(x) - \,^*f(x')| < \varepsilon/2$. Thus, $|g(y) - g(y')| \leq \varepsilon/2 < \varepsilon$. Uniqueness is left to the reader. □

Theorem 10.12 can be used to extend the exponential function $f(x) = a^x$, $a > 0$ in R, defined on the rationals Q to the reals $R = \bar{Q}$. The function a^x, $x \in Q$, satisfies the following properties.

10.13 Properties of Exponents If a and b are positive reals and q and r are rational then

    (i) $1^q = 1$,
    (ii) $a^q a^r = a^{q+r}$, $a^{-q} = 1/a^q$,
    (iii) $(a^q)^r = a^{qr}$,
    (iv) $a^q b^q = (ab)^q$,
    (v) $a < b$ and $q > 0$ implies $a^q < b^q$,
    (vi) $1 < a$ and $q < r$ implies $a^q < a^r$,
    (vii) $a \geq 0$ and $q \geq 1$ implies $(a + 1)^q \geq aq + 1$.

The useful inequality (vii) follows by noting that, for $x \geq 0$, $(x + 1)^q - qx - 1$ has a minimum at $x = 0$. Properties (i) through (vi) are obvious.

To extend $f(x) = a^x$, $a > 0$, $x \in Q$, to R we need only show that f is uniformly continuous on bounded subsets of Q. That is, we need the following lemma.

10.14 Lemma If $a > 0$ in R, then $a^p \simeq a^q$ if $p \simeq q$ in $*Q \cap G(0)$.

Proof: We may suppose that $p > q$ and $a \geq 1$ [if $0 < a < 1$ consider $a^q = (1/a)^{-q}$]. Let $b = a^{p-q} - 1$; we must show that $b \simeq 0$. By transfer from 10.13(vi), $b \geq 0$, and, by transfer from 10.13(vii),

$$(10.6) \qquad a = (b + 1)^{1/(p-q)} \geq b/(p - q) + 1 \geq 1,$$

so $b/(p - q)$ is a finite number ρ, and hence $b = (p - q)\rho \simeq 0$. $\square$

This argument is due to Keisler [23]. It is easy to show that properties 10.13 are satisfied by the extension $g(x)$, $x \in R$, of $f(x) = a^x$, $x \in Q$. For example, $g(y + y') \simeq *f(q + q') = *f(q)*f(q') \simeq g(y)g(y')$ if $q \simeq y$, $q' \simeq y'$, and $q, q' \in *Q$; this establishes the first part of 10.13(ii) for g since g is real-valued.

Most of the results in this section can be extended to functions f of n variables defined on subsets of R^n simply by using the definition of nearness for points in $*R^n$ introduced in the previous section. The details are left to the reader.

Exercises I.10

1. Prove parts (b)–(d) of Proposition 10.1.
2. Prove Proposition 10.2.
3. Prove Theorem 10.3.
4. Complete the proof of Theorem 10.7 by showing that for each $d \in [a, b]$ there is a $k \in *N$ with $0 \leq k \leq n$ such that $d \simeq x_{n,k}$.

5. Prove that if f is uniformly continuous on a bounded set $B \subset R$, then $*f(x)$ is finite for each $x \in *B$.

6. Prove uniqueness in Theorem 10.12.

7. Show that there are infinite rational numbers p and q with $p \simeq q$ such that $2^p \not\simeq 2^q$. Where is the assumption that $p, q \in G(0)$ used in the proof of Lemma 10.14?

8. Let

$$f(x) = \begin{cases} \sin(1/x), & 0 < x \le 1, \\ 0, & x = 0 \end{cases}$$

   (a) Show that $f(x)$ is not continuous on $[0, 1]$.
   (b) Show that the function $xf(x)$ is uniformly continuous on $[0, 1]$.

9. Show that the function $f(x) = x^2$ on $(0, \infty)$ is continuous but not uniformly continuous.

10. Show that $\lim_{x \to a} f(x) = L$ iff for each sequence $\langle s_n \rangle$ with $\lim_{n \to \infty} s_n = a$ and $s_n \neq a$, $n \in N$, we have $\lim_{n \to \infty} f(s_n) = L$.

11. Prove that if f is uniformly continuous on R and $\langle s_n \rangle$ is a Cauchy sequence then $\langle f(s_n) \rangle$ is a Cauchy sequence.

12. Suppose that f is continuous on R and satisfies $\lim_{x \to \infty} f(x) = \lim_{x \to -\infty} f(x) = 0$. Prove that f is uniformly continuous.

13. Suppose that f is defined on a compact set A in R. Prove that f is continuous iff the graph $\{(x, f(x)) \in R^2 : x \in A\}$ of f is compact.

14. Show that if the function f is continuous on the set A then the zero set $\{x \in A : f(x) = 0\}$ of f is closed.

15. Suppose that the function f on the closed bounded interval $[a, b]$ is monotone [e.g., $x < y$ implies $f(x) \le f(y)$] and that for any r between $f(a)$ and $f(b)$ there is an x_0 such that $f(x_0) = r$. Prove that f is continuous on $[a, b]$. (Hint: Proceed by contradiction.)

16. (Hyperreal Intermediate Value Theorem) Suppose that f is continuous on the closed bounded interval $[a, b]$. If $d \in *R$ satisfies $*f(x) < d < *f(y)$ for $x, y \in *[a, b]$ then there is some $c \in *[a, b]$ with $x < c < y$ or $y < c < x$ such that $*f(c) = d$.

17. Prove that if f is continuous on $[a, b]$ and $*f(x)$ is real for all $x \in *[a, b]$ then f is a constant.

18. (Hyperreal Extreme Value Theorem) Suppose that f is continuous on the closed bounded interval $[a, b]$. Then $*f$ has a maximum on $*[a, b]$; i.e., there is a $c \in *[a, b]$ so that $*f(c) \ge *f(x)$ for all $x \in *[a, b]$.

19. Prove the Intermediate Value Theorem for $[0, 1] \times [0, 1]$. That is, let f be continuous on $[0, 1] \times [0, 1]$ and assume that $f(0, 0) < d < f(1, 1)$. Show that there is a point $(x_0, y_0) \in [0, 1] \times [0, 1]$ with $f(x_0, y_0) = d$.

20. Prove the Extreme Value Theorem for $[0, 1] \times [0, 1]$.

I.11 Differentiation

The theory of differentiation can now be developed easily using the results of the previous section.

11.1 Proposition Let f be defined at $a \in R$. The derivative $f'(a)$ of f at a exists iff for any infinitesimal $h \neq 0$

(i) $*f(a + h)$ is defined,
(ii) $[*f(a + h) - f(a)]/h$ is finite,
(iii) $\mathrm{st}([*f(a + h) - f(a)]/h)$ is independent of the choice of h.

In this case, $f'(a) = \mathrm{st}([*f(a + h) - f(a)]/h)$. The right-hand (left-hand) derivative of f at a exists iff (i)–(iii) hold for any infinitesimal $h > 0$ ($h < 0$), in which case that derivative equals $\mathrm{st}([*f(a + h) - f(a)]/h)$.

Proof: Immediate from Proposition 10.1. □

11.2 Proposition Let f be defined on $[a, b]$. The following statements are equivalent:

(i) f' exists and is continuous on $[a, b]$, where $f'(a)$ is the right-hand derivative at a and $f'(b)$ is the left-hand derivative at b.
(ii) For all x, y, x', y', in $*[a, b]$ with $x \simeq x' \simeq y \simeq y'$ and $x \neq y$ and $x' \neq y'$,

$$\frac{*f(x) - *f(y)}{x - y} \simeq \frac{*f(x') - *f(y')}{x' - y'} \in G(0).$$

If (ii) holds, then $f'(\mathrm{st}(x)) = \mathrm{st}([*f(x) - *f(y)]/(x - y))$.

Proof: If (i) holds and, in $*[a, b]$, $x \simeq y$, $x < y$, $\mathrm{st}(x) = c \in [a, b]$, then by the transfer of the mean value theorem there is an x_0 with $x < x_0 < y$ such that $[*f(x) - *f(y)]/(x - y) = *f'(x_0)$. (How is a Skolem function used here?) Since f' is continuous, $*f'(x_0) \simeq f'(c)$, whence (ii) follows. Assume that (ii) holds. If $c = x = x' \in [a, b]$, then $f'(c)$ exists by Proposition 11.1. Using a Skolem function and the transfer principle, we can obtain for each $x \in *[a, b]$ and positive infinitesimal ε a positive infinitesimal δ such that when $y \in *[a, b]$ and $0 < |x - y| < \delta$, $|*f'(x) - [*f(x) - *f(y)]/(x - y)| < \varepsilon$. It follows from (ii) that if $x \simeq x'$ in $*[a, b]$ then $*f'(x) \simeq *f'(x')$; i.e., f' is uniformly continuous on $[a, b]$ by Proposition 10.8. □

11.3 Examples

1. If $f(x) = 2x^2 + 3x$ then

$$\frac{*f(x + h) - f(x)}{h} = \frac{2(x + h)^2 + 3(x + h) - 2x^2 - 3x}{h}$$

$$= \frac{4xh + 2h^2 + 3h}{h}$$

$$= 4x + 3 + 2h$$

$$\simeq 4x + 3$$

for all $h \simeq 0$, $h \neq 0$, and hence $f'(x) = 4x + 3$.

2. Starting with the definition $e = \lim_{x \to \infty} (1 + 1/x)^x$, we show that $de^x/dx = e^x$. If $f(x) = e^x$ then $[*f(x + h) - f(x)]/h = e^x(e^h - 1)/h$, and we need to show that $(e^h - 1)/h \simeq 1$ if $h \simeq 0$. If $b = (e^h - 1)/h$ then $e^h = 1 + bh$. If $h \simeq 0$, $h > 0$, $e^h \simeq 1$ by the continuity of e^x (which we assume here) and so $bh \simeq 0$ and $1/bh$ is infinite. Then

$$e = \lim_{x \to \infty} \left(1 + \frac{1}{x}\right)^x \simeq (1 + bh)^{1/bh} = (e^h)^{1/bh} = e^{1/b}.$$

Hence $b \simeq 1$, and $[*f(x + h) - f(x)]/h \simeq e^x$ if $h > 0$. A similar argument works for $h < 0$, showing that $f'(x) = e^x$ (this argument is due to Keisler [23]).

11.4 Theorem If f is differentiable at $x \in (a, b)$, then f is continuous at x.

Proof: By proposition 10.1, $f(x + h) - f(x) \simeq f'(x)h$ for $h \simeq 0$, and so $f(x + h) \simeq f(x)$ for all $h \simeq 0$; i.e., f is continuous at x. □

11.5 Theorem If f, defined on (a, b), achieves a relative maximum or minimum at $x \in (a, b)$ and is differentiable at x, then $f'(x) = 0$.

Proof: Suppose that f achieves a relative minimum at x. Then, for all h sufficiently small and positive (negative), we have $[f(x + h) - f(x)]/h \geq 0$ (≤ 0). By transfer of the appropriate sentence, we see that $[*f(x + h) - f(x)]/h \geq 0$ (≤ 0) if $h \simeq 0$ and $h > 0$ ($h < 0$). Thus $f'(x) = 0$ from 11.1 and 6.7(iv). □

Rolle's theorem and the mean value theorem can be deduced in the standard way from this result and the extreme value theorem.

11.6 Theorem Let f and g be differentiable at x. Then $f + g$, fg, and [if $g(x) \neq 0$] f/g are differentiable at x, and

(i) $(f + g)'(x) = f'(x) + g'(x)$,
(ii) $(fg)'(x) = f'(x)g(x) + f(x)g'(x)$,
(iii) $(f/g)'(x) = [f'(x)g(x) - f(x)g'(x)]/g^2(x)$.

Proof: We prove (ii) and leave the remaining proofs to the reader. Let $h \simeq 0$. Then

$$\frac{f(x + h)g(x + h) - f(x)g(x)}{h}$$

$$= \frac{f(x + h)g(x + h) - f(x)g(x + h)}{h} + \frac{f(x)g(x + h) - f(x)g(x)}{h}$$

$$= g(x + h)\frac{f(x + h) - f(x)}{h} + f(x)\frac{g(x + h) - g(x)}{h}$$

$$\simeq g(x)f'(x) + f(x)g'(x)$$

by 11.1, 11.4 (applied to g), and 6.7. The result follows from Proposition 11.1. □

At this point it is natural to introduce differentials in the spirit of Leibniz. Denoting the nonzero infinitesimal h by Δx, we have

$$[{}^*f(x + \Delta x) - f(x)]/\Delta x \simeq f'(x)$$

if f is differentiable at x. We call

$$\Delta y = {}^*f(x + \Delta x) - f(x)$$

the *increment* of f at x corresponding to the increment Δx. The *differential* of f at x corresponding to Δx is defined to be $dy = f'(x)\,\Delta x$. Notice that $\varepsilon = \Delta y/\Delta x - f'(x)$ is infinitesimal, and so

$$(11.1) \qquad\qquad \Delta y = f'(x)\,\Delta x + \varepsilon\,\Delta x = dy + \varepsilon\,\Delta x.$$

11.7 Theorem (Chain Rule)   Let $h(t) = f(g(t))$ be the composite of f and g. If $g'(t)$ exists and $f'(g(t))$ exists [so that g is defined in an interval about t and f is defined in an interval about $g(t)$], then $h'(t)$ exists and $h'(t) = f'(g(t))g'(t)$.

Proof: Let $x = g(t)$ and $y = h(t) = f(x)$. By (11.1),

$$(11.2) \qquad\qquad \Delta y = f'(x)\,\Delta x + \varepsilon\,\Delta x, \qquad \varepsilon \simeq 0,$$

for any infinitesimal Δx. Setting $\Delta x = {}^*g(t + \Delta t) - g(t)$, where Δt is any non-zero infinitesimal, and dividing by Δt, we get $\Delta y/\Delta t = f'(x)(\Delta x/\Delta t) + \varepsilon(\Delta x/\Delta t)$. The result follows by taking standard parts. $\square$

11.8 Inverse Function Theorem Let f be continuous and strictly increasing (or decreasing) on (a, b) and let g be the inverse of f. If f is differentiable at $x \in (a, b)$ with $f'(x) \neq 0$, then g is differentiable at $y = f(x)$, and $g'(y) = 1/f'(x)$.

Proof: Let $\Delta y \simeq 0$, $\Delta y \neq 0$, and set $\Delta x = {}^*g(y + \Delta y) - g(y)$. Then Δx is infinitesimal and nonzero since g is continuous (why?) and one-to-one. Since $f'(x) \neq 0$,

$$(11.3) \qquad \frac{1}{f'(x)} \simeq \frac{\Delta x}{{}^*f(x + \Delta x) - f(x)} = \frac{\Delta x}{y + \Delta y - y} = \frac{\Delta x}{\Delta y}.$$

Since this is true for all nonzero infinitesimals Δy, $g'(y)$ exists and equals $1/f'(x)$. $\square$.

Partial derivatives of functions of several variables are defined as usual. For notational convenience, we confine ourselves to functions $z = f(x, y)$ of two variables; the extension to functions of n variables is obvious. The partial derivatives f_x and f_y are defined by $f_x(a, b) = g'(a)$ and $f_y(a, b) = h'(b)$, where $g(x) = f(x, b)$, $h(y) = f(a, y)$. Assuming that the partial derivatives exist, we define the *increment* Δz and *total differential dz* by

$$(11.4) \qquad\qquad \Delta z = {}^*f(a + \Delta x, y + \Delta y) - f(a, b)$$

and

$$(11.5) \qquad\qquad dz = f_x(a, b)\,\Delta x + f_y(a, b)\,\Delta y,$$

respectively, where Δx and Δy are arbitrary numbers in ${}^*R$. Note that both Δz and dz depend on a, b, Δx, and Δy. We say that f is *differentiable* at (a, b) if

$$(11.6) \qquad\qquad \Delta z = dz + \varepsilon\,\Delta x + \delta\,\Delta y$$

for any infinitesimals Δx and Δy and corresponding $\varepsilon \simeq 0$, and $\delta \simeq 0$.

11.9 Theorem If f_x and f_y are continuous at (a, b), then f is differentiable at (a, b).

Proof: If Δx and Δy are nonzero standard numbers, then

(11.7) $f(a + \Delta x, b + \Delta y) - f(a, b)$
$$= [f(a + \Delta x, b + \Delta y) - f(a + \Delta x, b)] + [f(a + \Delta x, b) - f(a, b)].$$

Using the mean value theorem, we have

(11.8)
$$f(a + \Delta x, b) - f(a, b) = f_x(u, b)\,\Delta x,$$
$$f(a + \Delta x, b + \Delta y) - f(a + \Delta x, b) = f_y(a + \Delta x, v)\,\Delta y,$$

where $|a - u| \leq \Delta x$, $|b - v| \leq \Delta y$. Hence

(11.9) $f(a + \Delta x, b + \Delta y) - f(a, b) = f_x(u, b)\,\Delta x + f_y(a + \Delta x, v)\,\Delta y.$

Since this equation is true for all standard Δx and Δy we have by transfer check; you must use Skolem functions) that for $\Delta x \simeq 0$, $\Delta y \simeq 0$,

(11.10) $\Delta z = {}^*f_x(u, b)\,\Delta x + {}^*f_y(a + \Delta x, v)\,\Delta y$

for u, $v \in {}^*R$ with $|a - u| \leq \Delta x$, $|b - v| \leq \Delta y$. The result follows since ${}^*f_x(u, b) \simeq f_x(a, b)$ and ${}^*f_y(a + \Delta x, v) \simeq f_y(a, b)$. □

Exercises I.11

1. Prove Theorem 11.6, parts (i) and (iii).
2. Why is the inverse function g in Theorem 11.8 continuous?
3. Use Proposition 11.2 to show that if f' exists then it is continuous on $[a, b]$ if and only if for each $x \in {}^*[a, b]$ and each Δx with $\Delta x \simeq 0$ and $x + \Delta x \in {}^*[a, b]$, we have $\Delta y = {}^*f(x + \Delta x) - {}^*f(x) = {}^*f'(x)\,\Delta x + \varepsilon\,\Delta x$, where $\varepsilon \simeq 0$. That is, at any $x \in {}^*[a, b]$, $\Delta y = dy + \varepsilon\,\Delta x$ with $\varepsilon \simeq 0$ when $\Delta x \simeq 0$.
4. Consider the example $f(x) = x^2 \sin(1/x)$, $x \neq 0$, $f(0) = 0$, to see what happens in Exercise 3 if f' exists but is not continuous.
5. (Darboux's Theorem) A function f on $[a, b]$ may possess a derivative f' on $[a, b]$ that is not continuous. Prove that if $f'(a) < c < f'(b)$ then $f'(x) = c$ for some x in (a, b). [Hint: (i) Prove that if the minimum of f is at a then $f'(a) \geq 0$, and if it is at b then $f'(b) \leq 0$. (ii) Suppose that $f'(a) < 0$ and $f'(b) > 0$; prove that $f'(x) = 0$ for some $x \in (a, b)$. (iii) Reduce the problem to (ii) by using an appropriate function.]
6. (Hyperreal Mean Value Theorem) Let f be differentiable on (a, b). Assuming the standard mean value theorem (i.e., if $x < y$ are points in (a, b) then there is a c, $x < c < y$, with $f'(c) = [f(y) - f(x)]/(y - x)$, show that if $x < y$ in $^*(a, b)$ then there is a $c \in {}^*(a, b)$, $x < c < y$, with ${}^*f'(c) = [{}^*f(y) - {}^*f(x)]/(y - x)$.
7. Let f be twice differentiable on (a, b). Prove that if $f'(c) = 0$ and $f''(c) < 0$ $[f''(c) > 0]$ for some $c \in (a, b)$ then f has a local maximum [minimum] at c. (Hint: Use Exercise 6.)

8. (Behrens [5]). A real-valued function f defined in a neighborhood of $c \in R$ is *uniformly differentiable* at c with derivative $f'(c)$ if, for each $\varepsilon > 0$ in R, there is a $\delta > 0$ in R so that

$$\left| \frac{f(x) - f(y)}{x - y} - f'(c) \right| < \varepsilon$$

for all $x, y \in (c - \delta, c + \delta)$.

(a) Show that f is uniformly differentiable at c iff there exists an $\alpha \in R$,

$$\alpha \simeq \frac{{}^*f(x) - {}^*f(y)}{x - y}$$

for all $x, y \in {}^*R$ with $x \simeq y \simeq c$ and $x \neq y$, and that in this case $f'(c) = \alpha$.

(b) Show that if f has a derivative on an open interval (a, b) containing c, then f' is continuous at c iff f is uniformly differentiable at c. [Hint: see the proof of Proposition 11.2].

(c) Give an example of a function f which is uniformly differentiable at a point c, but every neighborhood of c contains a point where f is not differentiable.

(d) Show that if f is uniformly differentiable at c then f is continuous on some neighborhood of c.

(e) Show that if f is increasing on an interval (a, b) and f is uniformly differentiable at $x \in (a, b)$ with $f'(x) \neq 0$, then the inverse function g is uniformly differentiable at $y = f(x)$ and $g'(y) = 1/f'(x)$.

I.12 Riemann Integration

Nonstandard analysis is a natural tool for developing the theory of Riemann integration on an interval $[a, b]$, and this section contains a few relevant results. We concentrate on integration of continuous functions on intervals $[a, b]$. The presentation in this section owes much to Keisler [23].

12.1 Definition Let f be a continuous function on $[a, b] \subset R$, $a < b$. A partition P of $[a, b]$ is a set $\{x_0, x_1, \ldots, x_n\}$, where $a = x_0 < x_1 < \cdots < x_{n-1} < x_n = b$. The upper, lower, and ordinary Riemann sums $\bar{S}_a^b(f, P)$, $\underline{S}_a^b(f, P)$, and $S_a^b(f, P)$ of f with respect to P on $[a, b]$ are defined by

$$\bar{S}_a^b(f, P) = \sum M_i \Delta x_i (1 \leq i \leq n),$$
$$\underline{S}_a^b(f, P) = \sum m_i \Delta x_i (1 \leq i \leq n),$$

and

$$s_a^b(f, P) = \sum f(x_{i-1}) \Delta x_i (1 \le i \le n),$$

where M_i and m_i are the maximum and minimum of f on $[x_{i-1}, x_i]$ and $\Delta x_i = x_i - x_{i-1}$, $1 \le i \le n$. If P is given by setting $x_k = a + k \Delta x$, $0 \le k \le n - 1$, where Δx is a fixed positive number and n is the greatest integer for which $a + (n - 1) \Delta x < b$, then we write $\bar{S}_a^b(f, \Delta x)$, $\underline{S}_a^b(f, \Delta x)$, and $S_a^b(f, \Delta x)$ for the upper, lower, and ordinary Riemann sums, and say that P is determined by Δx. Here, $\Delta x_n = b - x_{n-1} \le \Delta x$. If $a = b$, all Riemann sums are set equal to 0.

The partition P_2 is a *refinement* of P_1 if $P_1 \subseteq P_2$. It is easy to see that if P_2 is a refinement of P_1, then

$$\underline{S}_a^b(f, P_1) \le \underline{S}_a^b(f, P_2) \le S_a^b(f, P_2) \le \bar{S}_a^b(f, P_2) \le \bar{S}_a^b(f, P_1).$$

The *common refinement* P_3 of P_1 and P_2 is given by $P_3 = P_1 \cup P_2$. Since

$$\underline{S}_a^b(f, P_1) \le \underline{S}_a^b(f, P_3) \le \bar{S}_a^b(f, P_3) \le \bar{S}_a^b(f, P_2),$$

it follows that any lower Riemann sum is less than or equal to any upper Riemann sum.

12.2 Definition The function f on $[a, b]$ is said to be *Riemann integrable* on $[a, b]$ with *integral* $\int_a^b f(x) \, dx$ if (i) $\underline{S}_a^b(f, P) \le \int_a^b f(x) \, dx \le \bar{S}_a^b(f, P)$ for any partition P of $[a, b]$ and (ii) given any $\varepsilon > 0$ in R there is a partition P so that $\bar{S}_a^b(f, P) - \underline{S}_a^b(f, P) < \varepsilon$.

We now set out to show that a continuous function f is Riemann integrable. Although we do not have an extension of the set of partitions of $[a, b]$ in this chapter, we can fix f and extend the Riemann sums determined by positive numbers $\Delta x \in R$ to $\Delta x \in {}^*R$. In the following result, ${}^*\underline{S}_a^b(f, \cdot)$ and ${}^*\bar{S}_a^b(f, \cdot)$ denote the extensions to ${}^*R$ of such sums $\underline{S}_a^b(f, \cdot)$ and $\bar{S}_a^b(f, \cdot)$.

12.3 Proposition Let f be continuous on $[a, b]$, and let Δx be a positive infinitesimal in ${}^*R$. Then ${}^*\underline{S}_a^b(f, \Delta x) \simeq {}^*\bar{S}_a^b(f, \Delta x)$.

Proof: Given $\Delta x > 0$ in R,

$$\bar{S}_a^b(f, \Delta x) - \underline{S}_a^b(f, \Delta x) = \sum (M_i - m_i) \Delta x_i (1 \le i \le n)$$
$$\le \sum B \Delta x_i (1 \le i \le n) = B \sum \Delta x_i (1 \le i \le n) = B(b - a),$$

where $B = \max_{1 \le i \le n}(M_i - m_i)$. Thus to each $\Delta x \in R^+$ corresponds two points $\phi(\Delta x)$ and $\psi(\Delta x)$ on $[a, b]$ with $|\phi(\Delta x) - \psi(\Delta x)| \le \Delta x$ and

$$(12.1) \qquad \bar{S}_a^b(f, \Delta x) - \underline{S}_a^b(f, \Delta x) \le [f(\phi(\Delta x)) - f(\psi(\Delta x))](b - a).$$

For $\Delta x \simeq 0$ in $*R$ there is a $c \in [a, b]$ with $*\phi(\Delta x) \simeq c \simeq *\psi(\Delta x)$, and hence $*f(*\phi(\Delta x)) \simeq *f(*\psi(\Delta x))$ by the continuity of f at c. The result follows by transfer of (12.1). $\square$

12.4 Corollary Let f be continuous on $[a, b]$. Then f is Riemann integrable and $\int_a^b f(x)\,dx \simeq *S_a^b(f, \Delta x)$ for any infinitesimal Δx.

From Corollary 12.4 it follows that $\int_a^b f(x)\,dx = \lim_{\Delta x \to 0} S_a^b(f, \Delta x)$.

In the following we will write $S_a^b(f, \Delta x)$ and $*S_a^b(f, \Delta x)$ as $\sum_a^b f(x)\,\Delta x$ and $\sum_a^b *f(x)\,\Delta x$, respectively. By convention we set $\int_b^a f(x)\,dx = -\int_a^b f(x)\,dx$ and $\int_a^a f(x)\,dx = 0$.

12.5 Theorem Let f and g be continuous on $[a, b]$. Then

(i) $\int_a^b cf(x)\,dx = c \int_a^b f(x)\,dx$ for $c \in R$,
(ii) $\int_a^b [f(x) + g(x)]\,dx = \int_a^b f(x)\,dx + \int_a^b g(x)\,dx$,
(iii) $\int_a^b f(x)\,dx = \int_a^c f(x)\,dx + \int_c^b f(x)\,dx$ if $a \le c \le b$,
(iv) if $f(x) \le g(x)$ on $[a, b]$ then $\int_a^b f(x)\,dx \le \int_a^b g(x)\,dx$,
(v) if $m \le f(x) \le M$ on $[a, b]$ then $m(b - a) \le \int_a^b f(x)\,dx \le M(b - a)$.

Proof: We prove (iii) and (iv) and leave the remaining proofs to the reader.

(iii) For each natural number n, if $\Delta x = (c - a)/n > 0$ then $\sum_a^c f(x)\,\Delta x + \sum_c^b f(x)\,\Delta x = \sum_a^b f(x)\,\Delta x$. The result follows by taking standard parts of the terms in the transferred equality when $n \in *N_\infty$.

(iv) For each standard $\Delta x > 0$, $\sum_a^b f(x)\,\Delta x \le \sum_a^b g(x)\,\Delta x$. Thus by transfer $\sum_a^b *f(x)\,\Delta x \le \sum_a^b *g(x)\,\Delta x$, where $\Delta x > 0$ is infinitesimal. The result follows from Theorem 6.7(iv). $\square$

12.6 Theorem If f is continuous on $[a, b]$, then the function $F(x) = \int_a^x f(t)\,dt$, defined for $x \in [a, b]$, is differentiable. Moreover, $F'(c) = f(c)$ for each $c \in [a, b]$, where $F'(c)$ is the right- or left-hand derivative if $c = a$ or b.

Proof: Fix $c \in [a, b)$. For any standard $h \in (0, b - c)$ we have, using 12.5(iii) and (v), that $f(x_1)h \le F(c + h) - F(c) \le f(x_2)h$, where f has a minimum and maximum on $[c, c + h]$ at x_1 and x_2, respectively. Thus there are Skolem functions $\phi, \psi : (0, b - c) \to [c, c + h]$ so that $f(\phi(h))h \le F(c + h) - F(c) \le f(\psi(h))h$ for all $h \in (0, b - c)$. By transfer, $*f(*\phi(h))h \le *F(c + h) - F(c) \le$

$*f(*\psi(h))h$ for all $h \in *(0, b - c)$. In particular, if $h \simeq 0$ we have

$$*f(*\phi(h)) \le [*F(c + h) - F(c)]/h \le *f(*\psi(h)).$$

Now $*\phi(h) \simeq c$ and $*\psi(h) \simeq c$ if h is infinitesimal and so $*f(*\phi(h)) \simeq *f(*\psi(h)) \simeq f(c)$ by continuity of f at c. Therefore $[*F(c + h) - F(c)]/h \simeq f(c)$ if h is a positive infinitesimal. The argument is similar if h is a negative infinitesimal; the result follows from Proposition 11.1.   □

A result due to Keisler [23], which can be used to justify the definition via integrals of many quantities occurring in applications, is the following.

12.7 Infinite Sum Theorem If $f(x)$ is continuous on $[a, b]$ and $B(u, v)$ is a real-valued function of two variables $(u, v) \in [a, b] \times [a, b]$ satisfying

(a) $B(u, v) = B(u, w) + B(w, v)$ for $u \le w \le v$,
(b) for any infinitesimal subinterval $[x, x + \Delta x] \subseteq *[a, b]$, $*B(x, x + \Delta x) = *f(x)\Delta x + \varepsilon \Delta x$ with $\varepsilon \simeq 0$,

then $B(a, b) = \int_a^b f(x)\, dx$.

Proof: For $n \in N$ let $g(n)$ be the maximum of $[B(x_i, x_i + \Delta x) - f(x_i)\Delta x]/\Delta x$, where $x_k = a + k\Delta x$, $0 \le k < n$, and $\Delta x = (b - a)/n$. From (b), $*g(\omega) \simeq 0$ if $\omega \in *N_\infty$. From (a) and (b), $|B(a, b) - \sum_a^b f(x)\Delta x| \le g(n)(b - a)$ for each $n \in N$, and so, by transfer, $B(a, b) \simeq \sum_a^b *f(x)\Delta x = *S_a^b(f, \Delta x)$ for $n = \omega$.   □

12.8 Fundamental Theorem of Calculus If a function F has a continuous derivative f on $[a, b]$, then $\int_a^b f(x)\, dx = F(b) - F(a)$.

Proof: Let $B(u, v) = F(v) - F(u)$ in Theorem 12.7 and use Exercise I.11.3.
   □

The following calculation is a direct proof of Theorem 12.8. Let $\Delta x = (b - a)/\omega$, where $\omega \in *N_\infty$, $\Delta_k y = *F(a + k\Delta x) - *F(a + (k - 1)\Delta x)$, and $d_k y = *f(a + (k - 1)\Delta x)\Delta x$, $k = 1, 2, \ldots, \omega$. Then by Exercise I.11.3

$$F(b) - F(a) = \sum_{k=1}^{\omega} \Delta_k y$$

$$= \sum [*f(a + (k - 1)\Delta x) + \varepsilon_k]\Delta x(1 \le k \le \omega)$$

$$= \sum d_k y(1 \le k \le \omega) + \sum \varepsilon_k \Delta x(1 \le k \le \omega)$$

$$\simeq \int_a^b f(x)\, dx + \sum \varepsilon_k \Delta x(1 \le k \le \omega),$$

where $\varepsilon_k \simeq 0$ for each k. Since $|\sum \varepsilon_k \Delta x(1 \le k \le \omega)| \le \max_k|\varepsilon_k|(b - a) \simeq 0$, the result follows. The standard proof of 12.8 uses 12.6 to show that $F(x) = \int_a^x f(t)\, dt + F(a)$ for $x \in [a, b]$.

12.9 Example: Volume of Revolution Suppose that a volume V is obtained by revolving a region $R = \{\langle x, y \rangle \in R^2 : 0 \le x \le 1, 0 \le y \le f(x)\}$ about the x-axis, where $f(x)$ is continuous. To find a formula for V we let $B(u, v)$ be the volume obtained by revolving $R(u, v) = \{\langle x, y \rangle \in R^2 : u \le x \le v, 0 \le y \le f(x)\}$ about the x-axis and make the reasonable assumption that B satisfies (a) of 12.7. Also obvious is the fact that $\pi m^2 \Delta x \le B(x, x + \Delta x) \le \pi M^2 \Delta x$ if Δx is standard, $x \in [0, 1]$, and m and M are the minimum and maximum, respectively, of $f(x)$ in $[x, x + \Delta x]$. As in the proof of 12.6, for $\Delta x > 0$, $\Delta x \simeq 0$, we have $^*B(x, x + \Delta x) = \pi[^*f(x)]^2 + \varepsilon \Delta x$ where $\varepsilon \simeq 0$, and so $V = B(a, b) = \pi \int_a^b [f(x)]^2 \, dx$.

Exercises I.12

1. Prove Theorem 12.5, parts (i), (ii), and (v).
2. (Keisler [23]) An "approximate" average of a continuous function for an interval $[a, b]$ is given by $\sum_{k=0}^{n-1} f(a + k \Delta x)/n$, where $n \in N$ and $\Delta x = (b - a)/n$. What relationship does this have to the integral average $\int_a^b f(x) \, dx/(b - a)$?
3. Do Example 12.9 for the case in which the axis of rotation is the y-axis.
4. Prove Bliss's theorem: Let f and g be continuous functions on $[a, b]$. For each $\Delta x > 0$ and the corresponding partition P, let ϕ and ψ be Skolem functions such that, for $1 \le i \le n - 1$, $a + (i - 1) \Delta x \le \phi(i, \Delta x) \le a + i \Delta x$ and $a + (i - 1) \Delta x \le \psi(i, \Delta x) \le a + i \Delta x$ while $a + (n - 1) \Delta x \le \phi(n, \Delta x) \le b$ and $a + (n - 1) \Delta x \le \psi(n, \Delta x) \le b$. Let $S(\Delta x) = \sum_{i=1}^n f(\phi(i, \Delta x)) g(\psi(i, \Delta x)) \Delta x$. Show that $\lim_{\Delta x \to 0} S(\Delta x) = \int_a^b f(x) g(x) \, dx$.

I.13 Sequences of Functions

A sequence of functions on $A \subset R$ is a map $f : N \times A \to R$. As usual we denote $f(n, x)$ by $f_n(x)$ ($n \in N$, $x \in A$). We will use nonstandard analysis to study the convergence of such sequences.

13.1 Proposition The sequence $\langle f_n \rangle$, $f_n : A \to R$, $n \in N$, converges pointwise to the function $f : A \to R$ iff $^*f_n(x) \simeq f(x)$ for all $x \in A$ and all infinite $n \in {}^*N$.

Proof: The sequence $\langle f_n \rangle$ converges pointwise iff for each fixed $x \in A$ the sequence $\langle f_n(x) \rangle$ converges to $f(x)$. The result then follows from 8.1. □

13.2 Proposition The sequence $\langle f_n \rangle$, $f_n : A \to R$, converges uniformly to the function $f : A \to R$ iff $^*f_n(x) \simeq {}^*f(x)$ for all $x \in {}^*A$ and all infinite $n \in {}^*N$.

Proof: Recall that $\langle f_n \rangle$ converges uniformly to f iff, given $\varepsilon > 0$ in R, there exists a $k \in N$ so that $|f_n(x) - f(x)| < \varepsilon$ for all $x \in A$ if $n \geq k$. Suppose then that $\langle f_n \rangle$ converges uniformly to f and find the k corresponding to a specified $\varepsilon > 0$. Then the sentence

$$(13.1) \qquad (\forall n)(\forall x)[\underline{N}\langle n \rangle \wedge \underline{A}\langle x \rangle \wedge n \geq k \to |\underline{f}_n(x) - f(x)| < \varepsilon]$$

is true in $\mathscr{R}$. By transfer, $|{}^*f_n(x) - {}^*f(x)| < \varepsilon$ for all $n \in {}^*N$, $n \geq k$, and all $x \in {}^*A$. In particular, this is true for all infinite n, no matter what $\varepsilon > 0$ we choose. Hence ${}^*f_n(x) \simeq {}^*f(x)$ for all infinite $n \in {}^*N$ and all $x \in {}^*A$.

The converse is left to the reader. $\square$

13.3 Dini's Theorem Suppose that the sequence $\langle f_n \rangle$ of continuous functions on the compact set $A \subset R$ is monotone [i.e., $f_n(x) \leq f_m(x)$ or $f_n(x) \geq f_m(x)$ for all $n \geq m$, $x \in A$] and converges pointwise to the continuous function f. Then the convergence is uniform.

Proof: We may suppose that $f(x) = 0$, $x \in A$ (simply by considering the sequence $f_n - f$), and that f_n decreases (otherwise consider $-f_n$). By transfer we see that ${}^*f_n(x) \leq {}^*f_m(x)$ for all $n \geq m$ in ${}^*N$ and all $x \in {}^*A$. Fix $x \in {}^*A$. Since A is compact there is a $y \in A$, $y \simeq x$. Then, for each $n \in {}^*N_\infty$ and standard m, $0 \leq {}^*f_n(x) \leq {}^*f_m(x) \simeq f_m(y)$, and since $\lim_{m \to \infty} f_m(y) = 0$ it follows that ${}^*f_n(x) \simeq 0$. $\square$

13.4 Theorem If $\langle f_n \rangle$ converges uniformly to f on A, $a \in R$ is a limit point of A, and $\lim_{x \to a} f_n(x) = s_n$ exists for all $n \in N$, then $\langle s_n \rangle$ converges and $\lim_{x \to a} f(x) = \lim_{n \to \infty} s_n$.

Proof: Let $\varepsilon > 0$ in R be specified. Then there is a $k \in N$ so that $|f_n(x) - f(x)| < \varepsilon/4$, and hence $|f_n(x) - f_m(x)| < \varepsilon/2$ for all $x \in A$ and all $n, m \geq k$ by uniform convergence of $\langle f_n \rangle$ to f on A. By transfer as in 13.2, $|{}^*f_n(x) - {}^*f(x)| < \varepsilon/4$ and $|{}^*f_n(x) - {}^*f_m(x)| < \varepsilon/2$ for all $n, m \geq k$ and all $x \in {}^*A$. Since $s_n \simeq {}^*f_n(x)$ if $x \simeq a$, $x \in {}^*A$, we have $|s_n - s_m| \simeq |{}^*f_n(x) - {}^*f_m(x)| < \varepsilon$ if $n, m \geq k$, and so $\langle s_n \rangle$ is a Cauchy sequence and converges, say, to L. It follows (letting $x \simeq a$ and $n \geq k$) that

$$|{}^*f(x) - L| \leq |{}^*f(x) - {}^*f_n(x)| + |{}^*f_n(x) - s_n| + |s_n - L|$$
$$\leq \varepsilon/4 + \text{infinitesimal} + 2\varepsilon$$
$$< 3\varepsilon,$$

and hence ${}^*f(x) \simeq L$. $\square$

13.5 Corollary If the functions f_n are continuous on A and $\langle f_n \rangle$ converges uniformly to f on A, then f is continuous on A.

We end this section with a proof of the Arzelà–Ascoli theorem, a result which has many important applications in analysis. The theorem asserts that from a uniformly bounded, equicontinuous sequences $\langle f_n \rangle$ of functions on a closed bounded interval $[a, b] \subset R$ it is possible to select a subsequence which converges uniformly on $[a, b]$ to a continuous function f. That the result is not true for an arbitrary sequence of continuous functions is shown by the sequence in which $f_n(x) = x^n$ on $[0, 1]$. Here $\langle f_n \rangle$ actually converges point-wise (but not uniformly) to the discontinuous function

$$f(x) = \begin{cases} 0, & 0 \le x < 1, \\ 1, & x = 1. \end{cases}$$

13.6 Definition The sequence $\langle f_n \rangle$ of functions on $[a, b]$ is *uniformly bounded* if there exists an M so that $|f_n(x)| \le M$ for all $x \in [a, b]$ and all $n \in N$.

The sequence $\langle f_n \rangle$ of functions on $[a, b]$ is *equicontinuous* if, given $\varepsilon > 0$, there is a $\delta > 0$ (independent of x, y, and n) so that $|f_n(x) - f_n(y)| < \varepsilon$ for all $n \in N$ and all $x, y \in [a, b]$ such that $|x - y| < \delta$. (Each f_n, then, is uniformly continuous on $[a, b]$.)

13.7 Arzelà–Ascoli Theorem If $\langle f_n \rangle$ is a uniformly bounded and equicontin-uous sequence of functions on the closed and bounded interval $[a, b]$, then there is a subsequence $\langle f_{n_k} \rangle$ which converges uniformly to a continuous function f on $[a, b]$.

Proof: Let $\varepsilon > 0$ be given and find the corresponding $\delta > 0$ from the equi-continuity of the sequence. Then the sentence

$$(\forall n)(\forall x)(\forall y)[n \in \underline{N} \wedge x \in [a, b] \wedge y \in [a, b] \wedge |x - y| < \delta$$
$$\rightarrow |\underline{f}_n(x) - \underline{f}_n(y)| < \varepsilon]$$

is true. By transfer, for all $n \in {}^*N$ and all $x, y \in {}^*[a, b]$ such that $|x - y| < \delta$ we have $|{}^*f_n(x) - {}^*f_n(y)| < \varepsilon$. In particular, $|{}^*f_n(x) - {}^*f_n(y)| < \varepsilon$ if $x \simeq y$ for any $n \in {}^*N$. Since $\varepsilon > 0$ is arbitrary, we see that ${}^*f_n(x) \simeq {}^*f_n(y)$ for any $n \in {}^*N$ as long as $x \simeq y$.

Now let $n = \omega$ be a fixed infinite natural number. By an argument similar to that of the first paragraph we see that $|{}^*f_\omega(x)| \le M$ for any $x \in {}^*[a, b]$, so that ${}^*f_\omega(x)$ is near-standard for $x \in [a, b]$. Define $f(x) = {}^\circ({}^*f_\omega(x))$, $x \in [a, b]$. We claim that $f(x)$ is uniformly continuous. For let $\varepsilon > 0$ be given and find the $\delta > 0$ corresponding to $\varepsilon/2$ from equicontinuity. Then if $x, y \in [a, b]$

and $|x - y| < \delta$, we have

$$|f(x) - f(y)| \leq |f(x) - {}^*f_\omega(x)| + |{}^*f_\omega(x) - {}^*f_\omega(y)| + |{}^*f_\omega(y) - f(y)|.$$

The first and last terms on the right are infinitesimal by definition of f, and the middle term is $< \varepsilon/2$ by the argument of the first paragraph, and so $|f(x) - f(y)| < \varepsilon$.

Finally we show that a subsequence of $\langle f_n \rangle$ converges uniformly to f on $[a, b]$. To do this it suffices to show that for all $\varepsilon > 0$ and all $n \in N$ there is an $m > n$ so that $|f_m(x) - f(x)| < \varepsilon$ for all $x \in [a, b]$ (why?). Suppose this statement is not true. Then there exists an $\varepsilon_0 > 0$ and an $n_0 \in N$ so that for each $m > n_0$ we can find an $x \in [a, b]$ with $|f_m(x) - f(x)| \geq \varepsilon_0$. Thus there exists a Skolem function $\psi : \{n_0, n_0 + 1, \dots\} \to [a, b]$ so that the statement $(\forall m)[m \in N \wedge m \geq n_0 \to |f_m(\psi(m)) - f(\psi(m))| \geq \varepsilon_0]$ is true. By transfer, given $\omega \in {}^*N_\infty$, we have $\omega \geq n_0$, and so there exists an $x \in {}^*[a, b]$ [equal to ${}^*\psi(\omega)$] such that $|{}^*f_\omega(x) - {}^*f(x)| \geq \varepsilon_0$. But by compactness of $[a, b]$ and Robinson's theorem, this x is infinitesimally close to a $y \in [a, b]$, and so

$$|{}^*f_\omega(x) - {}^*f(x)| \leq |{}^*f_\omega(x) - {}^*f_\omega(y)| + |{}^*f_\omega(y) - f(y)| + |f(y) - {}^*f(x)|.$$

Each term on the right is infinitesimal, the last by the continuity of f. This contradiction proves the theorem.   □

A general form of the Arzelà–Ascoli theorem will be given in §III.8 of Chapter III.

Exercises I.13

1. Finish the proof of Proposition 13.2.
2. Prove Corollary 13.5 directly from Proposition 13.2.
3. Give an example to show that an equicontinuous sequence need not be uniformly bounded.
4. Let $\langle f_n \rangle$ be a sequence on $[a, b]$. Show that $\langle f_n \rangle$ is an equicontinuous sequence if and only if for any $n \in {}^*N$ and any pair $x, y \in {}^*[a, b]$ with $x \simeq y$ we have ${}^*f_n(x) \simeq {}^*f_n(y)$. (Hint: For the necessity see the proof of Theorem 13.7.)
5. Let $\langle f_n \rangle$ be a sequence of continuous functions on $[a, b]$ which converges uniformly to f. Show that $\lim_{n \to \infty} \int_a^b f_n(x)\,dx = \int_a^b f(x)\,dx$.

I.14   Two Applications to Differential Equations

As our first application we prove the Cauchy–Peano existence theorem for ordinary differential equations. A nonstandard proof was first presented by A. Robinson [40].

14.1 Cauchy–Peano Existence Theorem Let f be continuous and satisfy $|f(x, y)| \leq M$ on the rectangle $B = \{(x, y) \in R^2 : |x - x_0| \leq a, \; |y - y_0| \leq b\}$. Then there exists a function ϕ with continuous first derivative, defined on the closed interval $I = \{x \in R : |x - x_0| \leq c\}$, where $c = \min(a, bM^{-1})$, and satisfying $\phi(x_0) = y_0$ and $\phi'(x) = f(x, \phi(x))$ for $x \in I$.

Proof: We begin, as in [40], by constructing a family of polygonal approximations. It suffices to construct a solution on $[x_0, x_0 + c]$. Divide $[x_0, x_0 + c]$ into n equal parts by the points $x_k = x_0 + kc/n$, $0 \leq k \leq n$, and define ϕ_n by the equations

(14.1)
$$\phi_n(x_0) = y_0,$$
$$\phi_n(x) = \phi_n(x_k) + f(x_k, \phi_n(x_k))(x - x_k)$$

for $x_k < x \leq x_{k+1}$, $0 \leq k \leq n - 1$. For any $n \in N$, the graph of ϕ_n lies in B since $|f(x, y)| \leq M$. Moreover, $|\phi_n(x) - \phi_n(x')| \leq M|x - x'|$ for any $x, x' \in [x_0, x_0 + c]$. Thus the following statement is true in $\mathscr{R}$:

(14.2)   For all $n \in N$, $x, x' \in [x_0, x_0 + c]$, we have $|\phi_n(x) - y_0| \leq b$ and $|\phi_n(x) - \phi_n(x')| \leq M|x - x'|$.

By transfer, for all $n \in {}^*N$ and $x, x' \in {}^*[x_0, x_0 + c]$,

(14.3) $|{}^*\phi_n(x) - y_0| \leq b$

and

(14.4) $|{}^*\phi_n(x) - {}^*\phi_n(x')| \leq M|x - x'|.$

We now let $n = \omega \in {}^*N_\infty$ and note that ${}^*\phi_\omega(x)$ is finite for all $x \in {}^*[x_0, x_0 + c]$ by (14.3). We may therefore define the standard function ϕ on $[x, x_0 + c]$ by $\phi(x) = \mathrm{st}({}^*\phi_\omega(x))$. Now ϕ is continuous since, for standard x and x' in $[x_0, x_0 + c]$, $|\phi(x) - \phi(x')| \simeq |{}^*\phi_\omega(x) - {}^*\phi_\omega(x')| \leq M|x - x'|$ by (14.4). Therefore

(14.5) ${}^*\phi(y) \simeq \phi(\mathrm{st}(y)) \simeq {}^*\phi_\omega(\mathrm{st}(y)) \simeq {}^*\phi_\omega(y)$

if $y \in {}^*[x_0, x_0 + c]$. Since f is continuous and hence uniformly continuous on the closed bounded set B,

(14.6) ${}^*f(x, {}^*\phi(x)) \simeq {}^*f(x, {}^*\phi_\omega(x))$

for all $x \in {}^*[x_0, x_0 + c]$ (exercise). Now if $x \in [x_0, x_0 + c]$, then $x_k \leq x \leq x_{k+1}$ for some $k \in {}^*N$, $0 \leq k \leq \omega - 1$, whence $x_k \simeq x$ and so

$$\phi(x) \simeq {}^*\phi_\omega(x_k)$$
$$= y_0 + \sum_{i=0}^{k-1} [{}^*\phi_\omega(x_{i+1}) - {}^*\phi_\omega(x_i)]$$
$$= y_0 + \sum_{i=0}^{k-1} {}^*f(x_i, {}^*\phi_\omega(x_i))(x_{i+1} - x_i)$$

by transfer from (14.1). Thus

$$\phi(x) \simeq y_0 + \sum_{i=0}^{k-1} {}^*f(x_i, {}^*\phi(x_i))(x_{i+1} - x_i)$$

since

$$\left| \sum_{i=0}^{k-1} {}^*f(x_i, {}^*\phi_\omega(x_i))(x_{i+1} - x_i) - \sum_{i=0}^{k-1} {}^*f(x_i, {}^*\phi(x_i))(x_{i+1} - x_i) \right|$$
$$\leq c \max_{0 \leq i \leq k-1} |{}^*f(x_i, {}^*\phi_\omega(x_i)) - {}^*f(x_i, {}^*\phi(x_i))|,$$

which is infinitesimal by (14.6) (where have we used the transfer principle in this argument?). Since

$$\sum_{i=0}^{k-1} {}^*f(x_i, {}^*\phi(x_i))(x_{i+1} - x_i) \simeq {}^*S_{x_0}^x(f(x, \phi(x)), \Delta x),$$

where $\Delta x = c/\omega$, it follows from Corollary 12.4 that

$$\phi(x) = y_0 + \int_{x_0}^x f(t, \phi(t))\, dt.$$

Therefore ϕ has a continuous derivative and $\phi'(x) = f(x, \phi(x))$. □

The standard proof of this result [8] uses the Arzelà–Ascoli theorem, 13.7. The reader is referred to any standard text on differential equations for a discussion of a (Lipshitz) condition that ensures uniqueness of the solution.

Lastly we use nonstandard techniques to derive the wave equation for a vibrating string. We assume that the magnitude of the tension T and the density μ of the string are constant along the string. Given an infinitesimal segment of length Δs from P_1 to P_2 on the string as shown in the figure, the

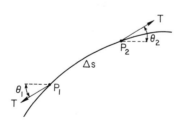

vertical force on the segment is $T({}^*\sin\theta_2 - {}^*\sin\theta_1)$ and its mass is $\mu\,\Delta s$. If the nearest standard point is x, and we are considering the vertical position y as a twice continuously differentiable function of x and t, then by Newton's law,

$$T({}^*\sin\theta_2 - {}^*\sin\theta_1) = \mu\,\Delta s(y_{tt}(x, t) + \varepsilon),$$

where $\varepsilon \simeq 0$, so that

$$\frac{T}{\mu}\left(\frac{{}^*\sin\theta_2 - {}^*\sin\theta_1}{\Delta s}\right) \simeq y_{tt}(x,t).$$

We want to show that $({}^*\sin\theta_2 - {}^*\sin\theta_1)/\Delta s \simeq y_{xx}(x,t)$. Often in deriving the wave equation the assumption that Δy is uniformly small is made, and then the expression $(\sin\theta_2 - \sin\theta_1)/\Delta s$ is replaced by $(\tan\theta_2 - \tan\theta_1)/\Delta x$, where $\Delta x^2 + \Delta y^2 = \Delta s^2$. Since Δs is infinitesimal, as is ${}^*\sin\theta - {}^*\sin\theta_1$, it is not clear why this replacement is justified. Let us instead fix t and consider small changes Δx_1 and Δx_2 at P_1 and P_2 resulting in changes $\Delta y_i, \Delta s_i$, $i = 1, 2$, respectively. We may take Δx_i, $i = 1, 2$, so small that

$$\Delta x_i/\Delta s_i = {}^*x_s(P_i,t) + \varepsilon_i,$$
$$\Delta y_i/\Delta s_i = {}^*\sin\theta_i + \varepsilon_{i+2},$$
$$\Delta y_i/\Delta x_i = {}^*y_x(P_i,t) + \varepsilon_{i+4}$$

for $i = 1, 2$, where $\varepsilon_j/\Delta s \simeq 0$ for $1 \le j \le 6$. Then, omitting the $*$, we have

$$\frac{\sin\theta_2 - \sin\theta_1}{\Delta s} \simeq \left(\frac{\Delta y_2}{\Delta x_2}\frac{\Delta x_2}{\Delta s_2} - \frac{\Delta y_1}{\Delta x_1}\frac{\Delta x_1}{\Delta s_1}\right)\Big/\Delta s$$

$$\simeq \frac{y_x(P_2,t)x_s(P_2,t) - y_x(P_1,t)x_s(P_1,t)}{\Delta s}.$$

Since we are assuming that $y(x,t)$ is twice continuously differentiable, it follows as in Proposition 11.2 that

$$\frac{\sin\theta_2 - \sin\theta_1}{\Delta s} \simeq [y_x(P_2,t)x_s(P_2,t) - y_x(P_1,t)x_s(P_2,t)$$

$$+ y_x(P_1,t)x_s(P_2,t) - y_x(P_1,t)x_s(P_1,t)](\Delta s)^{-1}$$

$$= \frac{y_x(P_2,t) - y_x(P_1,t)}{\Delta x}\frac{\Delta x}{\Delta s}x_s(P_2,t)$$

$$+ \frac{x_s(P_2,t) - x_s(P_1,t)}{\Delta s}y_x(P_1,t)$$

$$\simeq y_{xx}(x,t)[x_s(x,t)]^2 + x_{ss}(x,t)y_x(x,t).$$

The wave equation now results if y is "uniformly small" in the sense that $\partial x/\partial s$ can be taken to be 1 and $\partial^2 x/\partial s^2$ can be taken to be 0.

There are many potential applications of nonstandard analysis to differential equations. For example, for applications to singular perturbations, the reader is referred to the work of the Strasbourg group ([6] and the papers referenced there).

Exercises I.14

1. In the proof of Theorem 14.1, Eq. (14.6) states that $*f(x, *\phi(x)) \simeq *f(x, *\phi_\omega(x))$ for all $x \in *[x_0, x_0 + c]$. Show that this is correct.
2. Fill in the details in Theorem 14.1 on the use of the transfer principle to show that $\phi(x) = y_0 + \int_{x_0}^x f(t, \phi(t)) \, dt$.
3. Show that Theorem 14.1 goes through if we replace $f(x_k, \phi_n(x_k))$ in (14.1) by $M_{n,k}$ where $\min_{(x,y) \in A} f(x, y) \le M_{n,k} \le \max_{(x,y) \in A} f(x, y)$ and

$$A = [x_k, x_{k+1}] \times \left[\phi_n(x_k) - \frac{Mc}{n}, \phi_n(x_k) + \frac{Mc}{n} \right].$$

4. The conditions in Theorem 14.1 do not guarantee a unique solution ϕ to $\phi' = f(x, y)$. Use infinitesimal partitions and the idea in Exercise 3 to obtain the solutions $\phi(x) \equiv 0$ and $\phi(x) = x^3$ to the equation $\phi' = 3\phi^{2/3}$, $\phi(0) = 0$.
5. Generalize Theorem 14.1 to the vector situation. Let x denote a point in R and y denote a point in R^n. Let f be defined on $\{(x, y) \in R \times R^n : |x - x_0| \le a, \|y - y_0\| \le b\}$ where $x_0 \in R$, $y_0 \in R^n$, and $\|\cdot\|$ denotes the usual distance in R^n. Consider the system $\phi' = f(x, \phi)$, $\phi(x_0) = \phi_0$, where $\phi(x) = (\phi_1(x), \ldots, \phi_n(x))$ and $\phi'(x) = (\phi_1'(x), \ldots, \phi_n'(x))$. Find conditions on the vector function f so that a solution ϕ to this system exists in a certain interval about x_0.

I.15 Proof of the Transfer Principle

Recall that the only functions and relations that are in $*\mathscr{R}$ are extensions of standard functions and relations. We assume that each constant $\underline{c}$ in $L_{*\mathscr{R}}$ names an element of $*R$, and if $\underline{c}$ names an element of R then $\underline{c}$ is in $L_{\mathscr{R}}$. Recall Definitions 3.8, 4.1, and 5.2, which give the following inductive definition of a constant term which is interpretable in $*\mathscr{R}$:

(i) A constant $\underline{c}$ in $L_{*\mathscr{R}}$ is interpretable in $*\mathscr{R}$ and is interpreted as the element it names.

(ii) If $\underline{f}$ is the name of a function f of n variables on R and $\tau^1, \ldots, \tau^n$ are constant terms interpretable in $*\mathscr{R}$ as $r^1, \ldots, r^n$, respectively, and if the n-tuple $\langle r^1, \ldots, r^n \rangle$ is in the domain of the nonstandard extension $*f$ of f, then $*\underline{f}(\tau^1, \ldots, \tau^n)$ is a constant term interpretable in $*\mathscr{R}$ as $*f(r^1, \ldots, r^n)$.

We now want to associate with each constant term in $L_{*\mathscr{R}}$ a fixed sequence of constant terms in $L_{\mathscr{R}}$. We will denote the sequence for a constant term τ

by $\langle T_\tau(n) \rangle$ or just T_τ. A sequence T_τ is defined for all terms τ, interpretable or not, by the following inductive definition:

(α) For each $r \in {}^*R$ we choose a definite sequence $\langle r_n \rangle$ from R so that $r = [\langle r_n \rangle]$. If $r \in R$, we choose $r_n = r$ for all n. If $\underline{c}$ is a constant in $L_{*\mathscr{R}}$ that names r, we set $T_{\underline{c}}(n) = \underline{r}_n$, where $\underline{r}_n$ is a name in $L_{\mathscr{R}}$ of $r_n \in R$ for all $n \in N$. If $r \in R$, we set $T_{\underline{c}}(n) = \underline{c}$ for all n.

(β) If $\tau = {}^*\underline{f}(\tau^1, \ldots, \tau^k)$ where $\underline{f}$ is a name of the function f of k variables on R and the τ^i are constant terms in $L_{*\mathscr{R}}$, $1 \le i \le k$, then $T_\tau(n) = \underline{f}(T_{\tau^1}(n), \ldots, T_{\tau^k}(n))$.

Conditions (α) and (β) serve to define T_τ inductively for all constant terms τ in $L_{*\mathscr{R}}$. We are now able to prove a simple form of a theorem due to Łoś (pronounced "Wash").

15.1 Theorem

(A) If τ is a constant term in $L_{*\mathscr{R}}$ and $\langle r_n \rangle$ is a sequence of numbers in R, then τ is interpretable in ${}^*\mathscr{R}$ and names $[\langle r_n \rangle]$ iff $T_\tau(n)$ is almost everywhere (a.e.) interpretable in $\mathscr{R}$ and names r_n a.e. [i.e., for all n in a set U in $\mathscr{U}$, $T_\tau(n)$ is interpretable and names r_n].

(B) If $\tau^1, \ldots, \tau^k$ are constant terms in $L_{*\mathscr{R}}$ and ${}^*\underline{P}\langle \tau^1, \ldots, \tau^k \rangle$ is an atomic sentence in $L_{*\mathscr{R}}$, then ${}^*\underline{P}\langle \tau^1, \ldots, \tau^k \rangle$ holds in ${}^*\mathscr{R}$ iff $\underline{P}\langle T_{\tau^1}(n), \ldots, T_{\tau^k}(n) \rangle$ holds a.e. in $\mathscr{R}$.

Proof: (A) The proof is by induction on the complexity of the terms (as defined by 3.8 and 5.2).

(i) If $\tau = \underline{c}$, where $\underline{c}$ is a constant naming an element of ${}^*R$, then $\underline{c}$ names $[\langle r_n \rangle]$ iff $T_{\underline{c}}(n)$ a.e. names r_n by definition of $T_{\underline{c}}$ in (α).

(ii) Let $\tau = {}^*\underline{f}(\tau^1, \ldots, \tau^k)$, where $\underline{f}$ is a name of the function f of k variables and $\tau^1, \ldots, \tau^k$ are constant terms for which (A) is true; i.e., given j, $1 \le j \le k$, and a sequence $\langle r_n^j \rangle$, $r_n^j \in R$, τ^j is interpretable in ${}^*\mathscr{R}$ and names $[\langle r_n^j \rangle]$ iff $T_{\tau^j}(n)$ is a.e. interpretable and names r_n^j a.e. Let $\langle s_n \rangle$ be a sequence in R. Then the following statements are equivalent:

(a) The term $\tau = {}^*\underline{f}(\tau^1, \ldots, \tau^k)$ is interpretable in ${}^*\mathscr{R}$ and names $[\langle s_n \rangle]$.

(b) There exist elements $[\langle r_n^1 \rangle], \ldots, [\langle r_n^k \rangle]$ in ${}^*\mathscr{R}$ such that, for $1 \le j \le k$, τ^j is interpretable as $[\langle r_n^j \rangle]$, the k-tuple $\langle [\langle r_n^1 \rangle], \ldots, [\langle r_n^k \rangle] \rangle$ is in the domain of ${}^*f$, and ${}^*f([\langle r_n^1 \rangle], \ldots, [\langle r_n^k \rangle]) = [\langle s_n \rangle]$.

(c) There exist sequences $\langle r_n^1 \rangle, \ldots, \langle r_n^k \rangle$ in R and a set $U \in \mathscr{U}$ such that, for each $m \in U$, if $1 \le j \le k$ then $T_{\tau^j}(m)$ is interpretable as r_m^j, the k-tuple $\langle r_m^1, \ldots, r_m^k \rangle$ is in the domain of f, and $f(r_m^1, \ldots, r_m^k) = s_m$.

(d) $\underline{f}(T_{\tau^1}(n), \ldots, T_{\tau^k}(n))$ is a.e. interpretable as s_n in $\mathcal{R}$.

(e) $T_\tau(n)$ is a.e. interpretable in $\mathcal{R}$ as s_n.

Thus (A) is true by induction.

(B) To prove (B), let $\underline{P}$ be a name for the k-ary relation P on R, and let $\tau^1, \ldots, \tau^k$ be constant terms in $L_{*\mathcal{R}}$. Then the following are equivalent statements:

(a) $*\underline{P}\langle \tau^1, \ldots, \tau^k \rangle$ holds in $*\mathcal{R}$.

(b) There are elements $[\langle r_n^1 \rangle], \ldots, [\langle r_n^k \rangle]$ in $*R$ such that τ^j is interpretable as $[\langle r_n^j \rangle]$, $1 \leq j \leq k$, and the k-tuple $\langle [\langle r_n^1 \rangle], \ldots, [\langle r_n^k \rangle] \rangle$ is in $*P$.

(c) There are sequences $\langle r_n^1 \rangle, \ldots, \langle r_n^k \rangle$ in R and a set $U \in \mathcal{U}$ such that, for each $m \in U$, $T_{\tau^j}(m)$ is interpretable as r_m^j for $1 \leq j \leq k$ and the k-tuple $\langle r_m^1, \ldots, r_m^k \rangle$ is in P.

(d) $\underline{P}\langle T_{\tau^1}(n), \ldots, T_{\tau^k}(n) \rangle$ holds a.e. in $\mathcal{R}$.

This establishes (B). $\square$

We are now in a position to prove the transfer principle. If Φ is an atomic sentence which holds in $\mathcal{R}$ then 15.1 shows immediately that $*\Phi$ holds in $*\mathcal{R}$. Suppose that Φ is of the form

$$(\forall x_1) \cdots (\forall x_n) \left[\bigwedge_{i=1}^k \underline{P}_i \langle \tau_1^i, \ldots, \tau_{m_i}^i \rangle \rightarrow \bigwedge_{j=1}^l \underline{Q}_j \langle \sigma_1^j, \ldots, \sigma_{n_j}^j \rangle \right]$$

and Φ holds in $\mathcal{R}$. Let $*\tau_s^t$ and $*\sigma_s^t$ be the $*$-transforms of τ_s^t and σ_s^t and replace the variables $x_1, \ldots, x_n$ in $*\tau_s^t$ and $*\sigma_s^t$ with constant symbols $\underline{r}_1, \ldots, \underline{r}_n$ from $L_{*\mathcal{R}}$. Assume that with this replacement $*\underline{P}_i \langle *\tau_1^i, \ldots, *\tau_{m_i}^i \rangle$ holds in $*\mathcal{R}$ for each i, $1 \leq i \leq k$. Using 15.1, we see that there is a set $U \in \mathcal{U}$ such that if $n \in U$ then $\underline{P}_i \langle T_{\tau_1^i}(n), \ldots, T_{\tau_{m_i}^i}(n) \rangle$ holds in $\mathcal{R}$ for each i, $1 \leq i \leq k$. But then, since Φ holds in R, $\underline{Q}_j \langle T_{\sigma_1^j}(n), \ldots, T_{\sigma_{n_j}^j}(n) \rangle$ holds in $\mathcal{R}$ for $1 \leq j \leq l$ and $n \in U$. By 15.1 again, $*\underline{Q}_j \langle *\sigma_1^j, \ldots, *\sigma_{n_j}^j \rangle$ holds in $*\mathcal{R}$ for each j, $1 \leq j \leq l$, and we are through.

CHAPTER II

Nonstandard
Analysis on
Superstructures

In order to proceed to analysis more general than the calculus, we will need to consider mathematical systems which contain entities corresponding to sets of sets, sets of functions, and so on. For example, we might want to prove theorems involving the set of open subsets of R, or the set of all continuous functions on R. Such entities, regarded as objects in themselves, are not contained in any relational system based on R.

Beginning with a basic set X, we can construct a superstructure $V(X)$ which contains all of the entities normally encountered in the mathematics of X by successively taking subsets. This chapter is devoted to nonstandard analysis in this general setting. In particular, we consider mathematical logic for superstructures in §II.2, and the transfer principle in §II.3. The language presented is more general than that of Chapter I, and this will allow us to avoid Skolem functions and proofs by contradiction in applying the transfer principle in the rest of this book.

We generalize the ultrapower construction and *-mapping for *$\mathscr{R}$ to superstructures in §II.4, obtaining a superstructure $V(*X)$ and a map *: $V(X) \to V(*X)$. In §II.5 we show how to choose the ultrafilter in the construction of §II.4 to ensure that $V(*X)$ is an enlargement, a notion which is fundamental to nonstandard analysis as developed by Abraham Robinson.

The notions of internal and external entities and sentences are developed in §II.6. These notions are important in being able to recognize when a sentence Ψ about $V(*X)$ is of the form *Φ for some sentence Φ about $V(X)$. We will often use such a corresponding "downward transfer principle" in succeeding chapters. In §II.7 we present the permanence principle, which involves the idea of internal formulas, and is useful in many proofs.

Finally in §II.8 we survey the theory of κ-saturated superstructures, a concept which was introduced by W. A. J. Luxemburg [36] and is very important in some of the recent applications of nonstandard analysis.

II.1 Superstructures

In the succeeding chapters of this book, we will need to consider mathematical systems which contain entities corresponding to sets of sets, sets of functions, etc. Such sets, regarded as objects in themselves, are not contained in any relational system based on R; there are no names for them in the language of relational systems. More generally, we are led to work with a set X and all of the sets which can be obtained inductively from X in a finite number of steps by successively taking subsets of the preceding set, as indicated in the following definition. The resulting structure is called a *superstructure* over X. We will always assume that X contains the natural numbers N in order later to be able to define ordered n-tuples (Definition 1.2).

1.1 Definition Let X be a nonempty set containing at least the natural numbers N. The *power set* $\mathscr{P}(X)$ of X is the set of all subsets of X (including the empty set $\varnothing$). The nth *cumulative power set* $V_n(X)$ of X is defined recursively by

$$V_0(X) = X, \qquad V_{n+1}(X) = V_n(X) \cup \mathscr{P}(V_n(X)).$$

The *superstructure* over X is the set

$$V(X) = \bigcup_{n=0}^{\infty} V_n(X).$$

The elements of $V(X)$ are called *entities*, and the entities in X are also called *individuals*. The entities in $V_n(X) - V_{n-1}(X)$ are of *rank n*.

For example, let $X = N$, the set of natural numbers. Then some entities in $V_1(N)$ are 7, $\{7\}$, and the set $\{2, 4, 6, \ldots\}$ of even numbers. Similarly, some entities in $V_2(N)$ are 7, $\{1, 3, 5, \ldots\}$, and the set of all finite subsets of N.

As usual in set theory we use the symbol $\in$ to stand for "is an element of" and $\notin$ to stand for "is not an element of." Similarly, if $x, y \in V(X)$, we write $x \subseteq y$ if $z \in x$ implies $z \in y$; we write $x = y$ if $x \subseteq y$ and $y \subseteq x$ and write $x \neq y$ otherwise. In particular, we write $x \subset y$ if $x \subseteq y$ but $x \neq y$. Notice that an entity may simultaneously be a subset of, and an element of, another entity; in particular, $V_n(X) \subseteq V_{n+1}(X)$ and $V_n(X) \in V_{n+1}(X)$ for all n. We always assume that the individuals have no members; i.e., if $x \in X$ then $x \neq \varnothing$ and the statement $t \in x$ is false.

The choice of the basic set X is always somewhat arbitrary and depends on the context. If, for example, we want to study the real number system, and do not need to consider the manner of construction of each real number (as, for example, an equivalence class of Cauchy sequences of rational numbers), then we may take $X = R$. If, on the other hand, we want to study the real numbers as equivalence classes of Cauchy sequences of rationals, then we might take $X = Q$ (the rational numbers).

We next show how to describe relations and functions in the set theory of $V(X)$. The basic step is to define an ordered n-tuple set-theoretically, and the rest follows as in Definition I.2.1. We start with the definition of an ordered pair and make a distinction between ordered pairs and two-tuples.

1.2 Definition An ordered pair $\langle a, b \rangle$ is the set $\{\{a\}, \{a, b\}\}$. For $n \geq 2$, an ordered n-tuple $\langle x_1, \ldots, x_n \rangle$ of elements $x_1, x_2, \ldots, x_n$ is defined by $\langle x_1, \ldots, x_n \rangle = \{\langle 1, x_1 \rangle, \ldots, \langle n, x_n \rangle\}$, where for each $k \in N$, $1 \leq k \leq n$, $\langle k, x_k \rangle$ is an ordered pair. If $c_1, c_2, \ldots, c_n$ and c are sets, we define

$$c_1 \times c_2 \times \cdots \times c_n = \{\langle x_1, \ldots, x_n \rangle : x_i \in c_i \, (i = 1, \ldots, n)\}$$

and $c^n = c \times \cdots \times c$ (n factors).

For $n \geq 2$, an n-ary *relation* P on $c_1 \times c_2 \times \cdots \times c_n$ is a subset of $c_1 \times c_2 \times \cdots \times c_n$. P is a *relation in* $V(X)$ if each $c_i \in V_k(X)$ $(i = 1, \ldots, n)$ for some fixed integer k. If P is a 2-ary relation on $c_1 \times c_2$ we will call it a binary relation. In this case we define the *domain* and *range* of P by

$$\text{dom } P = \{x_1 \in c_1 : \text{there exists } x_2 \in c_2 \text{ such that } \langle x_1, x_2 \rangle \in P\}$$

and

$$\text{range } P = \{x_2 \in c_2 : \text{there exists } x_1 \in c_1 \text{ such that } \langle x_1, x_2 \rangle \in P\}.$$

Similarly, if $b \subseteq c_1$ we define the *image of* b *under* P by

$$P[b] = \{x_2 \in c_2 : \text{there exists } x_1 \in b \text{ such that } \langle x_1, x_2 \rangle \in P\},$$

and the *inverse image* of $b \subseteq c_2$ under P is the set

$$P^{-1}[b] = \{x_1 \in c_1 : \text{there exists } x_2 \in b \text{ with } \langle x_1, x_2 \rangle \in P\}.$$

If b is a singleton set, i.e., $b = \{x\}$, we will usually write $P[x]$ and $P^{-1}[x]$ for $P[b]$ and $P^{-1}[b]$.

A *function* f from a to b, which we denote by $f : a \to b$, is a subset of $a \times b$ (and hence a binary or 2-ary relation) such that, for each $x \in a$, there is exactly one $y \in b$ such that $\langle x, y \rangle \in f$. The element y is called the *image of* x and is denoted by $f(x)$. The set a is the *domain* of f, and we say that f is defined on a. If $f(x) = f(y)$ implies $x = y$ for all $x, y \in a$, we say that f is *one-to-one* (1–1) or *injective*. If range $f = b$ we say that f maps a *onto* b or that f is *surjective*. A function $g : c \to b$ is an *extension* of $f : a \to b$ (and f is the *re-striction* of g to a) if $c \supseteq a$ and $g(x) = f(x)$ for all $x \in a$; we write $f = g|a$ in this case. If $a \subseteq c_1 \times \cdots \times c_n$ we may say that f is a *function of n variables* and may write $f(x)$ as $f(x_1, \ldots, x_n)$, where $x = \langle x_1, \ldots, x_n \rangle$, $x_i \in c_i$. (Note that $f(x)$ will be different from $f[x]$; e.g., if $f(x) = x^2$ on R then $f(2) = 4$ and $f[2] = \{4\}$.)

The set-theoretic definition of ordered n-tuple in Definition 1.2 is justified by the following lemma, which expresses the definitive property of ordered n-tuples.

1.3 Lemma $\langle x_1, \ldots, x_n \rangle = \langle y_1, \ldots, y_n \rangle$ (set equality) iff $x_i = y_i$ $(i = 1, \ldots, n)$.

Proof: We prove the lemma for $n = 2$ and leave the rest of the proof to the reader.

It is immediate that if $x_1 = y_1$ and $x_2 = y_2$ then $\langle 1, x_1 \rangle = \langle 1, y_1 \rangle$ and $\langle 2, x_2 \rangle = \langle 2, y_2 \rangle$ so $\langle x_1, x_2 \rangle = \langle y_1, y_2 \rangle$.

Suppose, conversely, that $\langle x_1, x_2 \rangle = \langle y_1, y_2 \rangle$. Then

(1.1) $\{\{\{1\}, \{1, x_1\}\}, \{\{2\}, \{2, x_2\}\}\} = \{\{\{1\}, \{1, y_1\}\}, \{\{2\}, \{2, y_2\}\}\}$.

Suppose first that $x_1 = 1$. Then $\{\{1\}, \{1, x_1\}\} = \{\{1\}, \{1\}\} = \{\{1\}\}$ and so, from (1.1), $\{\{1\}, \{1, y_1\}\} = \{\{1\}\}$, so $\{1\} = \{1, y_1\}$ and hence $x_1 = y_1 = 1$. Suppose now that $x_1 \neq 1$. From (1.1), $\{\{1\}, \{1, x_1\}\} = \{\{1\}, \{1, y_1\}\}$, whence $\{1, x_1\} = \{1, y_1\}$ and so $x_1 = y_1$. Similarly $x_2 = y_2$. $\square$

From Definition 1.2 we see that n-ary relations on $V_k(X)$ and functions with domain and range in $V_k(x)$ for some $k \in N$ are entities in $V(X)$. Indeed, if $x, y \in V_k(x)$ then the ordered pair $\langle x, y \rangle \in V_{k+2}(X)$. Thus if $x_1, \ldots, x_n, x_{n+1} \in V_k(X)$ then $\langle x_1, \ldots, x_n \rangle \in V_{k+3}(X)$, and the 2-tuple $\langle \langle x_1, \ldots, x_n \rangle, x_{n+1} \rangle \in V_{k+6}$. Therefore, any relation P on a set $c_1 \times \cdots \times c_n$, $c_i \in V_k(X)$ $(i = 1, \ldots, n)$, is an element of $V_{k+4}(X)$, and a function of n variables on $c_1 \times \cdots \times c_n$ with range in $V_k(X)$ is in V_{k+7}; a function on just c_1 is again in $V_{k+4}(X)$. Thus superstructures are at least rich enough to contain entities corresponding to the usual relations and functions occurring in mathematical systems.

We conclude this section with some examples.

1.4 Examples

1. Let $X = R$ and let $\mathscr{I}$ be the set of all finite closed intervals in R; i.e., $I \in \mathscr{I}$ iff $I = \{x \in R : a \leq x \leq b, a, b \in R\} = [a, b]$. Then $\mathscr{I} \in V_2(R)$.

2. We define a relation P on $N \times \mathscr{I}$, where N is the set of natural numbers, by "$\langle n, y \rangle \in P$ iff $n \in y$." Clearly P is in $V_6(R)$ since N and $\mathscr{I}$ are in $V_2(R)$.

3. The relation μ defined on $\mathscr{I} \times R^+$ (where R^+ is the set of positive real numbers) defined by "$\langle y, r \rangle \in \mu$ iff $y = [a, b]$ and $r = |b - a|$" is a function on $\mathscr{I}$ which measures the length of each interval $I \in \mathscr{I}$; it is in $V_6(R)$.

Exercises II.1

1. Complete the proof of Lemma 1.3.
2. If $a \in V_k(X)$, $k \geq 1$, and $b \subseteq a$ then for what n is $b \in V_n(X)$?
3. If $a, b \in V_k(X) - V_0(X)$, $k \geq 1$, then for what n are $a \cup b$, $a \cap b$, and $a - b$ in $V_n(X)$?
4. Show that a relation P is in $V(X)$ iff dom P and range P are in $V(X)$.
5. Let $\mathscr{I}$ denote the collection of all finite closed intervals on the real line (i.e., sets of the form $[a, b]$). For each $I \in \mathscr{I}$ let $\mu(I) = b - a$ (i.e., the length of I). For which value of n is $\mu \in V_n(R)$?

II.2 Languages and Interpretation
for Superstructures

In this section we introduce a suitable language for superstructures and show how to interpret sentences in this language.

Let $V(X)$ be a given superstructure. The symbols of the language $\mathscr{L}_X$ for $V(X)$ consist of the following.

2.1 Connectives The symbols $\neg$, $\wedge$, $\vee$, $\rightarrow$, and $\leftrightarrow$, to be interpreted later as "not," "and," "or," "implies," and "if and only if," respectively.

2.2 Quantifiers The symbols $\forall$ and $\exists$, to be interpreted as "for all" and "there exists," respectively.

2.3 Parentheses The symbols $[\, , \,]$, $(\, , \,)$, and $\langle \, , \, \rangle$, to be used for bracketing.

2.4 Constant Symbols At least one symbol $\underline{a}$ for each element a of $V(X)$. For simplicity of notation we will identify a and its symbol $\underline{a}$. The context will clear up any possible confusion.

2.5 Variable Symbols A countable collection of symbols like $x, y, x_1, x_2, \ldots$, to be used as "variables."

2.6 Equality Symbol The symbol $=$, to be interpreted as "equals" [it denotes set-theoretic equality for elements of $V(X) - V_0(X)$].

2.7 Predicate Symbol The symbol $\in$, to be interpreted as "is an element of." We use the same symbol that was used informally in §II.1.

Notice that the language $\mathscr{L}_X$ is richer than the language $L_{\mathscr{S}}$ for a relational system $\mathscr{S}$ based on X in that $\mathscr{L}_X$ has the symbols $\neg$, $\vee$, $\leftrightarrow$, and also $\exists$. However, $\mathscr{L}_X$ is poorer in having no terms.

Sentences in $\mathscr{L}_X$ are built up inductively using the symbols just introduced. The basic building blocks are the atomic formulas, introduced in the following definition.

2.8 Definition A *formula* of $\mathscr{L}_X$ is built up inductively using the following rules:

(a) If $x_1, \ldots, x_n$, x, and y are either constants or variables, the expressions $x \in y$, $x = y$, $\langle x_1, x_2, \ldots, x_n \rangle \in y$, $\langle x_1, \ldots, x_n \rangle = y$, $\langle \langle x_1, \ldots, x_n \rangle, x \rangle \in y$, and $\langle \langle x_1, \ldots, x_n \rangle, x \rangle = y$ are formulas, called *atomic formulas*.

(b) If Φ and Ψ are formulas, then so are $\neg \Phi$, $\Phi \wedge \Psi$, $\Phi \vee \Psi$, $\Phi \rightarrow \Psi$, and $\Phi \leftrightarrow \Psi$.

(c) If x is a variable symbol, y is either a variable or a constant symbol, and Φ is a formula which does not already contain an expression of the form $(\forall x \in z)$ or $(\exists x \in z)$ (with the same variable symbol x), then $(\forall x \in y)\Phi$ and $(\exists x \in y)\Phi$ are formulas.

A variable *occurs in the scope of a quantifier* if whenever a variable x occurs in Φ, then x is contained in a formula Ψ which occurs in Φ in the form $(\forall x \in z)\Psi$ or $(\exists x \in z)\Psi$ (z may be either a variable or a constant); it is then said to be *bound*, and otherwise it is called *free*. A *sentence* is a formula in which all variables are bound.

For example, the expression $(\forall x \in b)[x \in y \wedge \langle y, a \rangle = b]$, where a and b are constants and x and y are variables, is a formula in $\mathscr{L}_X$ but not a sentence, since the variable y is free. The formula $(\forall x \in y)(\exists z \in c)[\langle x, z \rangle = c]$ is likewise not a sentence, but $(\exists y \in a)(\forall x \in y)(\exists z \in c)[\langle x, y \rangle = c \wedge \langle y, z \rangle = d]$ is a sentence.

We now indicate how to interpret a given sentence Φ in $\mathscr{L}_X$ in the superstructure $V(X)$. That is, we show how to decide whether Φ is true or false in $V(X)$.

2.9 Definition

(a) The atomic sentences $a \in b$, $\langle a_1, \ldots, a_n \rangle \in b$, $\langle \langle a_1, \ldots, a_n \rangle, c \rangle \in b$ and $a = b$, $\langle a_1, \ldots, a_n \rangle = b$, $\langle \langle a_1, \ldots, a_n \rangle, c \rangle = b$ are true (hold) in $V(X)$ if, respectively, the entity (corresponding to) a, $\langle a_1, \ldots, a_n \rangle$, or $\langle \langle a_1, \ldots, a_n \rangle, c \rangle$ is an element of, or identical to, b.

(b) If Φ and Ψ are sentences then

(i) $\neg\Phi$ is true if Φ is not true (does not hold),
(ii) $\Phi \wedge \Psi$ is true if both Φ and Ψ are true,
(iii) $\Phi \vee \Psi$ is true if at least one of Φ and Ψ is true,
(iv) $\Phi \rightarrow \Psi$ is true if either Ψ is true or Φ is not true,
(v) $\Phi \leftrightarrow \Psi$ is true if Φ and Ψ are either both true or both not true.

(c) Let $\Phi = \Phi(x)$ be a formula in which x is the only free variable, and b is a constant symbol. Then

(i) $(\forall x \in b)\Phi$ is true if, for all entities $a \in b$, when the symbol corresponding to the entity a is substituted for x in Φ, the resulting formula, which we denote by $\Phi(a)$, is true,
(ii) $(\exists x \in b)\Phi$ is true if there exists an entity $a \in b$ so that $\Phi(a)$ is true.

Theoretically, the induction scheme for interpretation of sentences implied by 2.9 could be rather involved. But it is nothing more than the obvious one consistent with the specified (and usual) interpretation of the logical symbols involved. For most sentences Φ in $\mathscr{L}_X$ which we will later encounter, it will be easy to check if Φ is true in $V(X)$. For example, let $\mathscr{L}_R$ be the language for $V(R)$, and let $\hat{S}$ be the ternary relation for sum (i.e., $\langle a, b, c \rangle \in \hat{S}$ if $a + b = c$), $\hat{P}$ be the ternary relation for product (i.e., $\langle a, b, c \rangle \in \hat{P}$ if $ab = c$), and R_0 be the set $R - \{0\}$ of nonzero reals. Then the sentence

(2.1) $(\forall x \in R_0)(\exists y \in R)[\langle x, y, 1 \rangle \in \hat{P}]$

is true in $V(R)$. For let $\Phi(x) = (\exists y \in R)[\langle x, y, 1 \rangle \in \hat{P}]$. Then $(\forall x \in R_0)\Phi$ is true if, for all nonzero $a \in R$, $(\exists y \in R)[\langle a, y, 1 \rangle \in \hat{P}]$ is true. This sentence is true if there is a number b so that $\langle a, b, 1 \rangle \in \hat{P}$. But this is true with $b = a^{-1}$. Thus the sentence states that "every nonzero real number has an inverse."

As in Chapter I, it is important to be able to translate an ordinary mathematical statement into a sentence in the language $\mathscr{L}_X$, since in the next section we will show how to write down the "transform" of such a sentence, which can then be interpreted in an appropriate "nonstandard" superstructure. As an example, consider the distributive law for the real numbers. To simplify matters we introduce the following notational convention.

2.10 Convention If f is the symbol for a function of n variables we may write $x_{n+1} = f(x_1, \ldots, x_n)$ for the atomic formula $\langle \langle x_1, \ldots, x_n \rangle, x_{n+1} \rangle \in f$. If f is a function of one variable, we may write $y = f(x)$ for $\langle x, y \rangle \in f$.

Noting that the ternary relations $\hat{S}$ and $\hat{P}$ define functions S and P of two variables [e.g., $S(a, b) = c$ if $\langle a, b, c \rangle \in \hat{S}$], we may express the distributive law by the sentence

(2.2) $(\forall x \in R)(\forall y \in R)(\forall z \in R)(\forall x_1 \in R) \cdots (\forall x_4 \in R)$
$$[[[S(y,z) = x_1] \wedge [P(x, x_1) = x_2]$$
$$\wedge [P(x, y) = x_3] \wedge [P(x, z) = x_4]]$$
$$\rightarrow [S(x_3, x_4) = x_2]].$$

Notice that since our language $\mathscr{L}_R$ does not have terms, the sentence (2.2) is somewhat more involved than the corresponding sentence

(2.3) $(\forall x)(\forall y)(\forall z)[R(x) \wedge R(y) \wedge R(z) \rightarrow E(P(x, S(y, z))), S(P(x, y), P(x, z))]$

in the language $L_{\mathscr{R}}$ of Chapter I. [Remember that, in $L_{\mathscr{R}}$, $x(y + z) = xy + xz$ is shorthand for the term $E(x(y + z), xy + xz)$, where E is the symbol for the equality relation.] However, it is still easy to check that (2.2) is true in $V(R)$.

For another example, consider the statement that $f: R \rightarrow R$ is continuous at the point $x = a$, or, more precisely, given $\varepsilon > 0$ there exists a $\delta > 0$ so that whenever $|x - a| < \delta$, $|f(x) - f(a)| < \varepsilon$. To translate this into a sentence in $\mathscr{L}_R$, let R_0^+ denote the entity in $V(R)$ which is the set of strictly positive real numbers. Let ρ be the function of two variables corresponding to distance (so that $\langle\langle x, y\rangle, z\rangle \in \rho$ iff $|x - y| = z$), and I be the binary relation of strict inequality (so that $\langle x, y\rangle \in I$ iff $x < y$). Then the corresponding sentence in $\mathscr{L}_R$ is

(2.4) $(\forall \varepsilon \in R_0^+)(\exists \delta \in R_0^+)(\forall x \in R)(\forall x_1 \in R)(\forall x_2 \in R)(\forall x_3 \in R)$
$$[[\rho(x, a) = x_1 \wedge \langle x_1, \delta\rangle \in I \wedge f(x) = x_2 \wedge f(a) = b$$
$$\wedge \rho(x_2, b) = x_3] \rightarrow [\langle x_3, \varepsilon\rangle \in I]].$$

Check that the interpretation of this sentence is true in $V(R)$ if f is continuous at $x = a$ and $f(a) = b$.

Since, with a little practice, the translation of ordinary mathematical statements in a given superstructure $V(X)$ into sentences in the language $\mathscr{L}_X$ will be routine, we adopt the following convention in the rest of this book.

2.11 Convention A sentence in $\mathscr{L}_X$ will often be written as a sentence in the language $L_{\mathscr{S}}$ of Chapter I, where $\mathscr{S}$ is a relational system over X, or even as a sentence in ordinary mathematical language, when the translation into a sentence in $\mathscr{L}_X$ is clear. We will also abbreviate $(\forall x_1 \in c) \cdots (\forall x_n \in c)$ by $(\forall x_1, \ldots, x_n \in c)$.

Thus, for example, the sentence in $\mathscr{L}_R$ which is equivalent to (2.4) is the sentence

(2.5) $(\forall \varepsilon \in R_0^+)(\exists \delta \in R_0^+)(\forall x \in R)[|x - a| < \delta \rightarrow |f(x) - f(a)| < \varepsilon].$

Similarly, the sentence in $\mathcal{L}_R$ which is equivalent to the semiformal sentence $(\forall x \in a)[x \subseteq b]$ is the sentence $(\forall x \in a)(\forall y \in x)[y \in b]$.

Exercises II.2

1. Write out the commutative and associative laws of addition for R in the form of sentence (2.2).
2. Write out a sentence in the form (2.4) which means that $\lim_{x \to a} f(x) = L$ in R.
3. Write out a sentence in the form (2.4) which means that the derivative $f'(a)$ exists and equals L.
4. Formulate a sentence in $\mathcal{L}_R$ which expresses the Archimedean property of the real number system (i.e., for each $x \in R$ there is an $m \in N$ so that $m \geq x$).
5. Write sentences in $\mathcal{L}_R$ expressing the fact that a collection $\mathcal{F}$ of subsets of R is a filter.
6. Let X be any set. Write sentences in $\mathcal{L}_X$ which express the facts that a function $f: A \to B$ is surjective (i.e., onto) and injective (i.e., one-to-one) respectively.

II.3 Monomorphisms between Superstructures:
The Transfer Principle

In §I.5 we stated that the relational systems $\mathcal{R}$ and $*\mathcal{R}$ are connected by a transfer principle. To be precise, we stated that, with the $*$-mapping and the associated $*$-transform of simple sentences defined in §I.5, if Φ is any simple sentence which is true in $\mathcal{R}$ then $*\Phi$ is true in $*\mathcal{R}$. In this section we generalize this relationship to superstructures. The basic properties of the new mapping $*$, which was introduced by Robinson and Zakon [45, 48], are abstracted in the notion of a monomorphism. In the next section we show that with each superstructure $V(X)$ one can associate a superstructure $V(*X)$ and a monomorphism $*: V(X) \to V(*X)$.

Let X and Y be two sets of individuals with associated superstructures $V(X)$ and $V(Y)$ and languages $\mathcal{L}_X$ and $\mathcal{L}_Y$, respectively. We will again assume that there is at least one constant symbol in $\mathcal{L}_X$ and $\mathcal{L}_Y$ for each entity in $V(X)$ and $V(Y)$, respectively, and identify the constant symbols with the corresponding entities. The context should settle any possible confusion. A constant symbol in $\mathcal{L}_X$ names something in $V(X)$ and the same is true for constant symbols in $\mathcal{L}_Y$.

Now let $*\colon V(X) \to V(Y)$ be a one-to-one mapping (injection). For $a \in V(X)$ we write $*(a) = *a$. We assume that for each $a \in V(X)$ the symbol $*a$ is in $\mathscr{L}_Y$ and names $*(a)$.

3.1 Definition If Φ is a formula (or sentence) in $\mathscr{L}_X$, the $*$-transform $*\Phi$ of Φ is the formula (or sentence) in $\mathscr{L}_Y$ obtained from Φ by replacing each constant symbol c in Φ with the symbol $*c$ in $\mathscr{L}_Y$ associated with the entity $*(c)$.

For example, given the set R of real numbers, we assume that a superstructure $V(*R)$ over a set $*R$ and a monomorphism $*\colon V(R) \to V(*R)$ exist (this will be established in the next section). Then the $*$-transform of the sentence (2.1) of the last section is the sentence

(3.1) $$(\forall x \in *R_0)(\exists y \in *R)[\langle x, y, *1 \rangle \in *\hat{P}],$$

and the $*$-transform of (2.4) is

(3.2) $$(\forall \varepsilon \in *R_0^+)(\exists \delta \in *R_0^+)(\forall x \in *R)(\forall x_1 \in *R)(\forall x_2 \in *R)(\forall x_3 \in *R)$$
$$[[*\rho(x, *a) = x_1 \wedge \langle x_1, \delta \rangle \in *I \wedge *f(x) = x_2 \wedge *f(*a) = *b$$
$$\wedge *\rho(x_2, *b) = x_3] \to [\langle x_3, \varepsilon \rangle \in *I]].$$

3.2 Definition The injection $*\colon V(X) \to V(Y)$ is called a *monomorphism* if

 (i) $*(\varnothing) = \varnothing$, where $\varnothing$ is the empty set,
 (ii) $a \in X$ implies $*a \in Y$, and $n \in N$ implies $*n = n$ (recall that $N \subseteq X$ and $N \subseteq Y$ by assumption),
 (iii) $a \in V_{n+1}(X) - V_n(X)$ implies $*a \in V_{n+1}(Y) - V_n(Y)$, $n \geq 0$,
 (iv) if $a \in *V_n(X)$, $n \geq 1$, and $b \in a$, then $b \in *V_{n-1}(X)$,
 (v) (*transfer principle*) for any sentence Φ in $\mathscr{L}_X$, Φ holds in $V(X)$ iff $*\Phi$ holds in $V(Y)$.

Property (iv) is called *strictness* by Zakon [48]. We will later interpret it to say that elements of "internal" sets are internal.

Because of (ii) we may, and will, assume that X is actually a subset of Y and $*a = a$ for $a \in X$ [this is the analogue of Convention 2.5(c) of Chapter I].

The transfer principle as stated is redundant in that if from Φ holding in $V(X)$ one can conclude that $*\Phi$ holds in $V(Y)$, then when $\neg\Phi$ holds in $V(X)$, $*(\neg\Phi)$, i.e., $\neg(*\Phi)$, holds in $V(Y)$. The principle that $*\Phi$ holding in $V(Y)$ implies that Φ holds in $V(X)$ will sometimes be called the *downward transfer principle*.

We now suppose that $*\colon V(X) \to V(Y)$ is a monomorphism and collect together some elementary results that follow easily from the transfer principle; the proofs are good illustrations of the use of that principle.

3.3 Theorem

(a) Let $a, b, a_1, \ldots, a_n$ be fixed entities in $V(X)$. Then

(i) $*\{a_1, \ldots, a_n\} = \{*a_1, \ldots, *a_n\}$,
(ii) $*\langle a_1, \ldots, a_n \rangle = \langle *a_1, \ldots, *a_n \rangle$,
(iii) $a \in b$ iff $*a \in *b$,
(iv) $a = b$ iff $*a = *b$,
(v) $a \subseteq b$ iff $*a \subseteq *b$,
(vi) $*(\bigcup_{i=1}^n a_i) = \bigcup_{i=1}^n *a_i$, $*(\bigcap_{i=1}^n a_i) = \bigcap_{i=1}^n *a_i$,
(vii) $*(a_1 \times a_2 \times \cdots \times a_n) = *a_1 \times \cdots \times *a_n$.

(b) If P is a relation on $a_1 \times \cdots \times a_n$ then $*P$ is a relation on $*a_1 \times \cdots \times *a_n$, and, for $n = 2$, $*(\text{dom } P) = \text{dom } *P$ and $*(\text{range } P) = \text{range } *P$.

(c) If f is a mapping from a into b then $*f$ is a mapping from $*a$ into $*b$, and $*[f(c)] = *f(*c)$ for each $c \in a$. Also f is one-to-one iff $*f$ is one-to-one.

Proof: (a)(i) Let $b = \{a_1, \ldots, a_n\}$ and transform the sentence $(\forall x \in b)$ $[x = a_1 \vee x = a_2 \vee \cdots \vee x = a_n]$, as well as the sentences $a_1 \in b, \ldots, a_n \in b$.

(a)(ii) Exercise 1.

(a)(iii) Clear.

(a)(iv) Clear.

(a)(v) The sentence $(\forall x \in a)[x \in b]$ is true in $V(X)$ iff its $*$-transform $(\forall x \in *a)[x \in *b]$ is true in $V(Y)$. The interpretation of the latter sentence is that $*a \subseteq *b$.

(a)(vi) We show that $*(a \cup b) = *a \cup *b$; it then follows by induction that $*(\bigcup_{i=1}^n a_i) = \bigcup_{i=1}^n *a_i$. The proof that $*(\bigcap_{i=1}^n a_i) = \bigcap_{i=1}^n *a_i$ is similar (Exercise 2). Let $c = a \cup b$. The sentence $(\forall x \in c)[x \in a \vee x \in b]$ is true in $V(X)$, so its $*$-transform $(\forall x \in *c)[x \in *a \vee x \in *b]$ is true in $V(Y)$. The interpretation of the latter sentence is that $*(a \cup b) \subseteq *a \cup *b$. Similarly, the interpretation of the $*$-transforms of the sentences $(\forall x \in a)[x \in c]$ and $(\forall x \in b)[x \in c]$ shows that $*(a \cup b) \supseteq *a \cup *b$.

(a)(vii) We show that $*(a \times b) = *a \times *b$; the proof for $n > 2$ is similar. Interpretation of the $*$-transforms of the sentences $(\forall z \in (a \times b))(\exists x \in a)$ $(\exists y \in b)[\langle x, y \rangle = z]$ and $(\forall x \in a)(\forall y \in b)(\exists z \in (a \times b))[\langle x, y \rangle = z]$ shows that $*(a \times b) \subseteq *a \times *b$ and $*a \times *b \subseteq *(a \times b)$.

(b) That $*P$ is a relation on $*a_1 \times \cdots \times *a_n$ follows by interpretation of the $*$-transform of $(\forall x \in P)(\exists x_1 \in a_1) \cdots (\exists x_n \in a_n)[\langle x_1, \ldots, x_n \rangle = x]$. To show that, for $n = 2$, $*(\text{dom } P) \subseteq \text{dom } *P$, interpret the $*$-transform of the sentence $(\forall x \in \text{dom } P)(\exists y \in a_2)[\langle x, y \rangle \in P]$. The proof of the fact that $*(\text{dom } P) \supseteq \text{dom } *P$ is left to the reader (Exercise 3).

(c) $*f$ is a relation on $*a \times *b$ by (b). To show that $*f$ is a mapping, interpret the $*$-transform of the sentence $(\forall x \in a)(\forall y \in b)(\forall z \in b)[[\langle x, y \rangle \in f \wedge \langle x, z \rangle \in f] \to y = z]$, which is true in $V(X)$. The rest of the proof of (c) is left as Exercise 3. $\square$

The results in Theorem 3.3 are quite general in nature. To be more concrete we consider, as examples, the interpretation of the sentences (3.1) and (3.2).

Remember that the sentence (2.1) of which (3.1) is the $*$-transform holds in $V(R)$ because of the fact that there exists a multiplicative inverse of each nonzero element in the field $\mathscr{R}$, and (2.1) is a formal expression of that mathematical statement. Clearly (3.1) should be a formal expression of a similar fact about $V(*R)$. To see this, note that the ternary relation $\hat{P}$ defines a function P of two variables since the product of two real numbers is uniquely defined. By parts (b) and (c) of Theorem 3.3 we see that $*P$ is a function from $*R^2$ to $*R$. Thus for each $a, b \in *R$ the number $c \in *R$ such that $\langle \langle a, b \rangle, c \rangle \in *P$ is uniquely defined and is called the $*$-product of a and b. We denote c by $a \cdot b$ or ab. Now (3.1) is true by transfer in $V(*R)$ since (2.1) is true in $V(R)$, and its interpretation establishes the existence for each $a \neq 0$ in $*R$ of a number $y \in *R$ so that $a \cdot y = 1$. One can similarly show by transfer that y is unique.

Consider now the interpretation of (3.2). Proceeding as above, we see that (3.2) is equivalent to the ordinary mathematical statement "Given $\varepsilon > 0$ in $*R$ there is a $\delta > 0$ in $*R$ so that, for all $x \in *R$, $|x - a| < \delta$ implies $|*f(x) - *f(a)| < \varepsilon$." (The absolute value $|x|$ for $x \in *R$ is the extension of the usual absolute value in R.) Notice that here ε and δ are allowed to be any positive numbers in $*R$ (even infinitesimal). The function $*f$ will be said to be $*$-*continuous* at a if it satisfies (3.2), which will be the case, by transfer, if f is continuous at a.

In §I.2 we noted that if B was a subset of R then $*B$ was an extension of B (regarded as embedded in $*R$). This fact is again true in the present context. For if $b \in \mathscr{P}(X)$ and $a \in b$ then $*a \in *b$ by Theorem 3.3(a)(iii). But since $a \in X$ we have $*a = a$ and so $a \in *b$, and hence $b \subseteq *b$. One might expect that this fact is true in general, i.e., that $a \in b$ implies $a \in *b$ for any entities $a, b \in V(X)$, but in general Theorem 3.3(a)(iii) is the best we can do, as shown by the following example.

3.4 Example Let $\mathscr{I}$ denote the set of closed bounded intervals in R; each $I \in \mathscr{I}$ is of the form $I = \{x \in R : a \leq x \leq b, a, b \in R\} = [a, b]$. Then $\mathscr{I} \in V_2(R)$. Thus the following statements are true in $V(R)$:

$$(\forall x \in \mathscr{I})(\exists a, b \in R)(\forall y \in R)[a \leq y \leq b \leftrightarrow y \in x],$$
$$(\forall a, b \in R)(\exists x \in \mathscr{I})(\forall y \in R)[a \leq y \leq b \leftrightarrow y \in x].$$

By transfer, assuming a monomorphism $*: V(R) \to V(*R)$, we see that if $I \in *\mathscr{I}$ then there exist numbers $a, b \in *R$ so that $I = \{x \in *R : a \leq x \leq b\}$. Even if a and b are standard (i.e., in R), if $a \neq b$ such an interval is not identical to an interval in $\mathscr{I}$, since it contains non-standard reals between a and b. Thus $*\mathscr{I}$ contains the transform $*I = \{x \in *R : a \leq x \leq b\}$ of each standard interval $I = [a, b], a, b \in R$, and also all other intervals of the form $\{x \in *R : a \leq x \leq b\}$ where either a or b or both are non-standard. Notice, in particular, that $\mathscr{I}$ is not embedded in $*\mathscr{I}$; i.e., only singleton sets in $\mathscr{I}$ lie in $*\mathscr{I}$. This situation is indicative of what happens in general when one forms $*b$ for an entity b of rank higher than one.

The fact that the languages $\mathscr{L}_X$ and $\mathscr{L}_Y$ contain the existential quantifier $\exists$ allows alternative proofs of many of the results established in Chapter I. In particular, we may use $\exists$ to do the work done by Skolem functions in Chapter I. To illustrate, consider the following proof of the sufficiency of the condition in Proposition 8.1 of Chapter I, which states that if $\langle s_n \rangle$ is a standard sequence and $*s_n \simeq L \in R$ for all infinite n, then s_n converges to L. We present the proof in a hybrid of the languages $L_{\mathscr{R}}$ and $\mathscr{L}_R$. Translation into the language $\mathscr{L}_R$ is left to the interested reader.

Suppose then that $*s_n \simeq L$ for all infinite positive integers $n \in *N$. Let $\varepsilon > 0$ be a fixed *standard* real. Since $|*s_n - L|$ is infinitesimal for *all* infinite positive integers, the statement

$$(3.3) \qquad (\forall n \in *N)[n \geq \omega \to |L - *s_n| < \varepsilon]$$

is a sentence in $\mathscr{L}_{*R}$ which is true for any infinite positive integer ω. However, (3.3) is not the $*$-transform of a sentence in $\mathscr{L}_R$, since it involves the constant ω, which does not name the image $*(a)$ of an element $a \in V(R)$. But since (3.3) is true, the sentence

$$(3.4) \qquad (\exists m \in *N)(\forall n \in *N)[n \geq m \to |L - *s_n| < \varepsilon]$$

is also true in $V(*R)$ and is the $*$-transform of

$$(3.5) \qquad (\exists m \in N)(\forall n \in N)[n \geq m \to |L - s_n| < \varepsilon],$$

which is then true in $V(R)$ by virtue of the transfer principle. Since (3.5) is true for any $\varepsilon > 0$, we see that s_n converges to L.

3.5 Remark Comparison of this proof with that in Chapter I shows that we have avoided a proof by contradiction, and the use of Skolem functions. Another and more important aspect of this new technique of proof is that we construct a true sentence $*\Phi$ in $\mathscr{L}_{*X}$ which is the $*$-transform of a sentence Φ in $\mathscr{L}_X$, so Φ is true by transfer *down* and yields the desired result. In Chapter I we used the transfer principle only in the upward direction, i.e., from $L_{\mathscr{R}}$ to $L_{*\mathscr{R}}$. The construction of $*\Phi$ is often accomplished, as above, by writing down a sentence Ψ in $\mathscr{L}_{*X}$ which is true but involves entities, like the ω above, which do not occur in the $*$-transforms of sentences Φ in $\mathscr{L}_X$, and then appropriately adding the existential quantifier to convert Ψ to a sentence of the form $*\Phi$ for some Φ in $\mathscr{L}_X$. The proof above may seem surprising since we infer the existence of a standard integer m satisfying

$$(3.6) \qquad\qquad (\forall n \in N)[n \geq m \to |s - s_n| < \varepsilon]$$

from the existence of the infinite integer ω satisfying (3.3).

Since similar proofs will occur in the rest of this book it is important to be able to recognize when a sentence Ψ in $\mathscr{L}_{*X}$ is or is not of the form $*\Phi$ for some sentence Φ in $\mathscr{L}_X$. This question will be dealt with in §II.6.

Exercises II.3

1. Prove Theorem 3.3(a)(ii).
2. Show that $*(\bigcap_{i=1}^n a_i) = \bigcap_{i=1}^n *a_i$.
3. Finish the proof of parts (b) and (c) of Theorem 3.3.
4. Use the downward transfer principle to prove the sufficiency of the condition in Proposition 10.1(a) of Chapter I.
5. Use the downward transfer principle to prove the sufficiency of the condition in Proposition 10.8 of Chapter I.
6. Use the transfer principle to show that the set N of standard natural numbers is not an element of $*\mathscr{P}(N)$.
7. Let $*: V(X) \to V(Y)$ be a monomorphism. Show that if $f \in V(X)$ maps a onto b then $*f$ maps $*a$ onto $*b$.

*II.4 The Ultrapower Construction
for Superstructures

In this section we show how to generalize the construction of $*R$ in Chapter I by constructing, for any superstructure $V(X)$, a superstructure $V(*X)$ on an appropriate set $*X$ and a monomorphism $*: V(X) \to V(*X)$.

We begin with an ultrafilter $\mathscr{U}$ on an index set I (see the Appendix); both I and $\mathscr{U}$ will be fixed in the construction of $V(*X)$, but in later sections we will choose them to have additional properties. Now let $V(X) = \bigcup_{n=0}^\infty V_n(X)$ be a given superstructure.

4.1 Definition Let S be an entity in $V(X)$. The set of all maps $a: I \to S$ is denoted by $\prod S$; we write $a(i) = a_i$ for $i \in I$. The maps a and b in $\prod S$ are *equivalent* (with respect to $\mathscr{U}$), and we write $a =_{\mathscr{U}} b$ iff $\{i \in I : a_i = b_i\} \in \mathscr{U}$ (the equality is set-theoretic except when $S \subseteq X$, in which case it is identity). If $a =_{\mathscr{U}} b$ we say that $a_i = b_i$ almost everywhere (a.e.). The relation $=_{\mathscr{U}}$ is an equivalence relation on $\prod S$. The set of associated equivalence classes is denoted by $\prod_{\mathscr{U}} S$, and is called the *ultrapower* of S (with respect to $\mathscr{U}$). The equivalence class in $\prod_{\mathscr{U}} S$ containing $a \in \prod S$ is denoted by $[a]$. Let $V_{-1}(X) = \varnothing$, the empty set. The *bounded ultrapower* of $V(X)$ is the set

$$\prod_{\mathscr{U}}^0 V(X) = \bigcup_{n=0}^{\infty} \prod_{\mathscr{U}} [V_n(X) - V_{n-1}(X)].$$

We define the map $e \colon V(X) \to \prod_{\mathscr{U}}^0 V(X)$ by $e(a) = [\bar{a}]$, where $\bar{a}_i = a$ for all $i \in I$.

The proof that $=_{\mathscr{U}}$ is an equivalence relation on $\prod S$ is similar to the proof of Lemma 1.4 of Chapter I and is left as an exercise. We see immediately from Definitions 1.3 and 1.5 of Chapter I that $*R = \prod_{\mathscr{U}} R$, where $\mathscr{U}$ is the ultrafilter of §I.1. The map e is a generalization of the map $* \colon R \to *R$ of Definition 1.9 of Chapter I. $\prod_{\mathscr{U}}^0 V(X)$ is called a bounded ultrapower since, for each $[a] \in \prod_{\mathscr{U}}^0 V(X)$, $a_i \in V_k(X) - V_{k-1}(X)$, $i \in I$, for some fixed $k \in N$; thus, there is a uniform upper bound to the rank of a_i, $i \in I$.

We now want to construct from $\prod_{\mathscr{U}}^0 V(X)$ a superstructure $V(*X)$ over a set $*X$, and an associated mapping $M \colon \prod_{\mathscr{U}}^0 V(X) \to V(*X)$. We will finally define the mapping $* \colon V(X) \to V(*X)$ as the composition of e and M, and show that $*$ is a monomorphism. In the literature, M is called a *Mostowski collapsing function*.

First we must define $*X$. In analogy with the definition of $*R$ we put

$$(4.1) \qquad *X = \prod_{\mathscr{U}} X = \prod_{\mathscr{U}} V_0(X).$$

Now $V(*X)$ is completely determined and we proceed to the definition of $M \colon \prod_{\mathscr{U}}^0 V(X) \to V(*X)$.

We define M successively on $\prod_{\mathscr{U}} [V_n(X) - V_{n-1}(X)]$ by induction. By (4.1) $\prod_{\mathscr{U}} V_0(X) = *X$, and by definition $V_0(*X) = *X$, and so we define M to be the identity on $*X$, i.e.,

$$(4.2) \qquad M(a) = a, \qquad a \in \prod_{\mathscr{U}} V_0(X) = *X.$$

For higher levels we need the following definition.

4.2 Definition If $[a], [b] \in \prod_{\mathscr{U}}^0 V(X)$, then $[a] \in_{\mathscr{U}} [b]$ iff $\{i \in I \mid a_i \in b_i\} \in \mathscr{U}$.

The reader should check that $\in_{\mathscr{U}}$ is well defined (exercise).

To motivate the definition of M on $\prod_{\mathscr{U}} [V_1(X) - V_0(X)]$ we let $X = R$ and recall from §I.2 the definition of $*A$, where A is a subset of R. By Definition 2.2 of Chapter I (with $I = N$), $*A$ consists of those elements $[a]$ of

*R for which $\{i \in I : a_i \in A\} \in \mathcal{U}$. For our more general situation, the subset A is mapped by e to the element $e(A) = [\bar{A}]$ in $\prod_{\mathcal{U}} V_1(R)$. Note that $[\bar{A}]$ is not a subset of *R. We want *A to be a subset of *R and to consist of precisely those elements $[a] \in {}^*R$ for which $[a] \in_{\mathcal{U}} [\bar{A}]$. Since * will be the composition of e and M, it follows that we should put

$$M([\bar{A}]) = \{[a] \in {}^*R : [a] \in_{\mathcal{U}} [\bar{A}]\}$$
$$= \{M([a]) \in V_0({}^*R) : [a] \in \prod_{\mathcal{U}} V_0(R) \text{ and } [a] \in_{\mathcal{U}} [\bar{A}]\}.$$

The general definition is now clear.

4.3 Definition We define $M : \prod_{\mathcal{U}}^0 V(X) \to V(^*X)$ inductively by

$$M([b]) = [b] \qquad \text{for} \quad [b] \in \prod_{\mathcal{U}} V_0(X),$$

$$M([b]) = \{M([a]) : [a] \in \bigcup_{k=0}^{n-1} \prod_{\mathcal{U}} [V_k(X) - V_{k-1}(X)] \text{ and } [a] \in_{\mathcal{U}} [b]\}$$

$$\text{for} \quad [b] \in \prod_{\mathcal{U}} [V_n(X) - V_{n-1}(X)], \quad n \geq 1.$$

The important properties of M and e are collected together in the following result.

4.4 Lemma

   (i) e and M are one-to-one maps; i.e., $a = b$ iff $e(a) = e(b)$, and $[a] = [b]$ iff $M([a]) = M([b])$.

   (ii) e maps X into *X; M maps *X onto *X.

   (iii) e maps $V_{n+1}(X) - V_n(X)$ into $\prod_{\mathcal{U}} [V_{n+1}(X) - V_n(X)]$; M maps $\prod_{\mathcal{U}} [V_{n+1}(X) - V_n(X)]$ into $V_{n+1}(^*X) - V_n(^*X)$.

   (iv) $a \in b$ iff $e(a) \in_{\mathcal{U}} e(b)$; $[a] \in_{\mathcal{U}} [b]$ iff $M([a]) \in M([b])$.

   (v) $e(X) = [\bar{X}]$ and $M([\bar{X}]) = {}^*X$.

   (vi) Let $[a], [b] \in \prod_{\mathcal{U}}^0 V(X)$ and put $c_i = \{a_i, b_i\}$, $i \in I$. Then $[c] \in \prod_{\mathcal{U}}^0 V(X)$ and $M([c]) = \{M([a]), M([b])\}$. Similar statements hold with $\{\ \}$ replaced by $\langle\ \rangle$ and $=$ replaced by $\in$, and also for three or more terms.

   (vii) If $[b] \in_{\mathcal{U}} e(a)$, $a \in V_n(X) - V_{n-1}(X)$, then $[b] \in_{\mathcal{U}} e(V_{n-1}(X))$.

Proof: We leave the proof of (ii)–(v) and (vii) as exercises.

   (i) To show e is one-to-one let $a \neq b \in V(X)$. Then $e(a) \neq e(b)$ since $\bar{a}_i \neq \bar{b}_i$ for all $i \in I$, and $\varnothing \notin \mathcal{U}$. To show M is one-to-one, we consider only the case that $[a]$ and $[b]$ are in $\prod_{\mathcal{U}}^0 V(X) - {}^*X$, and $[a] \neq [b]$ in $\prod_{\mathcal{U}}^0 V(X)$. Let $U_a = \{i \in I : \text{there exists } v_i \in a_i \text{ with } v_i \notin b_i\}$ and $U_b = \{i \in I : \text{there exists } v_i \in b_i \text{ with } v_i \notin a_i\}$. If neither U_a nor U_b is in $\mathcal{U}$, then $I - (U_a \cup U_b)$ is in $\mathcal{U}$ and $a_i = b_i$

for almost all $i \in I$. But this is impossible. Assume, therefore, that $U_a \in \mathcal{U}$. Choose $v_i \in a_i - b_i$ for each $i \in U_a$ and let v_i be a fixed v_{i_0} otherwise. Then $M([v]) \in M([a])$ and $M([v]) \notin M([b])$. The rest is left to the reader.

(vi) We prove the first statement and leave the rest to the reader.

Now $M([c]) = \{M([y]) : y_i \in \{a_i, b_i\} \text{ a.e.}\}$. If $y_i \in \{a_i, b_i\}$ a.e., let $A = \{i \in I : y_i = a_i\}$ and $B = \{i \in I : y_i = b_i\}$. Then $A \cup B \in \mathcal{U}$, and so either $A \in \mathcal{U}$ or $B \in \mathcal{U}$ since $\mathcal{U}$ is an ultrafilter. Thus

$$M([c]) = \{M([y]) : y_i = a_i \text{ a.e.}\} \cup \{M([y]) : y_i = b_i \text{ a.e.}\}$$
$$= \{M([a]), M([b])\}. \quad \square$$

With $*$ defined as the composition of e and M, we now show that $*: V(X) \to V(*X)$ is a monomorphism. To do so we need the following fundamental result; in the proof we use the axiom of choice.

4.5 Theorem (Łoś) If $\Phi(x_1, \ldots, x_n)$ is a formula in $\mathcal{L}_X$ with $x_1, \ldots, x_n$ its only free variables, and $[a_1], \ldots, [a_n] \in \prod_{\mathcal{U}}^0 V(X)$, then $*\Phi(M([a_1]), \ldots, M([a_n]))$ is true in $V(*X)$ iff

$$\{i \in I : \Phi(a_1(i), \ldots, a_n(i)) \text{ is true}\} \in \mathcal{U}.$$

Proof: 1. We first establish the result when Φ is an atomic formula. If Φ is of the form $x \in y$ or $x = y$, where x and y are either constants or variables, the result is immediate from 4.4(i) and 4.4(iv). The result for Φ of the form $\langle x_1, \ldots, x_n \rangle \in x_{n+1}$, $\langle x_1, \ldots, x_n \rangle = x_{n+1}$, $\langle \langle x_1, \ldots, x_n \rangle, x \rangle \in x_{n+1}$, and $\langle \langle x_1, \ldots, x_n \rangle, x \rangle = x_{n+1}$ can be proved by induction using 4.4(vi) (Exercise 4).

2. Suppose now that the theorem has been established for the formulas $\Phi(x_1, \ldots, x_n)$ and $\Psi(x_1, \ldots, x_n)$. We would like to prove it for the formulas $\neg\Phi$, $\Phi \wedge \Psi$, $\Phi \vee \Psi$, and $\Phi \to \Psi$. We do so for the first two and leave the proofs for the last two as exercises (Exercise 4); recall, however, that $\Phi \vee \Psi$ is equivalent to $\neg[(\neg\Phi) \wedge (\neg\Psi)]$.

(i) For $\neg\Phi$ note that the following are equivalent:

$$*(\neg\Phi)(M([a_1]), \ldots, M([a_n])) \text{ is true};$$
$$\neg *\Phi(M([a_1]), \ldots, M([a_n])) \text{ is true};$$
$$\{i \in I : \Phi(a_1(i), \ldots, a_n(i)) \text{ is true}\} \notin \mathcal{U};$$
$$\{i \in I : \neg\Phi(a_1(i), \ldots, a_n(i)) \text{ is true}\} \in \mathcal{U}$$

(since $\mathcal{U}$ is an ultrafilter).

(ii) For $\Phi \wedge \Psi$ note that the following are equivalent:

$*(\Phi \wedge \Psi)(M([a_1]), \ldots, M([a_n]))$ is true;

$*\Phi(M([a_1]), \ldots, M([a_n])) \wedge *\Psi(M([a_1]), \ldots, M([a_n]))$ is true;

$\{i \in I : \Phi(a_1(i), \ldots, a_n(i))$ is true$\} \in \mathcal{U}$,   and

$\{i \in I : \Psi(a_1(i), \ldots, a_n(i))$ is true$\} \in \mathcal{U}$;

$\{i \in I : \Phi(a_1(i), \ldots, a_n(i))$ is true$\}$

   $\cap \{i \in I : \Psi(a_1(i), \ldots, a_n(i))$ is true$\} \in \mathcal{U}$ (since $\mathcal{U}$ is a filter);

$\{i \in I : (\Phi \wedge \Psi)(a_1(i), \ldots, a_n(i))$ is true$\} \in \mathcal{U}$.

3. Suppose the result is true for a formula of the form $\Phi(x_1, \ldots, x_n, y)$. We want to show it is true for formulas of the form $(\exists y \in c)\Phi$, $(\exists y \in z)\Phi$, $(\forall y \in c)\Phi$, and $(\forall y \in z)\Phi$, where c is a constant and z is a variable. We consider the case $(\exists y \in c)\Phi$ and leave the case $(\exists y \in z)\Phi$ to the reader (Exercise 4). For the quantifier $\forall$, replace $(\forall y \in c)\Phi$ with $\neg(\exists y \in c)\neg\Phi$ and $(\forall y \in z)\Phi$ with $\neg(\exists y \in z)\neg\Phi$.

Suppose $*(\exists y \in c)\Phi(M([a_1]), \ldots, M([a_n]), y)$ holds in $V(*X)$, i.e.,

$$(\exists y \in *c)*\Phi(M([a_1]), \ldots, M([a_n]), y)$$

holds in $V(*X)$. Thus we can find $M([a]) \in V(*X)$ so that

$$(M([a]) \in *c) \wedge \Phi(M([a_1]), \ldots, M([a_n]), M([a]))$$

holds in $V(*X)$. Using step 2, this is equivalent to

$$\{i \in I : a(i) \in c \wedge \Phi(a_1(i), \ldots, a_n(i), a(i)) \text{ is true}\} \in \mathcal{U}.$$

Hence also the larger set $\{i \in I : (\exists y \in c)\Phi(a_1(i), \ldots, a_n(i), y)$ is true$\}$ is in $\mathcal{U}$.

Conversely, let $\{i \in I : (\exists y \in c)\Phi(a_1(i), \ldots, a_n(i), y)$ is true$\} = U$ belong to $\mathcal{U}$. Then, for each $i \in U$, we can use the axiom of choice to choose some $a(i) \in c$ and for $i \in I - U$ put $a(i) = d \in c$, where d is a fixed element of c, so that $\{i \in I : a(i) \in c \wedge \Phi(a_1(i), \ldots, a_n(i), a(i))\} \in \mathcal{U}$. Changing $a(i)$ on the complement of a set in $\mathcal{U}$ if necessary, we may assume that $a(i) \in V_n(X) - V_{n-1}(X)$ for some $n \in N$ and all i (Exercise 6). Now the map $a: I \to c$ defines $[a] \in \prod^0_{\mathcal{U}} V(X)$ and the steps of the previous paragraph can be retraced, yielding the result.

4. The general result now follows by induction based on Definition 2.9 (Exercise 4). $\square$

4.6 Theorem The map $*: V(X) \to V(*X)$ defined by $* = M \circ e$ is a monomorphism.

Proof: We prove (v) of Definition 3.2 and leave the remaining proofs as exercises. Let Φ be a sentence in $\mathscr{L}_X$. Then Φ has no free variables, so $*\Phi$ is true in $V(*X)$ iff $\{i \in I : \Phi$ is true$\} \in \mathscr{U}$ by Theorem 4.5. But the set $\{i \in I : \Phi$ is true$\}$ is either I [if Φ is true in $V(X)$] or $\varnothing$ [if Φ is not true in $V(X)$], so $*\Phi$ is true if and only if Φ is true. $\square$

Whenever nonstandard analysis is applied in any concrete situation in the rest of this book, we will start with a superstructure $V(S)$ based on a suitable set S, and then use a superstructure $V(*S)$ and a monomorphism $*: V(S) \to V(*S)$ constructed with an ultrafilter $\mathscr{U}$ as in this section. Usually the monomorphism will not be mentioned explicitly, but we will always choose $\mathscr{U}$ in such a way that $V(*S)$ has a special property, that of being an enlargement. This will guarantee that $*S$ is large enough to contain "infinite" entities. We turn to this question in the next section.

Exercises II.4

1. Prove that $=_{\mathscr{U}}$ is an equivalence relation on $\prod S$.
2. Show that the relation $\in_{\mathscr{U}}$ of Definition 4.2 is well defined.
3. Finish the proof of Lemma 4.4.
4. Finish the proof of Theorem 4.5.
5. Finish the proof of Theorem 4.6.
6. Show that if $a(i) \in V_n(X)$ for a fixed n and all $i \in I$, then, for some $k \le n$ and all $i \in U$ for some $U \in \mathscr{U}$, $a(i) \in V_k(X) - V_{k-1}(X)$.

II.5 Hyperfinite Sets, Enlargements, and Concurrent Relations

In §I.1 we showed that $*R$ was strictly larger than R (regarded as embedded in $*R$) by exhibiting elements like $[\langle 1, 2, 3, \ldots \rangle]$ in $*R$ which were not equal to any element of R. The demonstration involved the fact that the ultrafilter $\mathscr{U}$ on N was free, i.e., it contained the cofinite filter $\mathscr{F}_N$. In the general case it is interesting to determine the conditions under which $*X$ is strictly larger than X. It should be recalled that, by assumption, X contains N and hence is infinite. The following result shows that $*X = X$ and hence $V(*X) = V(X)$ when $\mathscr{U}$ is a principal (nonfree) ultrafilter on I; thus we get nothing new in this case.

5.1 Lemma If $\mathscr{U}$ is a principal ultrafilter on I then $*X$ (as constructed in §II.4) equals X (regarded as embedded in $*X$).

Proof: A principal ultrafilter $\mathscr{U}$ is generated by a single element $i_0 \in I$; i.e., $\mathscr{U}$ consists of all sets $U \subseteq I$ which contain i_0 (see the Appendix). If $[a] \in *X$

and $a_{i_0} = a_0$ then $[a] =_\mathcal{U} [\bar{a}]$, where $\bar{a}_i = a_0$ for all $i \in I$. Thus $[a] \in X$, where X is regarded as embedded in $*X$. □

We will next show how to choose an index set and an ultrafilter of subsets of the index set so that the $*X$ constructed as in §II.4 is strictly larger than X, and so that $V(*X)$ has other desirable properties; the most important is that of being an enlargement.

We begin by introducing the notion of a hyperfinite or $*$-finite set.

5.2 Definition If $A \in V_n(X) - V_0(X)$ for some n, we denote by $\mathscr{P}_F(A)$ the set of all finite subsets of A. $\mathscr{P}_F(A)$ is in $V(X)$, and we call the image $*\mathscr{P}_F(A) \in V(*X)$ (with respect to a monomorphism $*$) the set of *hyperfinite* or *$*$-finite* subsets of $*A$. The set of all hyperfinite subsets is the set $\bigcup_{n=1}^{\infty} *\mathscr{P}_F(V_n(X))$.

Any elementary mathematical result that holds for finite sets extends to a similar result for hyperfinite sets by the transfer principle. An example of a hyperfinite set is the set $J \subset *N$ of positive integers less than some $j \in *N$. To see that J is hyperfinite consider the collection $\mathscr{I} \subset \mathscr{P}_F(N)$ of all finite subsets of N of the form $\{1, 2, \ldots, j\}$ for some $j \in N$ (a set of this form is called an *initial segment*). Then $*\mathscr{I} \subset *\mathscr{P}_F(N)$ contains sets of the form $\{n \in *N : n \leq j\}$ for some $j \in *N$. The following result shows that these hyperfinite sets are in some sense the prototype.

5.3 Theorem If $B \in *V_k(X)$, $k \in N$, is a hyperfinite set, then there is an initial segment $J = \{n \in *N : n \leq j\}$ for some $j \in *N$ and a one-to-one, onto mapping $f : J \to B$ in $*V_{k+4}(X)$.

Proof: Suppose $B \in *\mathscr{P}_F(A)$, where $A \in V_n(X)$, $n \geq 1$. Now the following statement (in semiformal language) is true in $V(X)$:

$$(\forall B \in \mathscr{P}_F(V_n(X)))(\exists j \in N)(\exists f \in V_{n+4}(X))$$
$$[f \text{ maps } J \text{ one-to-one onto } B, \text{ where } J = \{n \in N : n \leq j\}]$$

[the reader should check that the sentence in square brackets can be translated into a sentence in $\mathscr{L}_X$ (exercise)]. The result follows by transfer. □

Because of Theorem 5.3 we will often write a hyperfinite set B as $B = \{b_1, b_2, \ldots, b_j\}$, where $b_k = f(k)$, $k \in J$, and f is the function of the theorem. It should be noted that the dots in this representation cover somewhat more ground than they do in the standard case, and that this representation is really an abbreviation of the setup in Theorem 5.3. Hyperfinite sets are an important tool in nonstandard analysis by virtue of the fact that many standard mathematical structures can be "approximated" by hyperfinite structures in a natural way. We will illustrate this fact later in this section.

5.4 Definition Entities in $V(X)$, and entities which are of the form $*b$ for some $b \in V(X)$, are called *standard*; all others are called *non-standard*.

5.5 Examples

1. Each individual in $X \subseteq *X$ is standard.
2. In Example 3.4 the intervals $I \in *\mathscr{I}$ of the form $I = \{x \in *R : a \le x \le b\}$, where $a < b$, are themselves standard entities even though they contain non-standard numbers. An interval $\{x \in *R : \alpha \le x \le \beta\}$, where $0 < \alpha < \beta$ and α and β are infinitesimal, is a non-standard entity.

5.6 Definition The superstructure $V(*X)$ [with respect to a monomorphism $* : V(X) \to V(*X)$] is called an *enlargement* of $V(X)$ if for each set $A \in V(X)$ there is a set $B \in *\mathscr{P}_F(A)$ such that $*a \in B$ for each $a \in A$, i.e., B contains the standard entities in $*A$.

We have already seen that a hyperfinite set of the form $\{n \in *N : 1 \le n \le j\}$, where $j \in *N_\infty$, contains every standard natural number. Definition 5.6 is a generalization for arbitrary sets in $V(X)$.

We will now show that for a given superstructure $V(X)$ it is possible to choose an index set J and a free ultrafilter $\mathscr{V}$ on J so that the associated superstructure $V(*X)$, constructed as in §II.4 using J and $\mathscr{V}$, is an enlargement of $V(X)$. It will follow as a corollary, since X is infinite, that $*X$ is strictly larger than X. The proofs of Lemma 5.7 and Theorem 5.8 may be skipped on first reading of the chapter.

Let J be the set of all nonempty finite subsets of $V(X)$. It follows that $a \in J$ iff there is a $b \in V(X) - V_0(X)$ and $a \in \mathscr{P}_F(b) - \emptyset$ (why?). If $a \in J$ we define

$$J_a = \{b \in J : a \subseteq b\}.$$

5.7 Lemma The collection

$$\mathscr{F} = \{A \subseteq J : \text{there exists } a \in J \text{ such that } J_a \subseteq A\}$$

is a free filter on J.

Proof: It is easy to show that $\mathscr{F}$ is a filter. For example, if $A_1, A_2 \in \mathscr{F}$ there exist $a_1, a_2 \in J$ so that $A_i \supseteq J_{a_i}$ $(i = 1, 2)$. Since $A_1 \cap A_2 \supseteq J_{a_1} \cap J_{a_2} = J_{a_1 \cup a_2}$, $A_1 \cap A_2 \in \mathscr{F}$. The rest is left as an exercise.

To show $\mathscr{F}$ is free, let $a \in J$. Then there is an element $b \in J$ so that $a \cap b = \emptyset$. Since $a \notin J_b$, $J - \{a\} \supset J_b$, so $\mathscr{F}$ is free. $\square$

Now let $\mathscr{V}$ be an ultrafilter on J with $\mathscr{V} \supset \mathscr{F}$ (such an ultrafilter exists by Theorem A.5 of the Appendix).

5.8 Theorem If $V(*X)$ is constructed from $V(X)$ using $\mathscr{V}$ and J then it is an enlargement of $V(X)$.

Proof: Let A be a set in $V(X)$. We define a map $\Gamma: J \to \mathscr{P}_F(A)$ by $\Gamma_a = a \cap A$, and let $B = M([\Gamma])$. Then $B \in *\mathscr{P}_F(A)$. If $x \in A$ then $J_{\{x\}} = \{a \in J : x \in a\}$, so $\{a \in J : x \in a \cap A\} \in \mathscr{V}$. Thus $[\bar{x}] \in_{\mathscr{V}} [\Gamma]$ and so $*x \in B$.   □

Robinson's original definition of enlargement (see Theorem 5.10 below) made use of the notion of concurrent relation and was the cornerstone of his development of nonstandard analysis.

5.9 Definition A binary relation P is *concurrent* (finitely satisfiable) on $A \subseteq \operatorname{dom} P$ if for each finite set $\{x_1, \ldots, x_n\}$ in A there is a $y \in \operatorname{range} P$ so that $\langle x_i, y \rangle \in P$, $1 \leq i \leq n$. P is concurrent if it is concurrent on $\operatorname{dom} P$.

Examples of concurrent relations are the relation $\leq$ in N and $\subseteq$ in $\mathscr{P}_F(N)$.

5.10 Theorem The following are equivalent:

(i) $V(*X)$ is an enlargement of $V(X)$.
(ii) For each concurrent relation $P \in V(X)$ there is an element $b \in \operatorname{range} *P$ so that $\langle *x, b \rangle \in *P$ for all $x \in \operatorname{dom} P$.

Proof: (i) $\Rightarrow$ (ii): Let $B \in *\mathscr{P}_F(\operatorname{dom} P)$ be such that, for each $x \in \operatorname{dom} P$, $*x \in B$. Since the sentence $(\forall w \in \mathscr{P}_F(\operatorname{dom} P))(\exists y \in \operatorname{range} P)(\forall x \in w)[\langle x, y \rangle \in P]$ is true in $V(X)$ by concurrence of P, its $*$-transform is true in $V(*X)$. Thus there exists an element $b \in \operatorname{range} *P$ so that $\langle z, b \rangle \in *P$ for each $z \in B$, and in particular for each $*x$ with $x \in \operatorname{dom} P$.
(ii) $\Rightarrow$ (i): Exercise.   □

5.11 Corollary If $Y \in V(X)$ contains an infinite number of entities and $V(*X)$ is an enlargement, then $*Y$ contains entities which are not standard. In particular, if $A \subseteq X$ is infinite then $*A$ properly contains A.

Proof: The relation P on $Y \times Y$ defined by "$\langle a, b \rangle \in P$ iff $a \neq b$" is concurrent since Y is infinite. By 5.10(ii) there is a $b \in *Y$ such that $b \neq *x$ for all $x \in Y$.   □

Corollary 5.11 gives another proof of the existence, in an enlargement $V(*R)$ of $V(R)$, of non-standard numbers, but it holds in much more general situations.

5.12 Definition A set $\mathscr{S}$ of subsets of an entity $A \in V(X)$ is called *exhausting* if, for each finite subset $F \subseteq A$, there is an $S \in \mathscr{S}$ with $F \subseteq S$.

5.13 Proposition If $\mathscr{S}$ is an exhausting set of subsets of $A \in V(X)$ and $V(*X)$ is an enlargement, then there is a set $C \in *\mathscr{S}$ containing all the standard entities in $*A$.

Proof: Let B be a hyperfinite subset of $*A$ such that $*a \in B$ for each $a \in A$. Then there is a $C \in *\mathscr{S}$ with $B \subseteq C$. □

In spite of its simplicity, Proposition 5.13 turns out to be a very powerful tool in nonstandard analysis. The typical application runs as follows. Suppose A is an infinite set with some additional mathematical structure; for example, A could be an infinite graph, or a Hilbert space. Suppose further that A can be exhausted by a family $\mathscr{S}$ of substructures—finite subgraphs, finite-dimensional inner-product spaces, etc.—so that for each $S \in \mathscr{S}$ a certain result can be proved. One wants to establish a corresponding result for A. Using Proposition 5.13, we can find a set $C \in *\mathscr{S}$ containing all of the standard elements in $*A$, and by transfer the $*$-transform of the given result is true for C. The problem then is to show how the validity of the $*$-transform of the result on C induces the validity of the result on A. This last step can be quite difficult but is often easier than proving the result by standard methods. This method of proof was the basis of the first successful attack, by Bernstein and Robinson [7], on an invariant subspace problem in Hilbert space proposed by Smith and Halmos.

We illustrate the technique by proving a result in infinite graph theory due to de Bruijn and Erdös. (See also the related paper by Luxemburg [35].) The application indicates how nonstandard analysis is applicable in areas other than analysis. A *graph* $\langle A, E \rangle$ consists of a set A of *vertices* and a binary relation E on $A \times A$ which is symmetric (i.e., $\langle x, y \rangle \in E$ implies $\langle y, x \rangle \in E$). If $\langle x, y \rangle \in E$ we say that x and y are connected by an *edge*. $\langle A, E \rangle$ is *infinite* if A is infinite. $\langle A, E \rangle$ is *k-colorable* if there exists a map $f: A \to \{1, 2, \ldots, k\}$ (the set of "colors") such that if $\langle a, b \rangle \in E$ then $f(a) \neq f(b)$, i.e., no two vertices which are connected by an edge are given the same color. If $B \subseteq A$ then the *subgraph* $\langle B, E|B \rangle$ is defined by "$\langle x, y \rangle \in E|B$ iff $x, y \in B$ and $\langle x, y \rangle \in E$"; i.e., B inherits its edges from E.

5.14 Theorem (De Bruijn–Erdös [13]) If each finite subgraph of an infinite graph $\langle A, E \rangle$ is k-colorable, then $\langle A, E \rangle$ is k-colorable.

Proof: We work in the superstructure $V(A \cup N)$. Let $\mathscr{S}$ denote the set of all finite subsets of A (obviously exhausting). For each $F \in \mathscr{S}$ the graph $\langle F, E|F \rangle$ is k-colorable, so the following is true in $V(A \cup N)$:

$$(5.1) \qquad (\forall F \in \mathscr{S})(\exists f_F: F \to \{1, 2, \ldots, k\})(\forall x, y \in F)$$
$$[\langle x, y \rangle \in E \to f_F(x) \neq f_F(y)].$$

By the definition of enlargement, there exists a $B \in {}^*\mathscr{S}$ so that $B \supseteq A$. By transfer of (5.1) we see that there is a map (coloring) $f_B: B \to {}^*\{1, 2, \ldots, k\}$ $(= \{1, 2, \ldots, k\})$ so that if $\langle x, y \rangle \in {}^*E$ then $f_B(x) \neq f_B(y)$. We now restrict f_B to A to get a map $f: A \to \{1, 2, \ldots, k\}$. f is a coloring since it inherits the property "$\langle x, y \rangle \in E$ implies $f(x) \neq f(y)$" from f_B (check). $\square$

Intuitively, the proof of 5.14 given above is obvious; we have simply covered A by a $*$-finite and hence k-colorable graph B and then restricted the coloring. A similar technique can be used to give easy proofs of more intricate theorems in infinite graph theory.

In closing this section we note that the results of Chapter I for $*R$ remain valid for an enlargement of $V(R)$. To get more we need to consider the notions of internal and external entities in $V(*R)$; these are introduced in the next section.

Exercises II.5

1. Show that if j is infinite then $J = \{n \in N : n \leq j\} \in {}^*\mathscr{P}_F(N) - \mathscr{P}_F(*N)$.
2. Show that in general $*\mathscr{P}_F(A) \supseteq \mathscr{P}_F(*A)$ whereas $\mathscr{P}(*A) \supseteq *\mathscr{P}(A)$.
3. Check the translation into a sentence in $\mathscr{L}_X$ of the informal sentence in the proof of Theorem 5.3.
4. Show that the family $\mathscr{F}$ in Lemma 5.7 is a filter given that $A_1, A_2 \in \mathscr{F} \Rightarrow A_1 \cap A_2 \in \mathscr{F}$.
5. Prove that (ii) $\Rightarrow$ (i) in Theorem 5.10.
6. Show that if $\{O_\alpha : \alpha \in A\}$ is an open covering of a set $S \subset R$ but no finite subcollection covers S, then there is a $y \in {}^*S$ such that $y \not\approx x$ for all $x \in S$.
7. Give another proof of the existence of infinite natural numbers in an enlargement $V(*R)$ of $V(R)$ by using the concurrent relation $<$.
8. Let A be an entity in a superstructure $V(X) - X$ which is closed under finite unions; i.e., if $a_i \in A$ $(1 \leq i \leq n)$ then $\bigcup a_i (1 \leq i \leq n) \in A$. Show that if $V(*X)$ is an enlargement of $V(X)$, there is an element $b \in {}^*A$ so that $\bigcup *a(a \in A) \subseteq b$.
9. (Luxemburg) Let A be an entity of $V(X) - X$. The *intersection monad* of A is the set $\mu(A) = \bigcap *a(a \in A)$ in $V(*X)$. A has the *finite* intersection property (f.i.p.) if $a, b \in A$ implies $a \cap b \neq \varnothing$. Show that $V(*X)$ is an enlargement of $V(X)$ iff the intersection monad $\mu(A)$ of each A with the f.i.p. is nonempty.
10. (Luxemburg) Let $V(*X)$ be an enlargement of $V(X)$ and $\mathscr{F}$ be a filter in $V(X) - X$. Let $\mu(\mathscr{F})$ be the intersection monad (see Exercise 9). Show that if $B \in V(X)$ and $F \cap B \neq \varnothing$ for all $F \in \mathscr{F}$ then $\mu(\mathscr{F}) \cap *B \neq \varnothing$.
11. Show that if $\mathscr{F}$ is a filter in $V(X) - X$ and B is a set in $V(*X)$ such that $B \cap *F \neq \varnothing$ for all $F \in \mathscr{F}$ then it is not necessarily true that $\mu(\mathscr{F}) \cap B \neq \varnothing$, where $\mu(\mathscr{F})$ is the intersection monad of Exercise 9. [Hint: Let $X = N$,

$\mathscr{F}$ be the Fréchet filter on N (the collection of complements of finite sub-sets of N), and $B = N$.]

12. A standard result states (informally) that a set $A \in V(X)$ is finite iff every injective map $f: A \to A$ is surjective. Formalize this statement and so obtain a similar characterization for the $*$-finite sets.

13. Let $P \in V(X)$ be a binary relation and suppose that in some enlargement $V(*X)$ of $V(X)$ the following is true: for every $y \in$ range $*P$, there exists an $x \in$ dom P so that $\langle *x, y \rangle \in *P$. Show that there exists a finite set $\{x_1, \ldots, x_n\} \subseteq$ dom P such that for all $y \in$ range P, there is an i, $1 \le i \le n$, with $\langle x_i, y \rangle \in P$.

14. (Konig's lemma) Let $\langle S_n : n \in N \rangle$ be a sequence of mutually disjoint nonempty finite sets and let P be a binary relation on $\bigcup S_n (n \in N)$ such that, whenever $x \in S_{n+1}$ for some n, there exists a $y \in S_n$ such that $\langle y, x \rangle \in P$. Show that there exists an infinite sequence $\langle x_n : n \in N \rangle$ such that $x_n \in S_n$ and $\langle x_n, x_{n+1} \rangle \in P$ for $n \in N$.

15. (Total ordering) Let X be a nonempty set. A binary relation P on X is a *partial ordering* if the following holds: (a) P is reflexive, that is, $\langle x, x \rangle \in P$ for all $x \in X$; (b) P is antisymmetric, that is if $\langle x, y \rangle \in P$ and $\langle y, x \rangle \in P$ then $x = y$; (c) P is transitive, that is, if $\langle x, y \rangle \in P$ and $\langle y, z \rangle \in P$ then $\langle x, z \rangle \in P$. A partial ordering P on X is a *total ordering* if whenever $x, y \in X$ then either $\langle x, y \rangle \in P$ or $\langle y, x \rangle \in P$. Every finite set can be totally ordered. Assuming this result and the fact that enlargements exist, show that any set can be totally ordered.

16. (Rado's selection lemma)   Let $\{A_\lambda : \lambda \in \Lambda\}$ be a nonempty family of finite sets. A choice function over Λ is a function $\phi : \Lambda \to \bigcup A_\lambda (\lambda \in \Lambda)$ so that $\phi(\lambda) \in A_\lambda$ for each $\lambda \in \Lambda$. Let $\{A_\gamma : \gamma \in \Gamma\}$ be a nonempty family of finite sets. Assume that for each finite subset $F \subseteq \Gamma$ there is a choice function ϕ_F over F. Show that there exists a choice function ϕ over Γ so that for any finite set $F \subseteq \Gamma$, there exists a finite set $F' \supseteq F$ with $\phi(x) = \phi_{F'}(x)$ for all $x \in F$.

II.6 Internal and External Entities; Comprehensiveness

We noted in Remark 3.5 that a basic technique of proof in nonstandard analysis is to establish the validity of a sentence Φ in $\mathscr{L}_X$ by noticing that it is the downward $*$-transform of a sentence $*\Phi$ which is true in $\mathscr{L}_{*X}$. Thus it is particularly important to be able to recognize when a sentence Ψ in $\mathscr{L}_{*X}$ is of the form $*\Phi$ for some sentence Φ in $\mathscr{L}_X$. Notice that, for a sentence Φ in $\mathscr{L}_X$, $*\Phi$ uses only the names of standard objects and so $*\Phi$ involves only expressions like $(\forall x \in *a)\Psi$, $(\forall x \in y)\Psi$, $(\exists x \in *a)\Psi$, and $(\exists x \in y)\Psi$. Thus to check the truth of $*\Phi$ we need only look at elements b in $V(*X)$ which satisfy $b \in *a$ for some $a \in V(X)$. By 3.2(iv), if $c \in b$ and $b \in *a$, then $c \in *V_k(X)$ for some k. If $b \in *a$ for some $a \in V(X)$, we call it internal; otherwise we call b external (Definition 6.1).

A sentence Ψ in $\mathscr{L}_{*X}$ is *not* of the form $*\Phi$ if it contains names of external entities, i.e., is an external sentence. A common mistake in nonstandard arguments is to apply the transfer principle to external sentences Ψ in $\mathscr{L}_{*X}$. Thus it is important to be able to recognize external entities in $V(*X)$. We will learn in this section that R, N, Z, $*R_\infty$, $*N_\infty$, $*Z_\infty$, and $m(0)$ are external subsets of $*R$. Using these, we can construct many external functions and relations. For example, the characteristic function of an external set is an external function; the relation of nearness $\simeq$ is external. The properties of external entities cannot be obtained by transfer from those of $V(X)$. For example, it is true that any subset of N which is bounded below has a least element. However, this property is not true of $*N_\infty$, for if n were a least element in $*N_\infty$ then $n - 1$ would have to be finite, which is impossible.

In this section we first concentrate on internal entities and their properties and then present examples of external entities. The section ends with a discussion of comprehensiveness which involves internality.

6.1 Definition An entity $b \in V(*X)$ is called *internal* [with respect to $*$: $V(X) \to V(*X)$] if there exists an $a \in V(X)$ so that $b \in *a$; i.e., internal entities are elements of standard entities. An entity which is not internal is called *external*. Similarly, a sentence or formula Φ in $\mathscr{L}_X$ is called either *standard* or *internal* if the constants in Φ are names of standard or internal entities, respectively. A sentence which is not internal is called *external*.

6.2 Examples

1. All standard entities are internal (Exercise 1).
2. With $\mathscr{I}$ the set of closed and bounded intervals in R, every set $\{x : a \leq x \leq b, a, b \in *R\} \in *\mathscr{I}$ is internal; the standard $*$-intervals are those for which a and b are in R.
3. If $\mathscr{C}$ denotes the set of continuous functions on R, then each $f \in *\mathscr{C}$ is internal and is called a $*$-continuous function.
4. If P is concurrent, the element $b \in$ range $*P$ given by Theorem 5.10(ii) is internal.
5. The $*$-transform of any formula $\Phi \in \mathscr{L}_X$ is standard.
6. The sentence

$$(\forall \varepsilon > 0 \text{ in } *R)(\forall y \in *R)(\exists \delta > 0 \text{ in } *R)(\forall x \in *R)$$
$$[|x - y| < \delta \to |f(x) - f(y)| < \varepsilon],$$

where $f \in {}^*\mathscr{C}$ is internal, is an internal sentence and expresses the fact that f is *-continuous on $^*R$.

6.3 Theorem The set of all internal elements of $V(^*X)$ is the set $^*V(X) = \bigcup_{n=0}^{\infty} {}^*V_n(X)$.

Proof: If $b \in {}^*V(X)$ then $b \in {}^*V_n(X)$ for some natural number $n \geq 0$ and so b is internal since $V_n(X)$ is standard. Conversely, if b is internal then $b \in {}^*a$, where a is in $V_{n+1}(X) - V_n(X)$ for some $n \geq 1$, so $a \subseteq V_n(X)$. Thus $^*a \subseteq {}^*V_n(X)$ and $b \in {}^*V_n(X)$. □

It is necessary to be able to recognize internal sets. In that regard the following result is very useful.

6.4 Theorem (Keisler's Internal Definition Principle [24]) Let $\Phi(x)$ be an internal formula in $\mathscr{L}_{*X}$ for which x is the only free variable, and let A be an internal set. Then $\{x \in A : \Phi(x) \text{ is true}\}$ is internal.

Proof: Let $c_1, \ldots, c_n$ be the constants in $\Phi(x)$; we write $\Phi(x) = \Phi(c_1, \ldots, c_n, x)$. Now $A, c_1, \ldots, c_n \in {}^*V_k(X)$ for some $k \in N$. Thus the sentence

$$(\forall x_1, \ldots, x_n, y \in V_k(X))(\exists z \in V_{k+1}(X))(\forall x \in V_k(X))$$
$$[x \in z \leftrightarrow [x \in y \wedge \Phi(x_1, \ldots, x_n, x)]]$$

in $\mathscr{L}_X$ holds in $V(X)$. Its interpretation in $V(^*X)$ says that $\{x \in A : \Phi(x) \text{ is true}\} \in {}^*V_{k+1}(X)$. □

6.5 Examples

1. The set Z_f of zeros of an internal $^*R$-valued function f in $V(^*R)$ is internal since $Z_f = \{x \in {}^*R : \langle x, 0 \rangle \in f\}$.

2. The characteristic function of an external set is external (Exercise 2).

A consequence of property 3.2(iv) of a monomorphism $*: V(X) \to V(^*X)$ is that any element of an internal entity is an internal entity. We use this fact in the proof of the following result.

6.6 Theorem If A and B are internal, then so are $A \cup B$, $A \cap B$, $A - B$, and $A \times B$.

Proof: We prove the result for $A \cup B$ and leave the remaining proofs as an exercise. Suppose $A, B \in {}^*V_{n+1}(X)$ and consider the following true statement in $V(X)$:

$$(\forall W, Y \in V_{n+1}(X))(\exists Z \in V_{n+1}(X))(\forall x \in V_n(X))[x \in Z \leftrightarrow x \in W \wedge x \in Y].$$

By transfer, there exists a set $C \in *V_{n+1}(X)$ having exactly the same elements from $*V_n(X)$ as $A \cup B$. But by 3.2(iv) all elements of A, B, and C are in $*V_n(X)$, and so $C = A \cup B$. $\square$

Having considered internal entities in some detail, we are now ready to demonstrate the existence of external entities. Recall that in Remark 7.8 of Chapter I we showed that there was no set $A \subset R$ so that $*A = R$. This fact is not sufficient to show that R is external in the sense of Definition 6.1; we would need to show that R was not an element in the $*$-transform of an element of $V(R)$. To show the existence of external subsets we use the following lemmas.

6.7 Lemma If $a \in V(X) - X$ then the internal entities in $\mathscr{P}(*a)$ consist exactly of the entities in $*\mathscr{P}(a)$.

Proof: Consider the following true statement in $V(X)$ with $n \geq 1$:

$$(\forall x \in V_n(X))[(\forall y \in x)[y \in a] \leftrightarrow x \in \mathscr{P}(a)]$$

[i.e., for all $x \in V_n(X)$, x is a subset of a if and only if $x \in \mathscr{P}(a)$]. Its $*$-transform says that, for all $x \in *V_n(X)$, x is a subset of $*a$ if and only if $x \in *\mathscr{P}(a)$. We see from Theorem 6.3 that if x is an internal set in $V(*X)$, i.e., $x \in *V(X)$, then $x \in *V_n(X)$ for some n. Such an x is a subset of $*a$ if and only if it is in $*\mathscr{P}(a)$. Thus $*V(X) \cap \mathscr{P}(*a) = *V(X) \cap *\mathscr{P}(a) = *\mathscr{P}(a)$. $\square$

As an example, we note that the internal subsets of $*N$ are exactly the members of $*\mathscr{P}(N)$.

6.8 Lemma Each nonempty internal subset of the hyperintegers $*Z$ which is bounded below (above) has a least (greatest) element.

Proof: If X is an internal nonempty subset of $*Z$ then $X \in *\mathscr{P}(Z)$ by Lemma 6.7. The result in the "bounded below" case now follows by transfer of the sentence

$$(\forall X \in \mathscr{P}(Z))[(\exists b \in Z)(\forall x \in X)[b \leq x] \wedge X \neq \varnothing$$
$$\rightarrow (\exists y \in X)(\forall x \in X)[y \leq x]],$$

which expresses the fact that each subset of Z which is bounded below has a least element. The "bounded above" case is similar. $\square$

6.9 Theorem In an enlargement $V(*R)$ of $V(R)$ the set $*N_\infty$ of infinite natural numbers is external.

Proof: Suppose that $*N_\infty \in \mathscr{P}(*N)$ is internal. Then by Lemma 6.8 there exists a least $b \in *N_\infty$. But then $b - 1 \in *N_\infty$ and $b - 1 < b$ (contradiction). $\square$

6.10 Corollary The sets R, N, Z, $*Z_\infty$ (the set of infinite integers), $*R_\infty$ (the set of infinite reals), and $m(0)$ (the set of infinitesimals) are external in an enlargement $V(*R)$ of $V(R)$.

Proof: Note that $*N_\infty = *N - N$. If N were internal then $*N_\infty$ would be internal by Theorem 6.6, contradicting Theorem 6.9.

Using the fact that the set of integers Z is external (exercise), we see that R is external, since otherwise $Z = R \cap *Z$ would be internal. Similarly $*Z_\infty$ and $*R_\infty$ are external.

To show that $m(0)$ is external, we note that $*R_\infty = \{x \in *R : (\exists y \in *R)$ $[\langle x, y \rangle \in P \wedge y \in m(0)]\}$, where P is defined by "$\langle x, y \rangle \in P$ if $y = 1/x$." If $m(0)$ is internal then so is $*R_\infty$ by Theorem 6.4 (contradiction). $\square$

Clearly, external entities and notions play a very important role in nonstandard analysis, as we see by noting the occurrence of the set of infinite natural numbers and the set of infinitesimals in many of the results of Chapter I. The reader might want to review some of the proofs in Chapter I to see just how external sets arise, and how the transfer principle is effective even though it involves only internal sets.

In many cases, external entities and notions are useful in recovering standard results from internal results. "Limiting" entities corresponding to "converging" families of entities in $V(X)$ can often be identified with internal entities in $*V(X)$, but to recover actual limiting entities in $V(X)$ usually involves some external operation (one which produces external entities). For example, consider Theorem 14.1 of Chapter I in this light. We constructed the solution $\phi(x)$ of the differential equation $\phi' = f(x, \phi)$ as the standard part, i.e., $\phi(x) = \text{st}(*\phi_\omega(x))$, of the internal function $*\phi_\omega(x)$. The operation of taking the standard part is an external operation. The solution is usually constructed as the limit of a subsequence of the polygonal sequence $\phi_n(x)$ by using the Arzelà–Ascoli theorem.

We end this section with a consideration of the notion of a comprehensive monomorphism; the special case of a denumerably comprehensive monomorphism will be used in Chapter IV.

6.11 Definition The monomorphism $*: V(X) \to V(*X)$ is *comprehensive* if, for any sets $C, D \in V(X)$ and any map $h: C \to *D$, there is an internal map $g: *C \to *D$ such that $g(*a) = h(a)$ for $a \in C$. The monomorphism is called *denumerably comprehensive* if the choice of C is restricted so that the cardinality of C is that of the natural numbers N.

6.12 Example Suppose that $*$ is comprehensive, and $\{A_n : n \in N\}$ is a sequence (in the ordinary sense) of entities in $*V_m(X)$ for some integer m. Then there is an internal sequence $\{B_n : n \in *N\}$ such that $A_n = B_n$ for all $n \in N$.

6.13 Theorem A monomorphism $*: V(X) \to V(*X)$, constructed as in §II.4, is comprehensive.

Proof: Let C, D, and h be as in Definition 6.11. Each element of $*D$ is of the form $M([b])$. Let $S(M[b])$ be a representative b from the equivalence class $[b]$. We may assume that $b_i \in D$ for all $i \in I$. For each $i \in I$, let k_i be the mapping from C to D given by

$$k_i = \{\langle a, S(h(a))(i)\rangle : a \in C\}$$

and let $[k]$ denote the equivalence class generated by the mapping $\{\langle i, k_i\rangle : i \in I\}$. We leave as an exercise the proof that $M([k])$ is an internal function from $*C$ to $*D$. If $M([a]) \in *C$ and $\{\langle i, a_i\rangle : i \in I\}$ is a representative from the equivalence class $[a]$, then the image of $M([a])$ under the mapping $M([k])$ is $M([b])$, where $b_i = S(h(a_i))(i)$ for $i \in I$. In particular, if $a_i = \alpha \in C$ for almost all $i \in I$, then $b_i = S(h(\alpha))(i)$ for all $i \in I$. Thus $M([k])$ extends h.   □

Exercises II.6

1. Show that all standard entities are internal.
2. Show that the characteristic function of an external set is external.
3. Finish the proof of Theorem 6.6.
4. Show that the sets Z, $*Z_\infty$, and $*R_\infty$ are external.
5. Show that $M([k])$ defined in the proof of Theorem 6.13 is an internal function from $*C$ to $*D$.
6. Show that if $\{x_n : n \in *N\}$ is an internal sequence and $|x_n| \leq 1/n$ for all $n \in N$ then, for some $k \in *N_\infty$, $|x_n| \leq 1/n$ for all $n \leq k$.
7. Let $\{A_n : n \in N\}$ be a sequence in the ordinary sense of internal subsets of $*R$ such that, for any $k \in N$, $\bigcap A_n(1 \leq n \leq k) \neq \varnothing$. Assume the monomorphism $*$ is denumerably comprehensive and show that $\bigcap A_n(n \in N) \neq \varnothing$.
8. (a) Show that every nonempty internal subset A of $*R$ with an upper bound has a least upper bound.
   (b) Show that every nonempty internal subset A of $*N$ has a minimal element.

9. Show that if P is an internal binary relation on $c_1 \times c_2$ then dom P and range P are internal. In particular, $c_1 \times c_2$ is internal if c_1 and c_2 are internal.

10. Show that st: $G(0) \to R$ is an external map.

11. Show that every internal subset of a $*$-finite set is $*$-finite.

12. Show that if a and b are internal then the set of all internal functions from a to b is internal.

13. (a) Let sup be the function in $V(R)$ which assigns, to each upper bounded set $E \subset R$, its supremum, sup E. The function sup can be extended to a function $*$sup defined on all internal subsets of $*R$ which are "$*$-bounded above." Characterize by a sentence in $\mathscr{L}_{*R}$ the collection $\mathscr{B}$ of sets which are $*$-bounded above, and then show that $*$sup $E \le *$sup F if $E \subseteq F$ are sets in $\mathscr{B}$.

(b) Show that if E is a finitely upper bounded external subset of $*R$, then $*$sup E may have no meaning in $*R$, but sup$\{{}^\circ r : r \in E\}$ has a meaning in R.

II.7 The Permanence Principle

In this section we present a principle with many applications called the permanence principle by Robinson and Lightstone [44] or Cauchy's principle by Stroyan and Luxemburg [46]. Throughout the section we suppose that X contains the set of reals R.

7.1 Theorem   (Permanence Principle)   Let $\Phi(x)$ be an internal formula in $\mathscr{L}_{*X}$ with x the only free variable.

(i) If $\Phi(x)$ holds for each $x \in N$ $(x \in *R^+ - *R_\infty)$, then there exists a $k \in *N_\infty$ so that $\Phi(b)$ holds for each $b \le k$ in $*N$ $(*R^+)$.

(ii) If $\Phi(x)$ holds for each $x \in *N_\infty$ $(*R_\infty^+)$, then there exists a $k \in N$ so that $\Phi(b)$ holds for each $b \ge k$ in $*N$ $(*R^+)$.

(iii) If $\Phi(x)$ holds for each infinitesimal x, then there is a standard $r > 0$ in R so that $\Phi(b)$ holds for all b with $|b| \le r$ in $*R$.

Proof: We prove the results in the case of the natural numbers N and leave the proofs for the real case (parentheses) of (i) and (ii) to the reader.

(i) Let $A = \{x \in *N : \neg \Phi(x) \text{ holds in } V(*X)\}$. Then A is internal by Theorem 6.4 (internal definition principle) and $A \subseteq *N_\infty$ by hypothesis. If $A = \varnothing$ we are through. Otherwise A is bounded below and hence has a least element l by Lemma 6.8; we may take $k = l - 1$.

(ii) Given the internal set A defined as in (i), $A \subseteq N$ and A is bounded above and hence has a largest element l by Lemma 6.8; we can take $k = l + 1$.

(iii) Let $A = \{x \in {}^*N - \{0\} : \Phi(y) \text{ holds for all } y \text{ with } |y| \leq 1/x\}$ and use (ii). $\square$

7.2 Corollary   (Spillover Principle)   Let A be an internal subset of ${}^*R$.

(i) If A contains all standard natural numbers then A contains an infinite natural number.

(ii) If A contains all infinite natural numbers then A contains a standard natural number.

(iii) If A contains the positive infinitesimals then A contains a standard positive real number.

Theorem 7.1 can be used to give yet another proof of the fact that if $\langle s_n : n \in N \rangle$ is a standard sequence and ${}^*s_n \simeq L \in R$ for all infinite n, then $\lim s_n = L$ (see §II.3). Let $\varepsilon > 0$ be a fixed number in R. Then $|{}^*s_n - L| < \varepsilon$ for all infinite n. Applying Theorem 7.1(ii) with $\Phi(b)$ the internal statement "$|{}^*s_b - L| < \varepsilon$", we see that there is a $k \in N$ so that $|{}^*s_b - L| < \varepsilon$ for all $b \geq k$ in ${}^*N$ and, in particular, $|s_b - L| < \varepsilon$ for all $b \geq k$ in N since ${}^*s_b = s_b$ if $b \in N$. This establishes the desired result.

The following result has many applications.

7.3 Theorem   (Robinson's Sequential Lemma)   Let $\langle s_n : n \in {}^*N \rangle$ be an internal ${}^*R$-valued sequence such that $s_n \simeq 0$ for each $n \in N$. Then there is an infinite natural number ω so that $s_n \simeq 0$ for all natural numbers $n \leq \omega$.

Proof: The sequence $\langle ns_n : n \in {}^*N \rangle$ is internal. Apply 7.1(i) with $\Phi(n)$ the internal formula "$|ns_n| \leq 1$" to obtain an $\omega \in {}^*N_\infty$ so that $|s_n| \leq 1/n$ if $n \leq \omega$. Thus $s_n \simeq 0$ if $n \in {}^*N_\infty$ and $n \leq \omega$, and so $s_n \simeq 0$ for all $n \leq \omega$. $\square$

One should beware of assertions similar to Theorem 7.3 which sound plausible but are not true. For example, it is not true that if $s_n \simeq 0$ for all infinite n then there exists a finite k so that $s_n \simeq 0$ for all $n \geq k$ as the example $s_n = 1/n$ shows.

As an application of Theorem 7.3 we give another proof of the fact (Corollary I.13.5) that if the sequence $\langle f_n(x) : n \in N \rangle$ of continuous real-valued functions on the interval $[a, b]$ converges uniformly then the limit $f(x)$ is continuous on $[a, b]$. Let $x_0 \in [a, b]$; we need to show that ${}^*f(x) \simeq f(x_0)$ if $x \simeq x_0$. But ${}^*f_n(x) \simeq {}^*f_n(x_0)$ for each $n \in N$, and so ${}^*f_\omega(x) \simeq {}^*f_\omega(x_0)$ for some infinite ω by Theorem 7.3. But ${}^*f_\omega(x) \simeq {}^*f(x)$ for all $x \in {}^*[a, b]$ by Proposition 13.2 of Chapter I, and we are through.

Robinson [41] applied Theorem 7.3 in a more significant context in giving a nonstandard construction for Banach limits of bounded sequences. Suppose $\langle s_n : n \in N \rangle$ is a bounded sequence, i.e., $|s_n| \leq M$ for some real $M > 0$. We would like to attach a "limit" to $\langle s_n \rangle$ even though it might not converge in the usual sense. For example, the sequence $t_n = (s_1 + s_2 + \cdots + s_n)/n$ ($n = 1, 2, \ldots$) of Cesàro means sometimes converges when $\langle s_n \rangle$ does not converge and defines a limit called the Cesàro sum of the sequence $\langle s_n \rangle$. Any generalized limit should satisfy the properties in the following definition.

7.4 Definition Let l_∞ denote the set of standard bounded sequences. A map $L : l_\infty \to R$ is called a *Banach limit* if

   (i) $L(a\sigma + b\tau) = aL(\sigma) + bL(\tau)$ $(a, b \in R, \sigma, \tau \in l_\infty)$,
   (ii) if $\sigma = \langle s_n | n \in N \rangle$ then $\liminf s_n \leq L(\sigma) \leq \limsup s_n$,
   (iii) if $\sigma = \langle s_n | n \in N \rangle$ and $\tau = \langle t_n | n \in N \rangle$, where $t_n = s_{n+1}$, then $L(\sigma) = L(\tau)$.

To obtain a Banach limit, we let $\sum_{i=1}^{\omega}$ for $\omega \in {}^*N_\infty$ extend the standard summation operators $\sum_{i=1}^{n}$, $n \in N$.

7.5 Theorem Fix $\omega \in {}^*N_\infty$, and let $L(\sigma) = {}^\circ((1/\omega) \sum_{i=1}^{\omega} {}^*s_i)$ for each $\sigma = \langle s_n : n \in N \rangle$ in l_∞. Then L is a Banach limit.

Proof: The mapping L clearly satisfies 7.4(i). Given $\sigma = \langle s_n : n \in N \rangle$, let $M = \sup\{|s_n| : n \in N\}$. For a given $m \in N$,

$$\left| \frac{1}{\omega - m} \sum_{i=m+1}^{\omega} {}^*s_i - \frac{1}{\omega} \sum_{i=1}^{\omega} {}^*s_i \right|$$

$$\leq \left| \frac{\omega}{\omega - m} \frac{1}{\omega} \sum_{i=m+1}^{\omega} {}^*s_i - \frac{1}{\omega} \sum_{i=m+1}^{\omega} {}^*s_i \right| + \left| \frac{1}{\omega} \sum_{i=m+1}^{\omega} {}^*s_i - \frac{1}{\omega} \sum_{i=1}^{\omega} {}^*s_i \right|$$

$$\leq \left(\frac{\omega}{\omega - m} - 1 \right) \frac{(\omega - m)M}{\omega} + \frac{mM}{\omega}$$

$$\simeq 0.$$

By Theorem 7.3, there is an $m \in {}^*N_\infty$ so that

(7.1) $$L(\sigma) \simeq \frac{1}{\omega - m} \sum_{n=m+1}^{\omega} {}^*s_n.$$

Fix $\varepsilon > 0$ in R. We see immediately from Definition 8.16 of Chapter I that for each $n \in {}^*N$ with $m + 1 \leq n \leq \omega$

$$\liminf s_n - \varepsilon < {}^*s_n < \limsup s_n + \varepsilon.$$

By the transfer of the usual properties of an average applied to (7.1),

$$\liminf s_n - \varepsilon \leq L(\sigma) \leq \limsup s_n + \varepsilon.$$

Since ε is arbitrary, we obtain 7.4(ii). The rest of the proof is left to the reader. $\square$

Exercises II.7

1. Prove the real case of (i) and (ii) of Theorem 7.1.
2. Assume that A is an internal set in $*N$ such that, for some infinite integer γ, if n is infinite and $n \leq \gamma$ in $*N$ then $n \in A$. Show that, for some finite $m \in N$, if $n \in N$ and $m \leq n$ then $n \in A$.
3. Prove that the mapping L of Theorem 7.5 satisfies property (iii) of Definition 7.4, i.e., L is invariant under finite translations.
4. Use the permanence principle to show that if f is a standard function and $|*f(x) - L| \simeq 0$ for all $x \simeq 1$ but $x \neq 1$, then $\lim_{x \to 1} f(x) = L$.
5. Let $\langle s_n : n \in *N \rangle$ be an internal $*R$-valued sequence, and suppose that there is an $M > 0$ in R so that $|s_n| \leq M$ for all $n \in N$. Show that there is an $\omega \in *N_\infty$ so that $|s_n| \leq M$ for all $n \leq \omega$ in $*N$.
6. Show that the assertion in Exercise 5 is not true if "$|s_n| \leq M$" is replaced by "s_n is finite."
7. A filter $\mathscr{F} \in V(X) - X$ has a *countable subbasis* if there is a countable family $\{A_i : i \in N\}$ of entities in $\mathscr{F}$ so that for each $F \in \mathscr{F}$ there is a sequence $i_1, \ldots, i_n$ with $\bigcap A_{i_k} (1 \leq k \leq n) \subseteq F$. Suppose that B is an internal set in $V(*X)$ and $\mathscr{F}$ has a countable subbasis. Show that if $B \cap *F = \varnothing$ for all $F \in \mathscr{F}$ then $B \cap \mu(\mathscr{F}) \neq \varnothing$, where $\mu(\mathscr{F})$ is the intersection monad of $\mathscr{F}$ introduced in Exercise II.5.9.
8. Let $*: V(R) \to V(*R)$ be comprehensive, and let $S = \{n_k : k \in N\}$ be a countable set contained in $*N_\infty$.

   (a) Show that S has a lower bound in $*N_\infty$. [Hint: Regard S as a sequence, i.e., a map $h: N \to *N$ with $h(k) = n_k$. Use comprehensiveness to extend h to an internal map $g: *N \to *N$ and apply the spillover principle to the set $A = \{m \in *N : g(k) > m \text{ for all } k \leq m\}$.] For decreasing sequences n_k this was presented by DuBois-Reymond and proved in our context by Robinson.

   (b) Use the transfer principle applied to g to show that S has an upper bound in $*N_\infty$.

9. Show that if f is an internal function on an internal set A in some superstructure $V(*X)$, and f is finite-valued, then there exists a *standard* $n \in N$ so that $|f(x)| \leq n$ for all $x \in X$. Give an example to show that the assertion is not necessarily true if f is not internal.

10. (∗-Convergence and S-Convergence)   An internal *R-valued sequence
$\langle s_n : n \in {}^*N \rangle$ is (i) ∗-convergent to $L \in {}^*R$ if for each $\varepsilon > 0$ in *R there is
an $m \in {}^*N$ so that $n > m$ implies $|s_n - L| < \varepsilon$, (ii) S-convergent to $L \in {}^*R$
if $s_n \simeq L$ for all $n \in {}^*N_\infty$.

   (a) Show that if $s_n = {}^*t_n$ where $\langle t_n \rangle$ is a standard sequence converging
to L, then $\langle s_n \rangle$ is ∗-convergent and S-convergent to L.

   (b) Show that there are internal sequences which are ∗-convergent but
not S-convergent and vice versa.

   (c) Show that if $\langle s_n \rangle$ is S-convergent to a finite $L \in {}^*R$ then there is an
$m \in N$ so that s_n is finite for $n \geq m$ and the standard sequence $\langle {}^\circ s_n : n \in N \rangle$
converges to $^\circ L$.

   (d) Show that if $s_n = {}^*t_n$, where $\langle t_n \rangle$ is a standard sequence, then $\langle s_n \rangle$
is S-convergent to a finite L iff there exists an infinite $\omega \in {}^*N_\infty$ so that
$^*s_n \simeq L$ for every $n \in {}^*N_\infty$ with $n \leq \omega$.

II.8 κ-Saturated Superstructures

   Theorem 7.5 of the last section is a good example of a result in which a
standard entity (a Banach limit) is obtained by performing a standardizing
operation on an internal entity [in this case, taking the standard part of the
internal sum $(1/\omega) \sum {}^*s_i (1 \leq i \leq \omega)$]. Similar applications of nonstandard
analysis often occur in more complicated circumstances, and sometimes the
internal structure in a given extension $V(^*X)$ of a superstructure $V(X)$ is not
rich enough to produce a desired result. A specific example arose from a result of Robinson, which was that if X is a metric space and B an internal
subset of *X in an enlargement $V(^*X)$, then the standard part of B is closed
(definitions and results will be presented in Chapter III). It was natural to
ask whether the result was still true if X was not metric. An example due
to H. J. Keisler showed that the answer was negative if $V(^*X)$ was only an
enlargement of $V(X)$ [36, Example 3.4.3]. Luxemburg [36, Theorem 3.4.2]
showed that the result does go through if $V(^*X)$ is large enough to satisfy
a generalization of the property of an enlargement, valid for internal concurrent binary relations on an appropriate set A in $V(^*X)$. $V(^*X)$ is called
κ-saturated, where κ is a cardinal number, if this generalization holds for all
sets A in $V(^*X)$ with the cardinality of $A < \kappa$ (Definition 8.1). It is not
necessary for the reader to be very knowledgeable about the theory of cardinal numbers for arbitrary sets in order to apply the theory. In a typical
application we will begin with an internal concurrent binary relation on
A—then we can assert that the results of the section will be applicable if
$V(^*X)$ is sufficiently large. Sufficiently large means that $V(^*X)$ is κ-saturated,

where $\kappa >$ card A, but this is irrelevant in the application as long as we are assured that κ-saturated structures exist (Theorem 8.2).

Let $V(X)$ be a given superstructure and $*: V(X) \rightarrow V(*X)$ a monomorphism. We write card A to denote the cardinality, in the standard sense, of a set A.

8.1 Definition $V(*X)$ is *κ-saturated* if, for each internal binary relation $P \in V(*X)$ which is concurrent (Definition 5.9) on some (not necessarily internal) set A in $V(*X)$ with card $A < \kappa$, there exists an element $y \in$ range P so that $\langle x, y \rangle \in P$ for all $x \in A$.

H. J. Keisler [21, 22] characterized those ultrafilters $\mathcal{U}$ such that the superstructure $V(*X)$ constructed from a given superstructure $V(X)$, using $\mathcal{U}$ as in §II.4, is κ-saturated; he called them κ-good ultrafilters. In [21] Keisler established the existence of κ-good ultrafilters on the assumption of the generalized continuum hypotheses. This assumption was subsequently removed by Kunen. Thus we have the following result.

8.2 Theorem Given any superstructure $V(X)$ and cardinal κ there is a κ-saturated superstructure $V(*X)$ and a monomorphism $*: V(X) \rightarrow V(*X)$.

For the proof of this and related results the interested reader is referred to the papers mentioned above and also to the book by Stroyan and Luxemburg [46], where the desired structures are constructed as limits of ultrapowers. In any applications it will not be necessary to know the details of the proof. It follows from Theorem 5.10 that if $\kappa >$ card $V(X)$ then $V(*X)$ is an enlargement.

In applying Theorem 8.2 it is important to note that the set A of Definition 8.1 need not be internal, although the binary relation P must be internal and so the elements of A are internal. For a successful application, however, we do need an upper bound on the cardinality of A which is *independent* of the particular construction of $V(*X)$. For example, suppose that P is the binary relation on $*R \times *\mathscr{P}_F(R)$ defined by "$\langle x, B \rangle \in P$ iff the $*$-finite set B contains x." Then P is concurrent on any subset $A \subseteq *R$. However, it is not possible to apply Theorem 8.2 and Definition 8.1 with $A = *R$; i.e., it is not possible to find a $*$-finite subset of $*R$ which contains all numbers of $*R$, no matter how large κ is. For then $*R$ itself would be a $*$-finite set and hence, by transfer down, R would be finite. The error occurs in trying to apply the result to the set $A = *R$ whose cardinality depends on the construction of the extension $V(*R)$ and is not fixed in advance.

In [36] Luxemburg developed a general theory of monads in enlargements and κ-saturated extensions. In the following we present several of his important results.

8.3 Definition Let $*: V(X) \to V(*X)$ be a monomorphism, and let A be an entity in $V(X)$. The (*intersection*) *monad* $\mu(A)$ *of* A (*with respect to* $*$) is the set

$$\mu(A) = \bigcap *a(a \in A).$$

Monads $\mu(A)$ are most important when A is a filter $\mathscr{F}$, i.e., when $\varnothing \notin \mathscr{F}$, F and G in $\mathscr{F}$ implies $F \cap G \in \mathscr{F}$, and $F \in \mathscr{F}$ and $G \supseteq F$ implies $G \in \mathscr{F}$. The next result generalizes the permanence principle.

8.4 Theorem (Luxemburg) Let $*: V(X) \to V(*X)$ be a monomorphism, and assume that $V(*X)$ is κ-saturated. Fix a filter $\mathscr{F} \in V(X)$ with card $\mathscr{F} < \kappa$; then

(a) given an internal set $B \in V(*X)$, if $*F \cap B \neq \varnothing$ for all $F \in \mathscr{F}$, then $\mu(\mathscr{F}) \cap B \neq \varnothing$,

(b) given an internal subset Λ of $*\mathscr{F}$ such that every standard element of $*\mathscr{F}$ is an element of Λ, there exists an element $E \in \Lambda$ such that $E \subseteq \mu(\mathscr{F})$,

(c) given an internal subset Λ of $*\mathscr{F}$ such that $E \in *\mathscr{F}$ and $E \subseteq \mu(\mathscr{F})$ implies $E \in \Lambda$, there exists an element $F \in \mathscr{F}$ such that $*F \in \Lambda$.

Proof: (a) Define an internal relation P, with domain $*\mathscr{F}$ and range contained in B, by "$\langle F, x \rangle \in P$ if $x \in B \cap F$." Then P is concurrent on the collection of standard elements of $*\mathscr{F}$, and this collection has the same cardinality as $\mathscr{F}$. Therefore there is a $y \in B$ so that $y \in B \cap *F$ for each $F \in \mathscr{F}$, i.e., $y \in \mu(\mathscr{F})$.

(b) Define an internal relation P, with domain $*\mathscr{F}$ and range contained in Λ, by "$\langle F, G \rangle \in P$ if $G \in \Lambda$ and $G \subseteq F$." Then P is concurrent on the collection of standard elements of $*\mathscr{F}$ (why?), so there is an $E \in \Lambda$ such that $E \subseteq *F$ for each $F \in \mathscr{F}$, i.e., $E \subseteq \mu(\mathscr{F})$.

(c) Let Λ satisfy the condition of (c). If Λ does not contain a standard element $*F \in *\mathscr{F}$ then the internal set $*\mathscr{F} - \Lambda \subset *\mathscr{F}$ contains all standard elements of $*\mathscr{F}$ and so by (b) there exists an element $E \in *\mathscr{F} - \Lambda$ with $E \subseteq \mu(\mathscr{F})$. But then $E \in \Lambda$ by the hypothesis on Λ (contradiction). $\square$

Several exercises in the preceding sections have dealt with situations in which, without saturation, the statement (a) of Theorem 8.4 may or may not hold. The results can be summarized as follows: The statement does not hold in general if B is not internal (Exercise II.5.11), but does hold if B is standard (Exercise II.5.10) or if B is internal and $\mathscr{F}$ has a countable basis (Exercise II.7.7). (See Theorem 8.6.) An example due to H. J. Keisler (see Example 2.7.4 in [36]) shows that the statement need not hold if B is internal but $V(*X)$ is only an ultrapower enlargement.

We note finally that an internal version of comprehensiveness holds in κ-saturated extensions.

8.5 Theorem Let $V(*X)$ be a κ-saturated extension of $V(X)$. Assume C is a (not necessarily internal) set of entities in $V_n(*X)$ for some $n \in N$ with card $C < \kappa$, and D is an internal set in $V(*X)$. For any mapping $\phi: C \to D$, there is an internal extension $\tilde{\phi}: \tilde{C} \to D$ of ϕ [i.e., $\tilde{C}$ is internal, contains C, and $\phi(a) = \tilde{\phi}(a)$ if $a \in C$]. If $C = \{*a : a \in C_0\}$ we may take $\tilde{C} = *C_0$.

Proof: Let P be the binary relation "$\langle \phi, \hat{\phi} \rangle \in P$ iff $\hat{\phi}$ is an extension of ϕ" [i.e., dom $\hat{\phi} \supseteq$ dom ϕ and $\hat{\phi}(a) = \phi(a)$ if $a \in$ dom ϕ] defined on the set of internal mappings with values in D. Let A be the set of all internal mappings $f_x: \{x\} \to \phi(x)$, $x \in C$. That is, each element of A is a set consisting of exactly one element from ϕ. Then card $A =$ card $C < \kappa$ and P is concurrent on A (check). Thus there exists an internal map $\tilde{\phi}$ with values in D which extends each f_x, $x \in C$, and so dom $\tilde{\phi} = \tilde{C} \supseteq C$ and $\tilde{\phi}(a) = \phi(a)$, $a \in C$. The rest is left as an exercise (Exercise 1). □

There is a converse of Theorem 8.5 when $\kappa = \aleph_1$, where $\aleph_1$ is the first cardinal number bigger than card N.

8.6 Theorem $V(*X)$ is a denumerably comprehensive extension of $V(X)$ (Definition 6.11) if and only if $V(*X)$ is $\aleph_1$-saturated.

Proof: Exercise. □

8.7 Corollary An extension $V(*X)$ constructed as in §II.4 is $\aleph_1$-saturated.

Proof: Follows from Theorems 6.13 and 8.6. □

Corollary 8.7 shows that assuming $\aleph_1$-saturation in an application of nonstandard analysis is not assuming very much. Later in this book we assume a stronger form of saturation (larger κ) only in the proof of Theorem 1.22 of Chapter III (which is not used afterward) and in the proofs of the last few results in §IV.3, where κ-saturation is used in a more significant way.

Exercises II.8

1. Show that if the set C in Theorem 8.5 has the form $\{*a : a \in C_0\}$ then one may take $\tilde{C} = *C_0$ in the conclusion of the theorem.
2. Prove Theorem 8.6.
3. Let $V(*R)$ be a κ-saturated extension of $V(R)$ with card $\mathscr{P}(R) < \kappa$. Let B be an internal subset of $*R$ and $\mathrm{st}(B) = \{x \in R : \text{there exists a } y \in B \text{ with } \mathrm{st}(y) = x\}$. Use Theorem 8.4(a) to show that $\mathrm{st}(B)$ is closed in R.

4. (Luxemburg [36])   Suppose that $V(*X)$ is a κ-saturated extension of $V(X)$ with $\kappa > \operatorname{card} V(X)$. Let $A \in V(X)$ contain an infinite number of elements. If $\Lambda \subset *(\mathscr{P}_F(A))$ is internal and moreover, $E \in \Lambda$ for every $*$-finite subset $E \subset *A$ with the property that $A = \{a \in V(X): *a \in E\}$, then there exists a finite subset $\{a_1, \ldots, a_n\} \subset A$ so that $\{*a_1, \ldots, *a_n\} \in \Lambda$. (Hint: Apply Theorem 8.4 to the Fréchet filter of A.)

CHAPTER III

Nonstandard Theory of Topological Spaces

In Chapter I we showed how the notion of continuity for real-valued functions of a real variable could be characterized in terms of the nonstandard concept of nearness [f is continuous at x if $*f(y) \simeq f(x)$ for all $y \simeq x$]. On the real line, nearness and the associated concept of monad are characterized in terms of the distance function, so that $x \simeq y$ if $|x - y| \simeq 0$. We also characterized open and closed sets in terms of monads. In this chapter we will show how these notions can be extended to more general settings.

In the standard development of topology one usually begins with a set X possessing a collection $\mathcal{T}$ of (open) subsets satisfying the abstract analogues of conditions (i) and (ii) of Theorem 9.2 in Chapter I. The pair $(X, \mathcal{T})$ is called a topological space. The notions of continuity can then be defined just in terms of the open sets; i.e., a function $f: X \to Y$ is continuous if $f^{-1}(V)$ is open in X for every set V which is open in Y. In the nonstandard theory developed here, we will show how the collection $\mathcal{T}$ on X can be used to characterize nearness and monad and so allow a simple development of the theory of topological spaces analogous to that of Chapter I.

One of the most useful results in the nonstandard development is a characterization of compact spaces (the analogues of closed bounded sets on the real line) due to Abraham Robinson. This development is presented in §III.2, with an elaboration in §III.7.

Sections III.3, III.4, and III.5 are devoted to the nonstandard theory of metric, normed, and inner-product spaces, which are of central importance in much of analysis. In §III.6 we show how one may begin with a standard metric space X and construct a (standard) metric space on the nonstandard set $*X$, leading to the so-called nonstandard hull of a metric space. This construction plays a central role in some recent applications of nonstandard analysis to the theory of Banach spaces by Henson and Moore (see [16] for a

109

review). The section ends with a discussion of some results in the theory of function spaces, and includes a generalization of the Arzelà–Ascoli theorem of Chapter I.

III.1 Basic Definitions and Results

A topological space is a pair $(X, \mathcal{T})$, where X is a set and $\mathcal{T}$ is a family of subsets of X satisfying the conditions in the following definition.

1.1 Definition A family $\mathcal{T}$ of subsets of X, called *open* sets, is a *topology* for X if

(a) $\emptyset, X \in \mathcal{T}$;

(b) $U, V \in \mathcal{T}$ implies $U \cap V \in \mathcal{T}$ (and thus every finite intersection of open sets is open),

(c) $U_i \in \mathcal{T}$ $(i \in I)$ implies $\bigcup U_i (i \in I) \in \mathcal{T}$, i.e., every arbitrary union of open sets is open.

Closed sets are complements of open sets. Often we call X rather than $(X, \mathcal{T})$ the topological space.

The usual family of open subsets of R, defined in the proof of Proposition 9.1 of Chapter I, is a topology for R (Theorem I.9.2). We will presently see that there are many topologies for R as for most sets.

With each topology we will associate corresponding notions of convergence and continuity, using only the open sets. In order to develop a nonstandard theory, we first generalize the notions of nearness and monad which were central to the work in Chapter I.

We begin with a few basic definitions.

1.2 Definition Let $(X, \mathcal{T})$ be a topological space. A set U is a *neighborhood* of a point $x \in X$ if U contains an open set V which contains x. The *neighborhood system* $\mathcal{N}_x$ of x is the set of all neighborhoods of x. We denote the system of *open neighborhoods* of $x \in X$ by $\mathcal{T}_x$. A collection $\mathcal{B} \subseteq \mathcal{T}$ is a *base* for $\mathcal{T}$ if each set in $\mathcal{T}$ is a union of sets in $\mathcal{B}$ or, equivalently, if for each $x \in X$ and each $U \in \mathcal{T}_x$ there is a $V \in \mathcal{T}_x \cap \mathcal{B}$ with $V \subseteq U$. (For example, open intervals form a base for the usual open sets in R.) A collection $\mathcal{B}$ is called a *subbase* for $\mathcal{T}$ if the collection of finite intersections of members of $\mathcal{B}$ is a base for $\mathcal{T}$. Similarly $\mathcal{B}_x \subseteq \mathcal{N}_x$ is a *(neighborhood) base* at x if for each $U \in \mathcal{N}_x$ there is a $V \in \mathcal{B}_x$ with $V \subseteq U$; $\mathcal{B}_x \subseteq \mathcal{N}_x$ is a *subbase* at x if the collection of finite intersections of members of $\mathscr{B}_x$ is a base at x. If $\mathscr{T}$ and $\mathscr{S}$ are topologies for X, then $\mathscr{T}$ is *weaker* than $\mathscr{S}$ (and $\mathscr{S}$ is *stronger* than $\mathscr{T}$) if $\mathscr{T} \subseteq \mathscr{S}$.

From now on we work in an enlargement $V(*S)$ of a superstructure $V(S)$, where $V(S)$ contains the standard space X under consideration, so $\mathscr{T} \in V(S)$ as well. In this section we will not use the fact that if $x \in X$ then x may contain elements. Therefore, we will write x instead of $*x$ for the nonstandard extension of x.

1.3 Definition The sets in $*\mathscr{T}$ are called *-open* subsets of $*X$. The *monad* of $x \in X$ is the subset $m(x) = \bigcap *U(U \in \mathscr{T}_x)$ of $*X$. A point $y \in *X$ is *near* $x \in X$, and x is the *standard part* of y, if $y \in m(x)$; then we write $y \simeq x$ and $x = \mathrm{st}(y)$. The set of *near-standard* points is the set $\mathrm{ns}(*X) = \bigcup m(x)(x \in X)$. A point $y \in *X$ is called *remote* if it is not near-standard.

An easy exercise shows that $m(x) = \bigcap *U(U \in \mathscr{N}_x)$.

1.4 Proposition If $\mathscr{B}_x$ is a local subbase at x, then $m(x) = \bigcap *U(U \in \mathscr{B}_x)$.

Proof: $\bigcap *U(U \in \mathscr{B}_x) \supseteq \bigcap *U(U \in \mathscr{N}_x)$ since $\mathscr{B}_x \subseteq \mathscr{N}_x$. On the other hand, for each $U \in \mathscr{N}_x$ there exist $V_i \in \mathscr{B}_x(1 \le i \le n)$ with $\bigcap V_i(1 \le i \le n) \subseteq U$, and so $\bigcap *V_i(1 \le i \le n) \subseteq *U$ by transfer. Hence $\bigcap *V(V \in \mathscr{B}_x) \subseteq \bigcap *U(U \in \mathscr{N}_x)$. $\square$

1.5 Examples

1. *Discrete topology.* $(X, \mathscr{T})$ is *discrete* if $\{x\}$ is open for each $x \in X$. In this case $m(x) = \{x\}$ for each $x \in X$.
2. *Trivial topology.* $(X, \mathscr{T})$ is *trivial* if $\mathscr{T} = \{\varnothing, X\}$. In this case $m(x) = *X$ for each $x \in X$.
3. *Usual topology on R.* The open sets in R as defined in §I.9 constitute a topology. The monads as defined here and in Definition 6.4 of Chapter I are identical [where we assume that $*\mathscr{R}$ and $V(*R)$ are obtained from the same ultrafilter]. This follows immediately from Proposition 1.4 since the set $\mathscr{B}_x$ of symmetric open intervals about x forms a local base by the definition of open set in R. A subbase for the topology is formed by intervals of the form $(-\infty, b)$, $(a, +\infty)$ with $a,b \in R$.
4. *Half-open interval topology on R.* Let $\mathscr{T}$ be the topology for R which has as base the set $\mathscr{B}$ of half-open intervals $[a, b) = \{x : a \le x < b\}$, where a and b are real. Here $m(x) = \{y \in *R : x \le y, x \simeq y\}$ (Exercise 1).

5. *Finite complement topology.* For simplicity let $X = N$ (any infinite set would do), and let $\mathscr{T}$ be the collection consisting of the empty set and those subsets of N whose complements are finite. It is an easy standard exercise to show that $\mathscr{T}$ is a topology. Here $m(x) = \{x\} \cup {}^*N_\infty$ (Exercise 1).

6. *Product topology.* Let $(X, \mathscr{T})$ and $(Y, \mathscr{S})$ be topological spaces. Then $X \times Y$ can be made into a topological space as follows: A set $W \subseteq X \times Y$ is open if to each $\langle x, y \rangle \in W$ there correspond sets $U \in \mathscr{T}_x$, $V \in \mathscr{S}_y$ so that $U \times V \subseteq W$; i.e., products of open sets form a base for the topology (check that this defines a topology). The resulting topology is called the *product topology* and is denoted by $\mathscr{T} \times \mathscr{S}$. If $m_\mathscr{T}$, $m_\mathscr{S}$, and m denote monads in $(X, \mathscr{T})$, $(Y, \mathscr{S})$, and $(X \times Y, \mathscr{T} \times \mathscr{S})$, respectively, then $m(\langle x, y \rangle) = m_\mathscr{T}(x) \times m_\mathscr{S}(y)$, $x \in X$, $y \in Y$ (Exercise 1).

The following facts should be noted in comparing the usual monads for R and monads in a general topological space $(X, \mathscr{T})$:

(a) The concept of nearness is derived from that of monad and not vice versa as in Definition 6.4 of Chapter I.

(b) We have defined monads only for standard points in ${}^*X$.

(c) Nearness is not in general an equivalence relation on ${}^*X$ [this is, of course, because of (b)].

The monad $m(x)$ always contains x. That $m(x)$ will in general contain points other than x follows from the following basic lemma, the proof of which requires that $V({}^*S)$ be an enlargement.

1.6 Proposition For each $x \in X$ there is a $*$-open set $V \in {}^*\mathscr{T}_x$ with $V \subseteq m(x)$.

Proof: The binary relation P on $\mathscr{T}_x \times \mathscr{T}_x$ defined by $P\langle U, V \rangle$ if $V \subseteq U$ is concurrent. For if $U_1, \ldots, U_n \in \mathscr{T}_x$ then $V = U_1 \cap \cdots \cap U_n$ satisfies $P\langle U_i, V \rangle$, $1 \leq i \leq n$. Since $V({}^*S)$ is an enlargement, Theorem 5.10 of Chapter II guarantees the existence of an element $V \in {}^*\mathscr{T}_x$, so that $V \subseteq {}^*U$ for all $U \in \mathscr{T}_x$ and hence $V \subseteq m(x)$. $\square$

1.7 Proposition Let A be a subset of X. Then

(i) A is open iff $m(x) \subseteq {}^*A$ for each $x \in A$,

(ii) A is closed iff $m(x) \cap {}^*A = \varnothing$ for each x in the complement A' of A.

Proof: (i) Suppose A is open and let $x \in A$. By definition there exists an open set $U \in \mathscr{T}_x$ with $U \subseteq A$. By transfer $m(x) \subseteq {}^*U \subseteq {}^*A$.

Conversely, suppose $m(x) \subseteq {}^*A$ for $x \in A$. By Proposition 1.6 there exists a $V \in {}^*\mathcal{T}_x$ with $V \subseteq m(x) \subseteq {}^*A$. Thus the internal sentence $(\exists V \in {}^*\mathcal{T}_x)[V \subseteq {}^*A]$ is true and so, by downward transfer, there exists a set $V \in \mathcal{T}_x$ with $V \subseteq A$. Thus A is open since $A = \bigcup V_x (x \in A)$.

(ii) This follows immediately from (i) and the definition of a closed set: A is closed if A' is open. □

1.8 Definition A point x is an *accumulation point* of the set $A \subseteq X$ if every open neighborhood of x contains points of A other than x. We let $\hat{A}$ denote the set of accumulation points of A; the set $\bar{A} = A \cup \hat{A}$ is the *closure* of A. A is *dense* in B if $\bar{A} = B$.

1.9 Proposition A point x is an accumulation point of A iff $m(x)$ contains a point $y \in {}^*A$ different from x.

Proof: If x is an accumulation point of A then the sentence $(\forall U \in \mathcal{T}_x)$ $(\exists y \in U \cap A)[y \neq x]$ is true for $V(X)$, and hence, by transfer, each $U \in {}^*\mathcal{T}_x$ contains a point $y \neq x$ in ${}^*A$. This is true, in particular, of the *-open set V of Proposition 1.6, and so there is a $y \in m(x) \cap {}^*A$ with $y \neq x$.

Conversely, suppose that $m(x)$ contains a point $y \neq x$ in ${}^*A$. Then, for a fixed $U \in \mathcal{T}_x$, ${}^*U$ contains a point $y \neq x$ in ${}^*A$. Thus the internal sentence $(\exists y \in {}^*(U \cap A))[y \neq x]$ is true, and it follows by downward transfer that there exists a $y \in U \cap A$ with $y \neq x$. □

1.10 Proposition The closure $\bar{A}$ of $A \subseteq X$ consists of those $x \in X$ for which $m(x) \cap {}^*A \neq \varnothing$. The closure of A is the smallest closed set containing A. Thus $A = \bar{A}$ if A is closed.

Proof: Exercise. □

Let $\mathcal{T}$ and $\mathcal{S}$ be two topologies for a set X with associated monads $m_{\mathcal{T}}(x)$ and $m_{\mathcal{S}}(x)$ $(x \in X)$. An easy exercise shows that $\mathcal{T}$ is weaker than $\mathcal{S}$ iff $m_{\mathcal{T}}(x) \supseteq m_{\mathcal{S}}(x)$ for each $x \in X$.

We noted in §I.6 that if x and y are distinct standard real numbers then $m(x) \cap m(y)$ is empty. Therefore, we say that R is a *Hausdorff space*. This property is not true in general for topological spaces. Properties of spaces which deal with the relationship between monads of distinct points are called *separation* properties. Some of the more important separation properties are presented next; the most important of these is the Hausdorff property.

1.11 Definition The space $(X, \mathcal{T})$ is

(a) T_0 if, for each pair x, y of distinct points in X, there is an open neighborhood of one not containing the other,

(b) T_1 if $\{x\}$ is closed for each $x \in X$,

(c) *Hausdorff* (or T_2) if whenever $x \neq y$ in X there are disjoint open neighborhoods U and V of x and y.

There are more separation properties (e.g., regularity and normality) which we will consider in the exercises.

1.12 Proposition The topological space $(X, \mathcal{T})$ is

(a) T_0 iff whenever $x, y \in X$ and both $x \in m(y)$ and $y \in m(x)$ then $x = y$,

(b) T_1 iff whenever $x, y \in X$ and $x \in m(y)$ then $x = y$,

(c) Hausdorff iff monads of distinct points in X are disjoint.

Proof: We prove (c) and leave the other proofs as exercises. Suppose $(X, \mathcal{T})$ is Hausdorff and $x, y \in X$ are distinct. Then there exist $U \in \mathcal{T}_x$, $V \in \mathcal{T}_y$ with $U \cap V = \varnothing$. Therefore, $*U \cap *V = \varnothing$, and since $m(x) \subseteq *U$ and $m(y) \subseteq *V$, we have $m(x) \cap m(y) = \varnothing$.

Conversely, if $m(x) \cap m(y) = \varnothing$ then by Proposition 1.6 there exist $U \in *\mathcal{T}_x$, $V \in *\mathcal{T}_y$ with $U \cap V = \varnothing$. By downward transfer of the appropriate sentence (check), there exist $U \in \mathcal{T}_x$, $V \in \mathcal{T}_y$ with $U \cap V = \varnothing$. $\square$

If $(X, \mathcal{T})$ is Hausdorff then there is only one standard point $\operatorname{st}(y)$ associated with each $y \in \operatorname{ns}(*X)$. It is defined by $\operatorname{st}(y) = x$, $y \in m(x)$. Thus for Hausdorff spaces we have a well-defined map $\operatorname{st}: \operatorname{ns}(*X) \to X$ called the *standard part map*, which has many applications (e.g., see §IV.3 below).

1.13 Examples

1. The discrete topology is Hausdorff, and every subset is both open and closed.

2. The trivial topology of a space with two or more points is not T_0.

3. The finite complement topology on N is T_1 but not Hausdorff by Proposition 1.12. Also a set is closed in the finite complement topology iff it is finite. For if A is finite then $*A = A$, and if $x \in A'$ then $m(x) \cap *A = (\{x\} \cup *N_\infty) \cap A = \varnothing$. On the other hand, if A is infinite then $*A \cap *N_\infty \neq \varnothing$ by 6.11 of Chapter I, and $m(x) \cap *A \neq \varnothing$ for any x.

So far we have used a topology $\mathcal{T}$ to define associated monads $m(x)$, $x \in X$. Conversely, it is possible to start with a collection $k(x)$, $x \in X$, of subsets of $*X$ with $x \in k(x)$, and define an associated family $\mathcal{T}$ as follows: $U \in \mathcal{T}$ if

$k(x) \subseteq *U$ for each $x \in U$. An easy exercise shows that $\mathcal{T}$ is a topology. If $\hat{k}(x)$ $(x \in X)$ are the monads of $\mathcal{T}$ then clearly $k(x) \subseteq \hat{k}(x)$ for all $x \in X$, but set equality does not necessarily hold (see Exercise 6). The sets $k(x)$ will be called *pseudomonads*; the concept will be used in §III.8.

Let $(X, \mathcal{T})$ and $(Y, \mathcal{S})$ be topological spaces with monads $m(x)$ $(x \in X)$ and $\bar{m}(y)$ $(y \in Y)$, respectively. To discuss continuity of mappings $f: X \to Y$ we work in an enlargement containing $*X$ and $*Y$ and thus $*\mathcal{T}$, $*\mathcal{S}$ and all mappings $*f: *X \to *Y$, etc. The symbol $\simeq$ will be used for the relation of nearness in both $(*X, *\mathcal{T})$ and $(*Y, *\mathcal{S})$; the context should clear up any ambiguities.

1.14 Definition The map $f: X \to Y$ is *continuous* at $x \in X$ if to each $V \in \mathcal{S}_{f(x)}$ there corresponds a $U \in \mathcal{T}_x$ with $f[U] \subseteq V$. f is *continuous* on X if it is continuous at each $x \in X$. A one-to-one map f from X onto Y is a *homeomorphism* if f and f^{-1} are continuous.

1.15 Proposition The map $f: X \to Y$ is continuous at $x \in X$ iff $*f(y) \simeq f(x)$ for each $y \simeq x$. That is, $*f[m(x)] \subseteq \bar{m}(f(x))$.

Proof: Suppose f is continuous at $x \in X$, and let V by any open neighborhood of $f(x)$. Find a corresponding $U \in \mathcal{T}_x$ from the definition of continuity so that $f[U] \subseteq V$. If $y \simeq x$ then $y \in *U$ by 1.7(i), so $*f(y) \in *V$ since $*f[*U] \subseteq *V$ by transfer. Thus $*f(y) \in *V$ for each $V \in \mathcal{S}_{f(x)}$, i.e., $*f(y) \simeq f(x)$. The converse is left to the reader. $\square$

Proposition 1.15 shows that for real-valued functions of a real variable, Definition 1.14 is equivalent to the ε-δ definition of continuity.

1.16 Theorem The map $f: X \to Y$ is continuous on X iff $f^{-1}[V] \in \mathcal{T}$ for each $V \in \mathcal{S}$.

Proof: Fix $x \in X$, suppose f is continuous, and let $V \in \mathcal{S}_{f(x)}$. Then $*f[m(x)] \subseteq \bar{m}(f(x)) \subseteq *V$ by continuity at x and the fact that V is open. It follows that $m(x) \subseteq *f^{-1}[*V] = *(f^{-1}[V])$ (check), and so $f^{-1}[V]$ is open by Proposition 1.7(i). The converse is left to the reader. $\square$

The reader will have noticed that the proofs of the results 1.6–1.16 are considerably simpler than the proofs of the corresponding results in §§I.9 and I.10. This is mainly because the richer language of Chapter II allows us to avoid proofs by contradiction which use Skolem functions.

If Y is a subset of the topological space $(X, \mathcal{T})$, then $\mathcal{T}$ induces a topology called the *relative topology* $\mathcal{T}_Y$ on Y. A subset $U \subseteq Y$ belongs to $\mathcal{T}_Y$ iff

$U = V \cap Y$ for some $V \in \mathcal{T}$. It is easy to see that the monads in $(Y, \mathcal{T}_Y)$ are given by $\hat{m}(y) = m(y) \cap {}^*Y$, $y \in Y$, where $m(y)$ is the monad of y in $(X, \mathcal{T})$ (check). The characterizations of relative openness, relative closedness, continuity, etc., are the obvious modifications of those we have just proved with $\hat{m}$ replacing m.

Next we define the important notion of a weak topology. Suppose that X is a set and $(X_i, \mathcal{T}^i)$ $(i \in I)$ is a family of topological spaces. We work in an enlargement containing ${}^*X$ and ${}^*Y$ where $Y = \bigcup X_i (i \in I)$. We let $m_i(y)$ $(i \in I, \, y \in X_i)$ denote the monads of y in $(X_i, \mathcal{T}^i)$. Let $\{\phi_i \colon X \to X_i \colon i \in I\}$ be a family of mappings.

1.17 Definition The *weak topology* $\mathcal{T}$ on X for the family $\{\phi_i \colon i \in I\}$ is the topology generated from the subbase $\mathcal{S}$ consisting of all inverse images of the form $\phi_i^{-1}[U]$, $U \in \mathcal{T}^i$; i.e., $\mathcal{T}$ consists of all sets obtained by taking arbitrary unions of finite intersections of sets in $\mathcal{S}$.

The weak topology is the weakest topology which makes all the maps ϕ_i continuous (Exercise 8).

1.18 Proposition If $m(x)$ $(x \in X)$ is a monad of the weak topology, then

$$m(x) = \{y \in {}^*X : {}^*\phi_i(y) \in m_i(\phi_i(x)) \text{ for all } i \in I\}.$$

Proof: Let the right-hand side of the equation be denoted by $k(x)$. If $x \in X$ then for $i \in I$ the sets $\phi_i^{-1}[U]$, $U \in \mathcal{T}^i_{\phi_i(x)}$, are open neighborhoods of x, so

$$m(x) \subseteq \bigcap \{y \in {}^*X : y \in \bigcap {}^*(\phi_i^{-1}[U])(U \in \mathcal{T}^i_{\phi_i(x)})\}(i \in I)$$

$$= \bigcap \{y \in {}^*X : y \in {}^*\phi_i^{-1}[\bigcap {}^*U(U \in \mathcal{T}^i_{\phi_i(x)})]\}(i \in I) = k(x).$$

On the other hand, if $V \in \mathcal{T}_x$ is a neighborhood in the base of $\mathcal{T}$ generated by the subbase $\mathcal{S}$, then V is a finite intersection of sets of the form $\phi_i^{-1}[U_i]$, $U_i \in \mathcal{T}^i_{\phi_i(x)}$. Clearly $k(x) \subseteq {}^*\phi_i^{-1}[{}^*U_i]$ for each $U_i \in \mathcal{T}^i_{\phi_i(x)}$ and so $k(x) \subseteq {}^*V$. It follows that $k(x) \subseteq m(x)$, and we are through. $\square$

1.19 Definition: The Product Topology Let $(X_i, \mathcal{T}^i)$ $(i \in I)$ be a family of topological spaces. Then the *product* $X = \prod X_i (i \in I)$ is defined to be the set of all mappings x on I with $x(i) \in X_i$ for $i \in I$. The *product topology* $\mathcal{T}$ for X is the weak topology generated by the mappings $\phi_i \colon X \to X_i$ defined by $\phi_i(x) = x(i)$.

To see what $*X$ is, note that each $x \in X$ is of the form $x: I \to \bigcup X_i (i \in I)$ with $x(i) \in X_i$. The $*$-transform of the collection $\{X_i : i \in I\}$ includes new sets X_i for $i \in *I - I$. Thus, by transfer, each $x \in *X$ is of the form $x: *I \to *[\bigcup X_i (i \in I)]$ with $x(*i) \in *X_i$ if $i \in I$, whereas if i is not standard, then $x(i) \in X_i$, but X_i need not be the extension of a standard set.

If $x \in X$, and $m(x)$ denotes the monad in $\mathcal{T}$, then by Proposition 1.18

$$m(x) = \{y \in *X : y(i) \in m_i(x(i)) \text{ for all standard } i \text{ in } *I\}.$$

That is, the monad is determined by just the standard indices in $*I$.

1.20 Theorem The topological product of Hausdorff spaces is Hausdorff.

Proof: Let $X = \prod X_i$, where the $(X_i, \mathcal{T}_i)$ are Hausdorff with monads $m_i(x)$. Let $\mathcal{T}$ be the product topology with monad $m(x)$. If $x, y \in X$ with $m(x) \cap m(y) \neq \emptyset$, let $z \in m(x) \cap m(y)$. Then $z(i) \in m_i(*x(i)) \cap m_i(*y(i))$ for each $i \in I$, and so $x(i) = y(i)$ for each $i \in I$ since $(X_i, \mathcal{T}_i)$ is Hausdorff, i.e., $x = y$. $\square$

We end this section with a result which is valid under the assumption that X is in $V(*S)$ for some S and $V(*S)$ is κ-saturated with $\kappa > \text{card } \mathcal{T}$. This result was mentioned at the beginning of §II.8 as a good example of the use of saturation in nonstandard analysis. It will be referred to again in §IV.3.

1.21 Definition Let $(X, \mathcal{T})$ be a topological space with monads $m(x)$, $x \in X$. The *standard part* st(A) of a set $A \subseteq *X$ is the set of all $x \in X$ for which there exists a $y \in A$ with $y \in m(x)$.

*1.22 Theorem Assume $X \in V(S)$ and $V(*S)$ is κ-saturated with $\kappa > \text{card } \mathcal{T}$. If $B \subseteq *X$ is internal then st(B) is closed.

Proof: Suppose z is an accumulation point of $^\circ B = \text{st}(B)$. If $U \in \mathcal{T}_z$ then there exists a point $x \in {}^\circ B$ with $x \in U$. Since $x \in {}^\circ B$ there exists a $y \in B$ with $y \in m(x)$, and hence $y \in *U$ since U is open. Thus $*U \cap B \neq \emptyset$ for all $U \in \mathcal{T}_z$. Since $V(*X)$ is κ-saturated with $\kappa > \text{card } \mathcal{T}_x$, we see from Theorem 8.4(a) of Chapter II that $\mu(\mathcal{T}_z) \cap B \neq \emptyset$, where $\mu(\mathcal{T}_z)$ is the intersection monad of the filter $\mathcal{T}_z$ (Definition 8.3 of Chapter II). Clearly $\mu(\mathcal{T}_z) = m(z)$, and so $z \in {}^\circ B$, and we are through. $\square$

Note that if $A \subseteq *X$ then st$(A) = \text{st}(A \cap \text{ns}(*X))$. Also note that Proposition 1.10 can be interpreted to say that $\bar{A} = \text{st}(*A \cap \text{ns}(*X))$, and so Theorem 1.22 is a generalization of Proposition 1.10. Theorem 1.22 was established for metric spaces by Robinson using an enlargement [42, Theorem 4.3.3], and in the general case (assuming saturation) by Luxemburg [36, Theorem 3.4.2]. An example due to Keisler shows that Theorem 1.22 is not true if $V(*S)$ is not κ-saturated with $\kappa >$ card $\mathcal{T}$ [36, Example 3.4.3].

Exercises III.1

1. Verify the statements in Examples 1.5.4–6.
2. Prove Proposition 1.10.
3. Prove (a) and (b) of Proposition 1.12.
4. Prove that a topology $\mathcal{T}$ is weaker than a topology $\mathcal{S}$ on X iff $m_{\mathcal{T}}(x) \supseteq m_{\mathcal{S}}(x)$ for each $x \in X$, where $m_{\mathcal{T}}$ and $m_{\mathcal{S}}$ denote the monads for $\mathcal{T}$ and $\mathcal{S}$, respectively.
5. A T_1 space is *normal* if for any two disjoint closed sets A and B there are disjoint open sets U and V with $A \subseteq U$ and $B \subseteq V$. A T_1 space is *regular* if the same condition holds for all A and B, where A is a point (actually a set consisting of a point) and B is a closed set. Give a nonstandard condition for regularity and normality.
6. (a) Let $k(x)$ be a subset of $*X$ for each $x \in X$. Define a collection $\mathcal{T}$ of subsets of X as follows: $U \in \mathcal{T}$ iff $k(x) \in *U$ for each $x \in U$. Show that $\mathcal{T}$ is a topology for X. Also show that if $\hat{k}(x)$ is the $\mathcal{T}$-monad of $x \in X$ then $k(x) \subseteq \hat{k}(x)$.

   (b) Fix an infinitesimal $\varepsilon > 0$ in $*R$ and for each $x \in R$ let $k(x)$ be the pseudomonad $\{y \in *R : |y - x| < \varepsilon\}$. Show that a set U is open in R in the usual sense if and only if, for each $x \in U$, $k(x) \subset *U$. Clearly $k(x) \subsetneqq m(x)$ for each $x \in R$.

   (c) Let X be any set. Let $\mathcal{B}_x$, $x \in X$, be a collection of subsets of X satisfying the following:

   (i) If $V \in \mathcal{B}_x$ then $x \in V$,
   (ii) If $V_1, V_2 \in \mathcal{B}_x$, there exists a $V \in \mathcal{B}_x$ with $V \subseteq V_1 \cap V_2$,
   (iii) If $y \in U \in \mathcal{B}_x$, then there is a $V \in \mathcal{B}_y$ with $V \subseteq U$.

   Use the sets $k(x) = \bigcap *U (U \in \mathcal{B}_x)$ to define a topology $\mathcal{T}$ as in 6(a). Show that $\mathcal{B}_x$ is a neighborhood base in $\mathcal{T}$ for each $x \in X$.
7. Finish the proof of Proposition 1.15.
8. Show that the weak topology is the weakest topology making the corresponding functions continuous. (See Definition 1.17.)
9. Let A be a subset of a topological space X. A point x is an *interior point* of A iff A is a neighborhood of x. The set of interior points of A is denoted by A°. A point x is a *boundary point* of A if x is not interior to A and not interior to A'. The set of boundary points of A is denoted by ∂A.

Show that

(a) $x \in A^\circ$ iff $m(x) \subseteq {}^*A$,
(b) $x \in \partial A$ iff $m(x) \cap {}^*A \neq \emptyset$ and $m(x) \cap {}^*A' \neq \emptyset$.

10. Let A be a subset of a topological space X. Use Exercise 9 and the text material to establish the following results:

   (a) $\partial A = \bar{A} \cap \bar{A}' = \bar{A} - A^\circ$,
   (b) $X - \partial A = A^\circ \cup (A')^\circ$,
   (c) $\bar{A} = A \cup \partial A$, $A^\circ = A - \partial A$,
   (d) A is closed iff $A \supseteq \partial A$,
   (e) A is open iff $A \cap \partial A = \emptyset$.

11. Let $(X \times Y, \mathcal{T} \times \mathcal{S})$ be the product of $(X, \mathcal{T})$ and $(Y, \mathcal{S})$. Show that if $A \subseteq X$ and $B \subseteq Y$ then

   (a) $\overline{A \times B} = \bar{A} \times \bar{B}$,
   (b) $(A \times B)^\circ = A^\circ \times B^\circ$,
   (c) $\partial(A \times B) = (\partial A \times \bar{B}) \cup (\bar{A} \times \partial B)$.

12. Let Y be a subset of $(X, \mathcal{T})$ with relative topology $\mathcal{T}_Y$. If $A \subseteq Y$ show that

   (a) A is $\mathcal{T}_Y$-closed iff it is the intersection of Y and a $\mathcal{T}$-closed set.
   (b) A point $y \in Y$ is a $\mathcal{T}_Y$-accumulation point of A iff it is a $\mathcal{T}$-accumulation point.

13. (a) Let $(X_1, \mathcal{T}_1)$, $(X_2, \mathcal{T}_2)$, and $(X_3, \mathcal{T}_3)$ be topological spaces. Show that a function $f: X_1 \to X_2$ is continuous iff, for each subset $A \subseteq X$, $f[\bar{A}] \subseteq \overline{f[A]}$.
   (b) Show that if $f: X_1 \to X_2$ and $g: X_2 \to X_3$ are continuous, then the composite function $h = g \circ f$ defined by $h(x) = g(f(x))$ for $x \in X$, is continuous.

14. Let $\mathcal{T}$ be the product topology on $X = \prod X_i (i \in I)$ where the $(X_i, \mathcal{T}_i)$ are topological spaces. If $A_i \subseteq X_i$ for each $i \in I$, show that $\overline{\prod A_i (i \in I)} = \prod \bar{A}_i (i \in I)$, so that the product of closed sets is closed.

15. (a) A sequence $\langle x_n : n \in N \rangle$ in a space $(X, \mathcal{T})$ converges to $x \in X$ if for every neighborhood U of x there is an m so that $x_n \in U$ if $n \geq m$. Show that $\langle x_n \rangle$ converges to x iff ${}^*x_\omega \in m(x)$ for all infinite ω.
   (b) Let $\langle x_n : n \in N \rangle$ be a sequence in $X = \prod X_i (i \in I)$, where the $(X_i, \mathcal{T}_i)$ are topological spaces. Show that $\langle x_n \rangle$ converges to $x \in X$ iff $\langle \phi_i(x_n) \rangle$ converges to $\phi_i(x)$ for each $i \in I$, where the ϕ_i are as in Definition 1.19.

16. Let $(X, \mathcal{T})$ and $(Y, \mathcal{S})$ be topological spaces with $(Y, \mathcal{S})$ being Hausdorff. Suppose that $f, g: X \to Y$ are continuous. Show that $\{x : f(x) = g(x)\}$ is closed.

III.2 Compactness

A cornerstone of topology is the notion of compactness, which is defined as follows.

2.1 Definition A collection $\mathscr{A} = \{A_i : i \in I\}$ of sets is a *cover* of (or covers) $A \subseteq X$ if $A \subseteq \bigcup A_i$ $(i \in I)$. A *subcover* of $\mathscr{A}$ is a subcollection of $\mathscr{A}$ which also covers A. A is a *compact* subset of a topological space $(X, \mathscr{T})$ if each open cover, that is, each cover of A by open sets U_i $(i \in I)$, contains a finite subcover.

Probably the most useful result in nonstandard analysis is the following pointwise characterization of compactness due to Robinson.

2.2 Robinson's Theorem Let $(X, \mathscr{T})$ be a topological space. Then $A \subseteq X$ is compact iff every $y \in {}^*A$ is near a standard point $x \in A$.

Proof: Suppose A is compact but that there is a point y which is not contained in the monad of any $x \in A$. Then each $x \in A$ possesses an open neighborhood U_x with $y \notin {}^*U_x$. The covering $\{U_x : x \in A\}$ of A has a finite subcovering $\{U_1, \ldots, U_n\}$; i.e., $U_1 \cup \cdots \cup U_n \supseteq A$. By transfer ${}^*U_1 \cup \cdots \cup {}^*U_n \supseteq {}^*A$. This contradicts the fact that $y \in {}^*A$ but $y \notin {}^*U_i$, $1 \le i \le n$.

Conversely, suppose that A is not compact. Then there is an open covering $\mathscr{A} = \{U_i : i \in I\}$ of A which has no finite subcover. The binary relation P on $\mathscr{A} \times A$ defined by $P\langle U, x \rangle$ iff $x \notin U$ is concurrent (check). By Theorem 5.10(ii) of Chapter II there is a point $y \in {}^*A$ with $y \notin {}^*U$ for all $U \in \mathscr{A}$. If $x \in A$ then $x \in U$ for some $U \in \mathscr{A}$, but $y \notin {}^*U$ so $y \notin m(x)$. $\square$

2.3 Examples

1. In the discrete topology the only compact subsets are finite.
2. All subsets in the trivial topology are compact.
3. In the finite complement topology for N, every subset A is compact. For if $A \ne \varnothing$ and $y \in {}^*A$ then either $y \in A$ or $y \in {}^*N_\infty$ (Corollary 7.6 of Chapter I). In the first case $y \in m(y)$, and in the second case $y \in m(x)$ for any $x \in N$ and, in particular, for some $x \in A$. Recall that a set must be finite to be closed in this topology, so there are compact subsets which are not closed in this non-Hausdorff topology.

We use Robinson's theorem to give proofs of the following standard results.

2.4 Theorem If X is compact in the topology $\mathscr{T}$ and $A \subseteq X$ is closed, then A is compact.

Proof: Let $y \in {}^*A$. Since X is compact there is an $x \in X$ with $y \in m(x)$, whence $x \in A$ by 1.7(ii), so A is compact. $\square$

2.5 Theorem If $(X, \mathscr{T})$ is Hausdorff and $A \subseteq X$ is compact, then A is closed.

Proof: Let $x \in A'$ and suppose that $y \in m(x)$, $y \in {}^*A$. Since A is compact, $y \in m(\tilde{x})$ for some $\tilde{x} \in A$, but then $m(x) \cap m(\tilde{x}) \neq \varnothing$, contradicting the fact that $(X, \mathscr{T})$ is Hausdorff. $\square$

2.6 Theorem If $(X, \mathscr{T})$ and $(Y, \mathscr{S})$ are topological spaces and $f: X \to Y$ is continuous, then $f[K]$ is compact for each compact $K \subseteq X$.

Proof: Exercise. $\square$

2.7 Theorem If $(X, \mathscr{T})$ is compact, $(Y, \mathscr{S})$ is Hausdorff, and $f: X \to Y$ is continuous, then

  (i) f is closed (i.e., takes closed sets onto closed sets),
  (ii) if f is one-to-one then it is a homeomorphism.

Proof: (i) Follows from 2.4–2.6.
  (ii) We may assume that $f[X] = Y$. We need only show that f is open (i.e., takes open sets onto open sets). But if U is open in X, then U' is closed. Since f is one-to-one, $f[U] = Y - f[U']$, which is open by (i). $\square$

The real power of Robinson's theorem is illustrated by the proofs of the following standard results. The standard proofs of these results as given in Kelley [20] are somewhat involved.

2.8 Tychonoff's Theorem If $(X_i, \mathscr{T}^i)\,(i \in I)$ are compact spaces and $X = \prod X_i (i \in I)$, then X is compact in the product topology $\mathscr{T}$.

Proof: Let $y \in {}^*X$. Then $y(i) \in {}^*X_i$ for (standard) $i \in I$ and so $y(i)$ is near a standard point $x_i \in X_i$ for each $i \in I$. That is, $y(i) \in m_i(x_i)$, where $m_i(x_i)$ denotes the monad of x_i in $(X_i, \mathscr{T}^i)$. By 1.19, $y \in m(x)$, where $m(x)$ is the monad in $\mathscr{T}$ of the point $x \in X$ defined by $x(i) = x_i$. $\square$

Tychonoff's theorem is used in many proofs in analysis. One can usually replace these standard proofs by simpler nonstandard ones which use Robinson's theorem directly (for example, see the proof of Alaoglu's theorem, 4.22, below).

*2.9 Alexander's Theorem If $\mathscr{S}$ is a subbase for the topology of $(X, \mathscr{T})$ and every cover of X by members of $\mathscr{S}$ has a finite subcover, then X is compact.

Proof: (Hirschfeld [18]) Suppose X is not compact. By 2.2 there exists a $y \in {}^{*}X$ which is not near-standard and so for each $x \in X$ there is an open set U_x with $x \in U_x$ and $y \notin {}^{*}U_x$. Since each U_x is a finite intersection of members V_i of $\mathscr{S}$, one of the ${}^{*}V_i$ must omit y, so we may as well assume that $U_x \in \mathscr{S}$ for each x. Then the covering $\{U_x : x \in X\}$ cannot have a finite subcover $U_1, \ldots, U_n$, for in that case ${}^{*}X = {}^{*}U_1 \cup \cdots \cup {}^{*}U_n$ and $y \in {}^{*}U_i$ for some i, $1 \leq i \leq n$ (contradiction). $\square$

Exercises III.2

1. Prove Theorem 2.6.
2. Let $(X, \mathscr{T})$ and $(Y, \mathscr{S})$ be topological spaces and suppose that $(Y, \mathscr{S})$ is compact Hausdorff. Show that $f : X \to Y$ is continuous iff the graph $G_f = \{(x, f(x)) \in X \times Y : x \in X\}$ of f is closed in $X \times Y$.
3. Let X have the topologies $\mathscr{T}$ and $\mathscr{S}$, and suppose that $(X, \mathscr{T})$ is compact Hausdorff. Show that (a) if $\mathscr{S}$ is strictly contained in $\mathscr{T}$ then $\mathscr{S}$ is not Hausdorff, (b) if $\mathscr{T}$ is strictly contained in $\mathscr{S}$ then $\mathscr{S}$ is not compact.
4. Show that if $(X, \mathscr{T})$ is compact then there is a hyperfinite set $F \subseteq {}^{*}X$ with $X \subseteq F \subseteq \text{ns}({}^{*}X)$ such that $X = \text{st}(F)$.
5. Suppose that $(X, \mathscr{T})$ is compact. Show that if $\langle A_n : n \in N \rangle$ is a sequence of nonempty closed subsets of X which is monotone, i.e., $A_1 \supseteq A_2 \supseteq \cdots$, then $\bigcap A_n (n \in N) \neq \varnothing$.
6. The following problem is derived from a result of A. Abian (see [1]): Let p_n be a sequence of polynomials and x_n a sequence of variables so that, for each n, $p_n = p_n(x_1, x_2, \ldots, x_n)$ is a function of the first n variables. Let I_n be a sequence of closed and bounded intervals in R. Assume that for each n there are values $a_i^n \in I_i$ for $1 \leq i \leq n$ such that, for each $i \leq n$, $p_i(a_1^n, a_2^n, \ldots, a_i^n) = 0$. Show that there are values $a_i \in I_i$ for $1 \leq i < \infty$ such that, for each $n \in N$, $p_n(a_1, a_2, \ldots, a_n) = 0$.
7. (Luxemburg [36]). Let $(X, \mathscr{T})$ be a regular Hausdorff space (see Exercise 1.5). If A is an internal set in a κ-saturated enlargement of $V(X)$ where $\kappa > \text{card } \mathscr{T}$, and $A \subseteq \text{ns}({}^{*}X)$, then $\text{st}(A) = \{x \in X : \text{there exists } y \in A \text{ with } x = \text{st}(y)\}$ is compact.

III.3 Metric Spaces

The most important topologies which occur in analysis are those associated with a metric or distance function. The corresponding spaces are called metric spaces.

3.1 Definition A *metric space* is a pair (X, d), where X is a set and d is a map from $X \times X$ into the nonnegative reals satisfying (for all $x, y, z \in X$)

(a) $d(x, y) = 0$ iff $x = y$,
(b) $d(x, y) = d(y, x)$,
(c) *(triangle inequality)* $d(x, z) \leq d(x, y) + d(y, z)$.

Each metric space (X, d) can be made into a topological space $(X, \mathcal{T}_d)$ by specifying that a set $U \in \mathcal{T}_d$ if, for each $x \in U$, there is an $\varepsilon > 0$ in R so that the *open ε-ball* $B_\varepsilon(x) = \{y \in X : d(x, y) < \varepsilon\} \subseteq U$. The resulting collection $\mathcal{T}_d$ is a topology (standard exercise). When the metric d and associated topology $\mathcal{T}_d$ are understood we simply call X rather than (X, d) or $(X, \mathcal{T}_d)$ the metric space. Note that the open ε-balls about a point $x \in X$ form a local base at x.

3.2 Examples

1. R is a metric space with the usual metric $d(x, y) = |x - y|$ for $x, y \in R$.
2. R is a metric space with the metric $\tilde{d}(x, y) = |x - y|/(1 + |x - y|)$ (check).
3. Let X be any set and define $d(x, y) = 1$ if $x \neq y$ and $d(x, y) = 0$ otherwise. It is easy to see that d is a metric and is called the *discrete metric*.
4. R^n is a metric space under each of the following metrics [where $x = (x_1, \ldots, x_n)$, $y = (y_1, \ldots, y_n)$]:

(α) $d_1(x, y) = \sum_{i=1}^n |x_i - y_i|$,
(β) $d_\infty(x, y) = \max\{|x_i - y_i| : 1 \leq i \leq n\}$.

Properties (a) and (b) of Definition 3.1 are trivial; to check property (c) for metric (α) we have [with $z = (z_1, \ldots, z_n)$]

$$d(x, z) = \sum_{i=1}^n |x_i - z_i| \leq \sum_{i=1}^n |x_i - y_i| + |y_i - z_i| = d(x, y) + d(y, z).$$

The triangle inequality (c) for metric (β) is left as an exercise.

5. Let l_∞ (also often denoted by l^∞) be the set of bounded sequences $x = \langle x_1, x_2, \ldots \rangle$. Then l_∞ is a metric space under the metric defined by $d_\infty(x, y) = \sup\{|x_i - y_i| : i \in N\}$ with $y = \langle y_1, y_2, \ldots \rangle$. Note that $d_\infty(x, y)$ is finite for any $x, y \in l_\infty$ since, for any i, $|x_i - y_i| \leq |x_i| + |y_i|$ and so

$$\sup\{|x_i - y_i| : i \in N\} \leq \sup\{|x_i| : i \in N\} + \sup\{|y_i| : i \in N\}.$$

To check the triangle inequality (c) we have [with $z = \langle z_1, z_2, \ldots \rangle$]

$$|x_i - z_i| \leq |x_i - y_i| + |y_i - z_i|$$
$$\leq \sup\{|x_i - y_i| : i \in N\} + \sup\{|y_i - z_i| : i \in N\}$$
$$= d_\infty(x, y) + d_\infty(y, z).$$

The result follows by taking sup over $i \in N$ on the left.

The nonstandard analysis of metric spaces will be carried out in an enlargement $V(*S)$ of a superstructure $V(S)$ that contains X. We always assume that S contains the set of real numbers R. In proving abstract theorems concerning a metric space (X, d) we will write x instead of $*x$ for an element of $*X$. In concrete examples, it might be important to investigate in more detail the structure of the elements of $*X$. For example, if $X = l_\infty$ then we could take $S = R$, in which case elements of l_∞ would appear as bounded real-valued functions on the integers. Often the set S in a particular example will not be specified; the reader should be able to fill in the details.

By transfer, the $*$-transform $*d$ of d satisfies the conditions of Definition 3.1 with $*d$ replacing d for all $x, y, z \in *X$.

3.3 Definition Let (X, d) be a metric space. Two points x and y in $*X$ are *near* if $*d(x, y) \simeq 0$. We write $x \simeq y$ if x and y are near and $x \not\simeq y$ otherwise. The *monad* of $x \in *X$ is the set $m(x) = \{y \in *X : y \simeq x\}$. Two points x, y in $*X$ are in the same *galaxy* if $*d(x, y)$ is finite. The *principal galaxy* of $*X$ is the one containing the standard points, and is denoted by fin($*X$). Points in fin($*X$) are called *finite*.

An easy exercise shows that for standard points $x \in X$ the monad $m(x)$ of Definition 3.3 coincides with the monad obtained from the associated topology $\mathscr{T}_d$. The metric monads, however, are defined for all points $x \in *X$. It is also easy to see that the relation $\simeq$ is an equivalence relation.

3.4 Examples

1. In the metric of Example 3.2.1, $x \simeq y$ iff $x - y$ is infinitesimal.
2. In each of the metrics on R^n defined in Example 3.2.4, $x \simeq y$ iff $x_i - y_i$ is infinitesimal for $1 \leq i \leq n$ (exercise).
3. Each element of $*l_\infty$ is an internal function $x : *N \to *R$, and we usually write $x(i) = x_i$ and $x = \langle x_i : i \in *N \rangle$. The standard elements in $*l_\infty$ are of the form $*y$, where $y = \langle y_i : i \in N \rangle$ is an element of l_∞. Each $x \in *l_\infty$ is $*$-bounded in the sense that there exists an $M \in *R$ (which could be infinite) so that $|x_i| \leq M$ for each $i \in *N$ (exercise). In passing, note that there are external

functions $z: {}^*N \to {}^*R$ which are also $*$-bounded; an example occurs when $z_i = 1$ for $i \in N$ and $z_i = 0$ for $i \in {}^*N_\infty$.

The real-valued function on $\mathscr{P}(N) \times l_\infty$ defined by $(A, x) \to \sup\{|x_i| : i \in A\}$ extends by transfer to a $*R$-valued function on $*\mathscr{P}(N) \times *l_\infty$. We again denote the value of this extended function by $\sup\{|x_i| : i \in A\}$, where $A \subseteq {}^*N$ is internal and $x \in *l_\infty$. Properties of the extended sup function can be obtained by transfer. For example, if A and B are internal subsets of $*N$ and $A \subseteq B$ then $\sup\{|x_i| : i \in A\} \leq \sup\{|x_i| : i \in B\}$. For each $x, y \in *l_\infty$ we have $*d_\infty(x, y) = \sup\{|x_i - y_i| : i \in {}^*N\}$.

The monads in $*l_\infty$ are easily characterized. We claim that if $x, y \in *l_\infty$ then $x \simeq y$ iff $x_i \simeq y_i$ for all $i \in {}^*N$. For suppose $x \simeq y$. Then, for any $i \in {}^*N$, $|x_i - y_i| \leq \sup\{|x_i - y_i| : i \in {}^*N\} \simeq 0$. The converse is left as an exercise.

The finite elements in $*l_\infty$ are those $x = \langle x_i : i \in {}^*N \rangle$ for which there exists a finite M (and hence even a standard M) in $*R$ so that $|x_i| \leq M$ for all $i \in {}^*N$. The value of M depends on x.

All of the results of 3.1 and 3.2 are available for the topological space $(X, \mathscr{T}_d)$ associated with a metric space (X, d). We concentrate in this section on some results which are special to metric spaces. The first few revolve around the notion of uniformity.

3.5 Definition Let (X, d) and $(Y, \bar{d})$ be two metric spaces and A a subset of X.

(a) A map $f: A \to Y$ is *uniformly continuous* on A if, given $\varepsilon > 0$ in R, there exists a $\delta > 0$ in R so that $\bar{d}(f(x), f(y)) < \varepsilon$ for all $x, y \in A$ for which $d(x, y) < \delta$.

(b) A sequence of maps $f_n: A \to Y$, $n \in N$, *converges uniformly* on A to $f: A \to Y$ if, given $\varepsilon > 0$, there exists a $k \in N$ so that $\bar{d}(f_n(x), f(x)) < \varepsilon$ for all $n \geq k$ in N and all $x \in A$.

In the following results, (X, d) and $(Y, \bar{d})$ are metric spaces and A is a subset of X. We use $\simeq$ to denote nearness in both $*X$ and $*Y$, letting the context settle any ambiguity.

3.6 Proposition The map $f: A \to Y$ is uniformly continuous on A iff $*f(x) \simeq *f(y)$ whenever $x, y \in *A$ and $x \simeq y$.

Proof: Let f be uniformly continuous on A. Find the $\delta > 0$ for a prescribed $\varepsilon > 0$ from 3.5(a). By transfer, $*\bar{d}(*f(x), *f(y)) < \varepsilon$ for all $x, y \in *A$ for which $*d(x, y) < \delta$. In particular, $*\bar{d}(*f(x), *f(y)) < \varepsilon$ for all $x, y \in *A$ for which $x \simeq y$. This is true for any $\varepsilon > 0$ in R, and so $*f(x) \simeq *f(y)$ for all $x, y \in *A$ for which $x \simeq y$.

Conversely, suppose $*f(x) \simeq *f(y)$ whenever $x, y \in *A$ and $x \simeq y$. Let $\varepsilon > 0$ in R be given. Then the internal sentence

$$(\exists \delta \in *R)[\delta > 0 \wedge (\forall x, y \in *A)[*d(x, y) < \delta \to *\bar{d}(*f(x), *f(y)) < \varepsilon]]$$

is true in $V(*S)$ (choose δ to be infinitesimal). That f is uniformly continuous follows by transfer to $V(S)$. $\square$

3.7 Proposition The sequence $f_n : A \to Y$ converges uniformly on A to $f : A \to Y$ iff $*f_n(x) \simeq *f(x)$ for all $n \in *N_\infty$ and all $x \in *A$.

Proof: Exercise. $\square$

3.8 Theorem If $f : A \to Y$ is continuous and A is compact, then f is uniformly continuous on A.

Proof: Let $x, y \in *A$ with $x \simeq y$. Then x and y are near a standard point $z \in A$ since A is compact, and $*f(x) \simeq f(z) \simeq *f(y)$ since f is continuous at z. The result follows from Proposition 3.6. $\square$

3.9 Theorem If $f_n : A \to Y$ is a sequence of continuous functions which converge uniformly on A to $f : A \to Y$, then f is continuous.

Proof: Let $x \in A$ and $y \in *A$ with $y \simeq x$. We need to show that $*f(y) \simeq f(x)$. Now $*f_n(y) \simeq *f_n(x)$ for each $n \in N$ and so, by Theorem 7.3 of Chapter II, $*f_\omega(y) \simeq *f_\omega(x)$ for some $\omega \in *N_\infty$. By Proposition 3.7, $*f_\omega(y) \simeq *f(y)$ and $*f_\omega(x) \simeq *f(x)$, so $*f(y) \simeq *f(x) = f(x)$. $\square$

Next we present the notion of a complete metric space. To do so we need the obvious generalizations of the definitions in §I.8.

3.10 Definition Let (X, d) be a metric space, and let $\langle s_n : n \in N \rangle$ be a sequence of points in X. Then

(i) $\langle s_n \rangle$ *converges* to s if, given $\varepsilon > 0$ in R, there is a $k \in N$ so that $d(s_n, s) < \varepsilon$ if $n \geq k$,

(ii) $\langle s_n \rangle$ is a *Cauchy sequence* if, given $\varepsilon > 0$ in R, there is a $k \in N$ so that $d(s_n, s_m) < \varepsilon$ if $n, m \geq k$,

(iii) s is a *limit point* of $\langle s_n \rangle$ if, for each $\varepsilon > 0$ in R and each $k \in N$, there is an $n > k$ so that $d(s_n, s) < \varepsilon$.

The reader will easily be able to prove that $\langle s_n \rangle$ converges to s iff $*s_n \simeq s$ for all $n \in *N_\infty$, $\langle s_n \rangle$ is a Cauchy sequence iff $*s_n \simeq *s_m$ for all $n, m \in *N_\infty$, and s is a limit point of $\langle s_n \rangle$ iff $*s_n \simeq s$ for some $n \in *N_\infty$.

3.11 Definition (X, d) is *complete* if each Cauchy sequence in X converges to a point in X.

3.12 Examples

1. The set R with the usual metric is complete by 8.5 of Chapter I.
2. Any set X with the discrete metric is complete.
3. R^n with each metric of Example 3.2.4 is complete. For example, let $\langle x^k \rangle$ be a Cauchy sequence in (R^n, d_1). Then for each i, $1 \le i \le n$, $|x_i^k - x_i^l| \le d_1(x^k, x^l)$. Thus, $\langle x_i^k \rangle$ is a Cauchy sequence for each i and so converges to a point x_i in R. The point $x = \langle x_1, \dots, x_n \rangle$ in R^n is the limit of x^k in R^n.

We now use nonstandard analysis to prove some abstract theorems on completeness. The nonstandard characterization of completeness requires the following notion.

3.13 Definition Let (X, d) be a metric space. A point $y \in *X$ is a *pre-near-standard* point if for every standard $\varepsilon > 0$ there is a standard $x \in X$ with $*d(x, y) < \varepsilon$.

3.14 Proposition A metric space (X, d) is complete iff every pre-near-standard point $y \in *X$ is near-standard.

Proof: Suppose (X, d) is complete. If y is pre-near-standard, find a sequence $s_n \in X$ so that $*d(y, s_n) < 1/n$. Then $\langle s_n \rangle$ is a Cauchy sequence with limit s and $y \simeq *s_n \simeq s$ if $n \in *N_\infty$.

Conversely, suppose every pre-near-standard point is near-standard, and let $\langle s_n \rangle$ be a Cauchy sequence. Given $\varepsilon > 0$, find the associated $k \in N$ from Definition 3.10. Then $d(*s_n, s_k) < \varepsilon$ if $n \in *N_\infty$. Thus $*s_n$ is pre-near-standard for every $n \in *N_\infty$ and each such $*s_n$ must be near-standard to the same $s \in X$ (check). The sequence $\langle s_n \rangle$ must converge to s. $\square$

3.15 Corollary A closed subset (A, d) of a complete metric space (X, d) is complete.

Proof: Let y be a pre-near-standard point in *A. Then $y \simeq x$ for some $x \in X$ since (X, d) is complete. But $x \in A$ by Proposition 1.10 since A is closed. □

Using this characterization, we will show that it is possible to adjoin "ideal" elements to a metric space (X, d) so that the result is a complete metric space in which (X, d) is densely embedded.

3.16 Definition Let (X, d) be a metric space. A metric space $(\hat{X}, \hat{d})$ is a *completion* of (X, d) if $(\hat{X}, \hat{d})$ is complete, there is an isometric embedding ϕ: $X \to \hat{X}$ [i.e., $d(x, y) = \hat{d}(\phi(x), \phi(y))$ for all $x, y \in X$, whence ϕ is one-to-one], and $\phi[X]$ is dense in $\hat{X}$.

3.17 Theorem Any metric space (X, d) has a completion $(\hat{X}, \hat{d})$.

Proof: We let X' be the pre-near-standard points in *X, and $\hat{X}$ be the equivalence classes of X' under the relation of nearness $\simeq$ (an equivalence relation); thus the elements of $\hat{X}$ are monads $m(x')$ of pre-near-standard points $x' \in$ *X. Also define $\hat{d}(m(x'), m(y')) = \text{st}(*d(x', y'))$ [note that *$d(x', y')$ is finite for any pre-near-standard points x', y']. This metric is independent of the pre-near-standard points chosen to represent the elements of $\hat{X}$, for if $x' \simeq x'_1$ and $y' \simeq y'_1$ then *$d(x', y') \simeq$ *$d(x'_1, y'_1)$ (Exercise 6).

The map $\phi: X \to \hat{X}$ defined by $\phi(x) = m(x)$ is obviously an isometric embedding. Also $\phi[X]$ is dense in $\hat{X}$. For if $m(x') \in \hat{X}$, where x' is pre-near-standard, then given $\varepsilon > 0$ there exists an $x \in X$ so that *$d(x', x) < \varepsilon$ and then $\hat{d}(m(x'), m(x)) = \text{st}(*d(x', x)) \leq \varepsilon$.

To show completeness, let $\langle m(x'_n) : n \in N \rangle$ be a Cauchy sequence in $(\hat{X}, \hat{d})$, with $x'_n \in X'$. Since each $x'_n \in X'$, there are elements $x_n \in X$ with *$d(x_n, x'_n) < 1/n$ for each $n \in N$. Given $\varepsilon > 0$ in R, there exists a $k \in N$ so that $\hat{d}(m(x'_n), m(x'_m)) < \varepsilon$ and hence *$d(x'_n, x'_m) < \varepsilon$ if $n, m \geq k$. Then $d(x_n, x_m) = $ *$d(x_n, x_m) \leq 2/n + \varepsilon$ if $m \geq n \geq k$ in N by the triangle inequality. Again by transfer, *$d(*x_n, *x_m) \leq 2/n + \varepsilon$ if $m \geq n \geq k$ in *N. In particular, if $\omega \in$ *N_∞, *$d(x_n, *x_\omega) \leq 2/n + \varepsilon$ if $n \geq k$, and so *x_ω is pre-near-standard. Therefore *$d(x'_n, *x_\omega) \leq$ *$d(x'_n, x_n) +$ *$d(x_n, *x_\omega) \leq 3/n + \varepsilon$ if $n \geq k$, yielding $\hat{d}(m(x'_n), m(*x_\omega)) \leq 3/n + \varepsilon$ if $n \geq k$. Thus $\langle m(x'_n) \rangle$ converges to $m(*x_\omega)$. □

As an example, note that the rationals Q form a metric space under the usual metric $d(x, y) = |x - y|$, $x, y \in Q$. The completion $(\hat{Q}, \hat{d})$ is isomorphic to the real metric space (R, d).

Recall that a subset of the real line is compact iff it is closed and bounded. In arbitrary metric spaces there is a similar relationship between compactness, completeness, and total boundedness, the last being a generalization of boundedness.

3.18 Definition A metric space (X, d) is *totally bounded* if, to each $\varepsilon > 0$ in R, there corresponds a finite covering $\{B_\varepsilon(x_i): 1 \leq i \leq n\}$ by open ε-balls [each $B_\varepsilon(x) = \{y \in X : d(x, y) < \varepsilon\}$].

3.19 Proposition A metric space (X, d) is totally bounded if every point of $*X$ is pre-near-standard.

Proof: Suppose (X, d) is totally bounded. Let $\varepsilon > 0$ be given and find the corresponding points x_i, $1 \leq i \leq n$, so that $X = \bigcup B_\varepsilon(x_i)(1 \leq i \leq n)$. By transfer, $*X = \bigcup *B_\varepsilon(x_i)(1 \leq i \leq n)$, and so every point of $*X$ is pre-near-standard. The converse is left to the reader. $\square$

3.20 Theorem A metric space (X, d) is compact iff it is complete and totally bounded.

Proof: Suppose (X, d) is compact. Then every point $y \in *X$ is near a point in X, so (X, d) is complete and totally bounded by 3.14 and 3.19, respectively.

Conversely, suppose (X, d) is complete and totally bounded. If $y \in *X$, then y is pre-near-standard by 3.19 and hence near-standard by 3.14. $\square$

One might expect that "totally bounded" may be replaced by "bounded" in this theorem, where boundedness is defined as follows.

3.21 Definition A set A in a metric space (X, d) is *bounded* if there is a point $x_0 \in X$ and a number M so that $d(x, x_0) \leq M$ for all $x \in A$.

Example 3.2.2 and the following example show that boundedness is not enough for Theorem 3.20.

3.22 Example Let $B_1 = \{x \in l_\infty : d_\infty(x, \theta) \leq 1\}$ be the "unit ball" in (l_∞, d_∞) where $\theta = \langle 0, 0, \ldots \rangle$. It is easy to see that B_1 is closed and hence is complete when regarded as a metric space with the metric induced by d_∞ (Exercises 8, 14). Also, B_1 is obviously bounded. Now consider the element $x = \langle x_i : i \in *N \rangle \in *B_1$ which is zero except at some infinite integer ω where $x_\omega = 1$. Then x is not near-standard. For if $x \simeq *y$ for some standard $y = \langle y_i : i \in N \rangle$ then $0 = x_i \simeq y_i$ for at least all $i \in N$, and so $y_i = 0$ for all $i \in N$. By transfer, $*y_i = 0$ for all $i \in *N$, and so $*y_\omega \not\simeq x_\omega$.

To end this section we consider another compactness criterion, which is especially important in applications. In many situations one can obtain a sequence $\langle x_n \rangle$ of points (in a given topological space X,) which has certain desirable properties, e.g., giving better and better approximate solutions to a set of equations. One would like to assert that a subsequence of the given sequence converges to a point in the space (in order, e.g., to produce an exact solution). Though the criterion of compactness in the sense of §III.2 is not always of help in constructing such a subsequence, if the assertion is nevertheless always true we call the space sequentially compact.

3.23 Definition A topological space X is *sequentially compact* if from each sequence $\langle x_n \rangle$ in X it is possible to select a subsequence which converges to a point $x \in X$.

It turns out that compactness is equivalent to sequential compactness in a metric space. Unfortunately this is not true in general topological spaces, as we shall see in §III.7.

3.24 Theorem A metric space (X, d) is compact iff it is sequentially compact.

Proof: (i) Suppose that (X, d) is compact and let $\langle x_n \rangle$ be a sequence in X. By Exercise 9 there is a point x_0 which is a limit point of $\langle x_n \rangle$. We will show that some subsequence of $\langle x_n \rangle$ converges to x_0. Consider the open ball $B_1 = \{x \in X : d(x, x_0) < 1\}$. Since x_0 is a limit point of $\langle x_n \rangle$ there is an $x_{n_1} \in B_1$. Similarly there is an x_{n_2} in $B_{1/2} = \{x \in X : d(x, x_0) < \frac{1}{2}\}$ with $n_2 > n_1$. Continuing this process inductively, we obtain a subsequence $\langle x_{n_k} \rangle$ with $x_{n_k} \in B_{1/k} = \{x \in X : d(x, x_0) < 1/k\}$; clearly $\langle x_{n_k} \rangle$ converges to x_0.
(ii) Suppose (X, d) is sequentially compact. Then it is obvious that (X, d) is complete, so that if (X, d) is not compact, it must not be totally bounded. Thus there exists some $\varepsilon > 0$ so that no finite collection $\{B_\varepsilon(y_i) : 1 \leq i \leq n\}$ covers X. Let $x_1 \in X$ be a given point. Then there is an x_2 with $d(x_1, x_2) \geq \varepsilon$. Similarly there is an x_3 with $d(x_1, x_3) \geq \varepsilon$ and $d(x_2, x_3) \geq \varepsilon$. Continuing in this way, we construct a sequence $\langle x_n \rangle$ with $d(x_n, x_m) \geq \varepsilon$ for any $n, m \in N$. Clearly $\langle x_n \rangle$ can have no convergent subsequence. $\square$

The procedure used in part (i) of the proof in going from a limit point to a convergent subsequence does not work in a general topological space. It uses in an essential way the fact that the neighborhood system of x has a countable base. A topological space is said to satisfy the *first axiom of countability* if the neighborhood system of each point has a countable base. Included in such spaces are the metric spaces. Clearly, a subset A in a metric or first countable space is closed iff A contains the limit of any convergent sequence in A.

Exercises III.3

1. Show that $d_\infty(x, y)$ satisfies the triangle inequality.
2. Show that for the metrics on R^n defined in Example 3.2.4, $x \simeq y$ iff $x_i \simeq y_i$ for $1 \le i \le n$.
3. Show that for each $x \in {}^*l_\infty$ there is an $M \in {}^*N$ such that $|x_i| \le M$ for all $n \in {}^*N$.
4. Prove that if x, y are internal sequences and $x_i \simeq y_i$ for all $i \in {}^*N$ then $\sup\{|x_i - y_i| : i \in {}^*N\} \simeq 0$.
5. Prove Proposition 3.7.
6. (a) Show that if a, b, c are points in a metric space (X, d) then $|d(a, c) - d(b, c)| \le d(a, b)$.
   (b) Show that if $x' \simeq x'_1$ and $y' \simeq y'_1$ in $(^*X, {}^*d)$ then $^*d(x', y') \simeq {}^*d(x'_1, y'_1)$.
7. Show that if (X, d) is a metric space and each point of $^*X$ is pre-near-standard then (X, d) is totally bounded.
8. Show that $B_1 = \{x \in l_\infty : d_\infty(x, \theta) \le 1\}$ is closed.
9. Show that a sequence in a compact metric space has a limit point.
10. Let $\langle x_n \rangle$ be a sequence in a compact metric space (X, d). Fix $\omega \in {}^*N_\infty$. Use the downward transfer principle and the fact that x_ω is near-standard to prove there is a subsequence x_{n_i} that converges to $\mathrm{st}(x_\omega)$.
11. Use Theorem 3.24 to prove Robinson's result: If (X, d) is a metric space and A is an internal set in X such that each $a \in A$ is near-standard, then $\mathrm{st}(A) = \{x \in X : \text{there exists an } a \in A \text{ with } x \simeq a\}$ is compact. (The generalization for regular topological spaces (Exercise 2.7) is due to Luxemburg [36].)
12. Prove that a Cauchy sequence in a metric space (X, d) is bounded.
13. Use Exercise 12 to show that (X, d) is complete if every finite point in $^*X$ is near-standard.
14. Show that (l_∞, d_∞) is complete.
15. ($\varepsilon\delta$-Continuity, *-Continuity, and S-Continuity) Let (X, d) be a metric space, A be a subset of $^*X$, and $f: A \to {}^*R$ be a function. We say that f is $\varepsilon\delta$-continuous (*-continuous) at $x \in A$ if, for each $\varepsilon > 0$ in R ($^*R$), there is a $\delta > 0$ in R ($^*R$) such that $|f(x) - f(y)| < \varepsilon$ if $y \in A$ and $^*d(x, y) < \delta$. We say that f is S-continuous at $x \in A$ if $f(y) \simeq f(x)$ for every $y \in A$ with $y \simeq x$.
    (a) $A = {}^*X$ and $f = {}^*g$, where $g: X \to R$. Show that if g is continuous at each $x \in X$, then f is *-continuous and S-continuous at each $x \in {}^*X$.

(b) Show that if f is $\varepsilon\delta$-continuous at $x \in A$ then f is S-continuous at $x \in A$ but not necessarily vice versa.

(c) Suppose that f is internal. Show that f is S-continuous at $x \in A$ iff f is $\varepsilon\delta$-continuous at x. (Hint: Use the spillover principle.)

(d) Show that there are internal functions f on $*R$ which are $*$-continuous but not S-continuous at zero and vice versa. (Hint: Look for examples on $X = R$ with the usual metric).

16. Let A be an internal set in $*X$ where (X, d) is a metric space and let $f: A \to *R$ be internal. Show that f is S-continuous at each point $x \in A$ iff, for every (standard) $\varepsilon > 0$ in R, there is a $\delta > 0$ in R such that $|f(x) - f(y)| < \varepsilon$ for all $x, y \in A$ for which $*d(x, y) < \delta$. (Hint: Again use the spillover principle.)

17. Let X be a compact metric space. Suppose that the internal function $f: *X \to *R$ is S-continuous at each point of $*X$ and finite at each $x \in X$. Let g be defined by $g(x) = {}^{\circ}f(x)$ for $x \in X$. Then g is continuous on X and $*g(x) \simeq f(x)$ for all $x \in *X$.

18. Two metrics on X are equivalent if they define the same topology. Show that the metrics d and d' are equivalent if there exist positive (nonzero) constants α and β in R so that $\alpha d(x, y) \le d'(x, y) \le \beta d(x, y)$ for all $x, y \in X$.

19. Let $I = [0, 1] \subset R$ and let X be the set of all continuous functions $f: I \to I$ such that $|f(x) - f(y)| \le |x - y|$. Define $d(f, g) = \sup\{|f(x) - g(x)|: x \in I\}$ for $f, g \in X$.

   (a) Show that (X, d) is a metric space.

   (b) Show that (X, d) is compact.

20. Use Robinson's theorem to show that the set of elements x of l_1 with $\|x\|_1 \le 1$ (the unit ball) is not compact.

21. (Lebesgue covering lemma). If $U_1, \ldots, U_n$ is an open covering of a compact metric space (X, d), then there is an $\varepsilon > 0$ in R such that the ε ball $B_\varepsilon(x)$ about any $x \in X$ is entirely contained in one of the sets U_i, $1 \le i \le n$.

III.4 Normed Vector Spaces
and Banach Spaces

The space R is not only a metric space with the usual metric; it is also equipped with operations of addition and multiplication, and the distance function $d(x, y) = |x - y|$ involves these operations. In this section we generalize this simple example. The metric spaces will have the additional structure of a vector space, and the metric will come from a generalization of the absolute value. Many theorems and exercises are standard.

As in §III.3, the nonstandard analysis will be carried out in an enlargement $V(*S)$ of a suitable superstructure $V(S)$. The choice of S will depend on the context and will not be mentioned explicitly.

4.1 Definition A (real)[†] *vector space* is a set X on which are defined operations of *vector addition* $(+)$ and *scalar multiplication* $(\cdot)$ (so that we form the sum $x + y$ of two vectors $x, y \in X$ and the *scalar multiple* $a \cdot x$ of the vector $x \in X$ by $a \in R$). These operations satisfy the following conditions (as usual we often omit the dot in scalar multiplication):

(i) $x + y = y + x$ for all $x, y \in X$.
(ii) $(x + y) + z = x + (y + z)$ for all $x, y, z \in X$.
(iii) There is a vector $\theta \in X$ called the *zero vector* so that $x + \theta = x$ for all $x \in X$.
(iv) $a(x + y) = ax + ay$ if $a \in R$ and $x, y \in X$.
(v) $(a + b)x = ax + bx$ if $a, b \in R$ and $x \in X$.
(vi) $a(bx) = (ab)x$ if $a, b \in R$ and $x \in X$.
(vii) $0 \cdot x = \theta$, $1 \cdot x = x$ for all $x \in X$.

We write $(-1)x = -x$, so that $x + (-x) = \theta$ by (v) and (vii). The set $Y \subseteq X$ is a (linear) *subspace* of X if $x, y \in Y$ and $a, b \in R$ imply $ax + by \in Y$.

An easy exercise shows that the element θ is unique. A subspace Y of a vector space X is itself a vector space with the inherited operations of addition and scalar multiplication.

4.2 Definition A *norm* on a vector space X is a nonnegative real-valued function $\| \ \|: X \to R$ satisfying

(a) $\|x\| = 0$ iff $x = \theta$,
(b) $\|x + y\| \le \|x\| + \|y\|$ (triangle inequality),
(c) $\|ax\| = |a| \|x\|$.

A normed vector space $(X, \| \ \|)$ is a metric space if we define the metric d by $d(x, y) = \|x - y\|$ (exercise). If the normed vector space is complete in this metric it is called a *Banach space*. A subspace $Y \subseteq X$ is *closed* if it is closed in the topology defined by the norm.

The reader should easily be able to prove that the norm function $\| \ \|: X \to R$ is continuous when X has the topology induced by d. Note also that a closed

---

[†] Much of this and the succeeding section obtains (with some obvious modifications) if the real numbers are replaced by complex numbers in the definition of vector space.

subspace of a Banach space is complete (Corollary 3.15) and hence a Banach space.

4.3 Examples

1. R^n can be made into a vector space in the following standard way: If $x = \langle x_1, \ldots, x_n \rangle$, $y = \langle y_1, \ldots, y_n \rangle$, and $a \in R$ we define $x + y = \langle x_1 + y_1, \ldots, x_n + y_n \rangle$, $ax = \langle ax_1, \ldots, ax_n \rangle$, and $\theta = \langle 0, 0, \ldots, 0 \rangle$. R^n is a normed space under each of the following definitions of a norm (exercise):

(a) $\|x\|_1 = \sum_{i=1}^{n} |x_i|$,
(b) $\|x\|_\infty = \sup\{|x_i| : 1 \le i \le n\}$

2. *The space l_1.* The space R^∞ of infinite sequences of real numbers is a vector space with the following definitions of addition and scalar multiplication: If $x = \langle x_1, x_2, \ldots \rangle$, $y = \langle y_1, y_2, \ldots \rangle$, and $a \in R$, we define $x + y = \langle x_1 + y_1, x_2 + y_2, \ldots \rangle$ and $ax = \langle ax_1, ax_2, \ldots \rangle$ (check). Let l_1 be the set of elements $x = \langle x_1, x_2, \ldots \rangle$ in R^∞ for which $\|x\|_1 = \sum_{i=1}^{\infty} |x_i|$ is finite. Then l_1 is a linear subspace of R^∞ and $\| \ \|_1$ is a norm on l_1 (regarded as a vector space). For example, to check the triangle inequality 4.2(b) and the fact that l_1 is closed under $+$, we have (with $x = \langle x_1, x_2, \ldots \rangle$ and $y = \langle y_1, y_2, \ldots \rangle$)

$$\sum_{i=1}^{n} |x_i + y_i| \le \sum_{i=1}^{n} |x_i| + \sum_{i=1}^{n} |y_i| \le \|x\|_1 + \|y\|_1,$$

and the results follow by taking the limit as $n \to \infty$ on the left. Properties 4.2(a) and 4.2(c) are immediate. Finally we show that l_1 is complete and so is a Banach space. Let $\langle x^k : k \in N \rangle$ be a Cauchy sequence in l_1 with $x^k = \langle x_1^k, x_2^k, \ldots \rangle$. Then given $\varepsilon > 0$ there is an $n \in N$ so that $\|x^k - x^l\|_1 \le \varepsilon$ if $k, l \ge n$. Since Cauchy sequences are bounded there exists a number A so that $\|x^k\|_1 \le A$ for all $k \in N$. Let ω be an infinite integer; by transfer we have $*\|x^\omega\|_1 \le A$. Now $|x_i^k| \le \|x^k\|_1$ for all k, and so by transfer $|x_i^\omega| \le A$. Let $x_i = \operatorname{st}(x_i^\omega)$. We will show that $x = \langle x_i \rangle \in l_1$ and $\langle x^k \rangle$ converges to x. For any k and L we have

$$\sum_{i=1}^{L} |x_i| \le \sum_{i=1}^{L} |x_i - x_i^k| + \|x^k\|_1,$$

and so by transfer

$$\sum_{i=1}^{L} |x_i| \le \sum_{i=1}^{L} |x_i - x_i^\omega| + *\|x^\omega\|_1 \le \text{infinitesimal} + A \le 2A.$$

This shows that $x \in l_1$. Finally, for any k, l, and L,

$$\sum_{i=1}^{L} |x_i - x_i^l| \le \sum_{i=1}^{L} |x_i - x_i^k| + \sum_{i=1}^{L} |x_i^k - x_i^l| \le \sum_{i=1}^{L} |x_i - x_i^k| + \|x^k - x^l\|_1.$$

By transfer, with $k = \omega$, we have

$$\sum_{i=1}^{L} |x_i - x_i^l| \leq \sum_{i=1}^{L} |x_i - x_i^\omega| + *\|x^\omega - x^l\|_1 \leq \text{infinitesimal} + *\|x^\omega - x^l\|_1.$$

The right-hand side is $\leq 2\varepsilon$ if $l \geq n$. Since this is true for any $L \in N$, we conclude that $\|x - x^l\|_1 \leq 2\varepsilon$ if $l \geq n$.

3. The space l_∞ is a Banach space under the norm defined by $\|x\|_\infty = \sup\{|x_i| : i \in N\}$, where $x = \langle x_1, x_2, \ldots \rangle$ (Exercise III.3.14).

4. *The space c_0.* The space c_0 consists of those $x = \langle x_i : i \in N \rangle \in l_\infty$ for which $\lim_{n \to \infty} x_n = 0$. It is easy to see that c_0 is a closed linear subspace of l_∞ and hence a Banach space.

5. *The spaces $B(S)$ and $C(S)$.* Let S be an arbitrary set. We denote by $B(S)$ the set of all bounded functions on S. Then $B(S)$ is a vector space with the usual definitions of addition and scalar multiplication of functions, that is, if $f, g \in B(S)$ and $a \in R$, we put $(f + g)(x) = f(x) + g(x)$ and $(af)(x) = af(x)$ for $x \in S$; we take θ to be the function that is identically zero. $B(S)$ is a Banach space under the norm defined by $\|f\|_\infty = \sup\{|f(x)| : x \in S\}$ (Exercise 3).

If S is a topological space we define $C(S)$ to be the subset of $B(S)$ consisting of continuous functions. Then $C(S)$ is a closed subspace of $B(S)$ (Exercise 4), and hence a Banach space.

Let $(X, \| \ \|)$ be a normed space. From now on we will follow the usual convention of denoting the $*$-transform of the norm $\| \ \|$ on $*X$ by $\| \ \|$ rather than $*\| \ \|$; the context will clear up any possible confusion. We see immediately that the (norm) *monad* of a point $x \in *X$ is the set $m(x) = \{y \in *X : \|y - x\| \simeq 0\}$. It is also almost immediate that $m(x) = \{y \in *X : y = x + z, z \in m(\theta)\}$, so that all monads are translates of the monad about zero (Exercise 5). The *finite* points in $*X$ (Definition 3.3) are those $x \in *X$ for which $\|x\|$ is finite.

Next we come to the basic notion of linear operator.

4.4 Definition Let X and Y be vector spaces. A map $T: X \to Y$ is called a *linear operator* if $T(ax + by) = aTx + bTy$ for all $a, b \in R$ and $x, y \in X$. The set of all such linear operators is denoted by $L(X, Y)$.

Let X and Y be normed vector spaces. (Since there is no possibility of confusion we denote the norms and zeros on both by $\| \ \|$ and θ, respectively.) A linear operator $T: X \to Y$ is *bounded* if the number $\|T\| = \sup\{\|Tx\| : \|x\| \leq 1\}$ is finite. This number is called the *norm* of T. Then $\|Tx\| \leq \|T\| \|x\|$ for all $x \in X$ (check). The set of all bounded linear operators $T: X \to Y$ is denoted by $B(X, Y)$.

If $Y = R$ (with the usual operations of addition and multiplication and usual norm) then a linear operator T is called a *linear functional*. In what follows, we will often write x and $T(x)$ for the nonstandard extensions $*x$ and $*T(x)$ of x and $T(x)$; $*x$ and $*T(x)$ may, however, have nonstandard elements.

4.5 Example Define a map $T: l_1 \to l_1$ as follows: if $x = \langle x_1, x_2, x_3, \ldots \rangle$ then $Tx = \langle 0, x_1, x_2, \ldots \rangle$. Then T is linear, one-to-one, and bounded (in fact $\|Tx\| = \|x\|$ for all $x \in l_1$). However, T does not map l_1 onto l_1.

4.6 Theorem (Robinson) Let $T \in L(X, Y)$, where X and Y are normed spaces. The following are equivalent:

 (i) T is bounded.
 (ii) $*T: *X \to *Y$ takes finite points to finite points.
 (iii) $*T$ takes the monad of θ into the monad of θ.
 (iv) $*T$ takes near-standard points to near-standard points. In fact, if $z \in *X$ is near $x \in X$ then $*Tz$ is near Tx.

Proof: (i) $\Rightarrow$ (ii): Suppose $\|Tx\| \leq M\|x\|$ for all $x \in X$. By transfer $\|*Tx\| \leq M\|x\|$ for all $x \in *X$ and (ii) follows.

(ii) $\Rightarrow$ (iii): Proceed by contradiction. Suppose $x \in m(\theta)$ but $\|*Tx\| \not\simeq 0$. Then the element $z = x/\|x\| \in *X$ is finite with norm 1 (here and in the following we use freely the transfers of the properties in Definitions 4.1, 4.2, and 4.4) but $*Tz = (1/\|x\|)*Tx$ is not finite since $\|x\| \simeq 0$ but $\|*Tx\| \not\simeq 0$.

(iii) $\Rightarrow$ (iv): Let $x \in X$ and $z \in m(x)$, so $x - z \in m(\theta)$. Then $*T(x - z) = Tx - *Tz \in m(\theta)$, so $*Tz$ is near Tx.

(iv) $\Rightarrow$ (i): Proceed by contradiction. If T is not bounded then there exists a sequence $\langle x_n \in X : n \in N \rangle$ so that $\|x_n\| = 1$ but $\|Tx_n\| > n$ for $n \in N$ (check). Then $\|*Tx_\omega\|$ is infinite for some infinite natural number ω. Now $z = x_\omega/\sqrt{\|*Tx_\omega\|}$ is near-standard since it belongs to $m(\theta)$, but $\|*Tz\| = \sqrt{\|*Tx_\omega\|}$ is not finite, so z cannot be near-standard. $\square$

It is easy to see that a linear operator is continuous if and only if it is continuous at θ (Exercise 6). Therefore we have the following result.

4.7 Corollary $T \in L(X, Y)$ is bounded iff it is continuous.

Proof: Use 4.6(iii) and 1.15. $\square$

4.8 Corollary If $T \in B(X, Y)$, then the null space $N(T) = \{x \in X : Tx = \theta\}$ is a closed linear subspace of X.

Proof: Exercise. $\square$

One of the most important results concerning bounded linear operators on Banach spaces is the uniform boundedness theorem. The proof is entirely standard.

4.9 Uniform Boundedness Theorem Let X be a Banach space, Y a normed vector space, and $\mathscr{F} \subset L(X, Y)$ a family of bounded linear operators. Suppose that for each $x \in X$ there is a constant M_x so that $\|Tx\| \leq M_x$ for all $T \in \mathscr{F}$. Then there is a constant M so that $\|T\| \leq M$ for all $T \in \mathscr{F}$, i.e., the operators in $\mathscr{F}$ are uniformly bounded.

Proof: Suppose that $T \in L(X, Y)$. Note that if $\|Tx\| \leq M$ for all x in the closed ball $\overline{B_\varepsilon(x_0)} = \{x \in X : \|x - x_0\| \leq \varepsilon\}$ then T is bounded and $\|T\| \leq 2M/\varepsilon$. The proof of this fact is left to the reader.

Now we proceed by contradiction. Let $x_0 \in X$ and $\varepsilon_0 > 0$ be given. Then there is an $x_1 \in B_{\varepsilon_0}(x_0)$ and a $T_1 \in \mathscr{F}$ so that $\|T_1 x_1\| > 1$. For otherwise $\|Tx\| \leq 1$ for all $x \in \overline{B_{\varepsilon_0}(x_0)}$ and all $T \in \mathscr{F}$, and then $\|T\| \leq 2/\varepsilon_0$ for all $T \in \mathscr{F}$ by the remark in the first paragraph. By continuity we can find an ε_1 with $0 < \varepsilon_1 < \frac{1}{2}$, and $B_{\varepsilon_0}(x_0) \supseteq B_{\varepsilon_1}(x_1)$ so that $\|T_1 x\| > 1$ for all $x \in \overline{B_{\varepsilon_1}(x_1)}$. Inductively we can find a sequence $\{B_{\varepsilon_n}(x_n) : n \in N\}$ with $B_{\varepsilon_n}(x_n) \supseteq B_{\varepsilon_{n+1}}(x_{n+1})$ and $\lim_{n \to \infty} \varepsilon_n = 0$, and a sequence $T_n \in \mathscr{F}$ so that $\|T_n x\| \geq n$ for all $x \in \overline{B_{\varepsilon_n}(x_n)}$. Now $\langle x_n \rangle$ is a Cauchy sequence since $\lim_{n \to \infty} \varepsilon_n = 0$. Let $x \in X$ be the limit of $\langle x_n \rangle$ (here we use the completeness of X). Then $x \in \overline{B_{\varepsilon_n}(x_n)}$, so $\|T_n x\| > n$ for all n, contradicting the assumption. $\square$

As a corollary we can prove the following result.

4.10 Theorem Let X be a Banach space and Y a normed vector space, and suppose that $\langle T_n : n \in N \rangle$ is a sequence in $B(X, Y)$ such that for each $x \in X$ there is an element y_x with $\lim_{n \to \infty} T_n x = y_x$ (limit in norm). Then the mapping T given by $Tx = y_x$ is in $B(X, Y)$.

Proof: An easy exercise shows that the map $T : X \to Y$ is linear. Since $\| \ \|$ is a continuous function, $\lim_{n \to \infty} \|T_n x\| = \|Tx\|$, and thus for each x there exists an M_x so that $\|T_n x\| \leq M_x$ for all n. By the uniform boundedness theorem there is an $M \in N$ with $\|T_n\| \leq M$ for all $n \in N$, so $\|Tx\| = \lim \|T_n x\| \leq M\|x\|$ and T is bounded. $\square$

Next we study an important class of bounded linear operators, the compact operators. These operators occur in many applications. There is an extensive analysis of equations in Banach spaces involving these operators; it is called the Fredholm theory.

4.11 Definition Let X and Y be normed vector spaces. An operator $T \in L(X, Y)$ is *compact* if $\overline{T[B]}$ is compact for every norm-bounded set $B \subset X$.

4.12 Theorem (Robinson) $T \in L(X, Y)$ is compact iff $^*T$ takes finite points to near-standard points.

Proof: Suppose T is compact and let $x \in {}^*X$ be finite, i.e., $\|x\| < M$ for some $M > 0$. The ball $B = \{x \in X : \|x\| \leq M\}$ is bounded and so $\overline{T[B]}$ is compact. Thus every point of $^*(\overline{T[B]}) = {}^*\overline{T[^*B]}$ is near-standard by Robinson's theorem, 2.2. Since $x \in {}^*B$ we conclude that $^*Tx$ is near-standard.

Conversely, suppose that $^*T$ maps finite points into near-standard points, and let B be a bounded set. By Theorem 2.4 we need only show that $T[B] \subseteq K$ for some compact set K. Let $K = \{y \in Y : y \simeq y'$ for some $y' \in {}^*(T[B])\} = \mathrm{st}(^*T[^*B])$. Then $T[B] \subseteq K$ and K is compact by Exercise III.3.11. $\square$

We see immediately from 4.6 and 4.12 that compact operators are bounded. Theorem 4.12 can be used to establish the compactness of many operators, as the following example shows.

4.13 Example: Integral Operators (Robinson [42, Theorem 7.1.7]) Let $T: C([0,1]) \to C([0,1])$ be defined by

$$Tf(x) = \int_0^1 K(x, y) f(y)\, dy,$$

where $K(x, y)$ is a continuous function on $[0,1] \times [0,1]$. The reader should check that T is a linear operator. To show that Tf is continuous notice that if $|f(x)| \leq M$ for all $x \in [0,1]$ then

$$(4.1) \qquad |Tf(x) - Tf(y)|$$

$$\leq \int_0^1 |K(x,t) - K(y,t)|\, |f(t)|\, dt$$

$$\leq M \max\{|K(x,t) - K(y,t)| : (x,t), (y,t) \in [0,1] \times [0,1]\},$$

and $\max|K(x,t) - K(y,t)|$ can be made as small as desired if $|x - y|$ is sufficiently small by the uniform continuity of $K(x,t)$. Also note that $|K(x,t)| \leq K$ for all $(x,t) \in [0,1] \times [0,1]$ for some constant K, and so, for any $x \in [0,1]$,

$$(4.2) \qquad\qquad |Tf(x)| \leq K \max\{|f(t)| : t \in [0,1]\}.$$

To show that T is compact we need to show that $^*Tf$ is near-standard for each finite f. Let $f \in {}^*C([0,1])$ be finite. This means that there is a finite standard M so that $|f(t)| \leq M$ for all $t \in {}^*[0,1]$.

From the transfer of (4.2) we see that $|{}^*Tf(x)| \le KM$ for all $x \in {}^*[0,1]$, i.e., ${}^*Tf$ is finite, and we may define a function ψ on $[0,1]$ by $\psi(x) = \mathrm{st}({}^*Tf(x))$, $x \in [0,1]$. To complete the proof we will show that ψ is continuous and ${}^*Tf$ is near ${}^*\psi$. From the transfer of (4.1) we have

$$|{}^*Tf(x) - {}^*Tf(y)|$$
$$\le M \max\{|{}^*K(x,t) - {}^*K(y,t)|:(x,t),(y,t) \text{ in } {}^*[0,1] \times {}^*[0,1]\}.$$

Thus ${}^*Tf(x) \simeq {}^*Tf(y)$ whenever $x, y \in {}^*[0,1]$ and $x \simeq y$ by the uniform continuity of $K(x,t)$ (Theorem 10.10 and Proposition 10.8 of Chapter I). Let $\varepsilon > 0$ be a fixed standard real, and let $D = \{\delta \in {}^*R, \delta > 0 : x, y \in {}^*[0,1]$ and $|x - y| < \delta$ implies $|{}^*Tf(x) - {}^*Tf(y)| < \varepsilon\}$. Then D contains all positive infinitesimals by the above remark, and so contains a standard $\delta > 0$ by Corollary 7.2(iii) of Chapter II. Now if $x, y \in [0,1]$ then

$$|\psi(x) - \psi(y)| \le |\psi(x) - {}^*Tf(x)| + |{}^*Tf(x) - {}^*Tf(y)| + |{}^*Tf(y) - \psi(y)|.$$

The first and last terms are infinitesimal, so that $|\psi(x) - \psi(y)| < 2\varepsilon$ if the δ is chosen as above; thus ψ is continuous. To show that ${}^*\psi$ is near ${}^*Tf$ notice that ${}^*\psi(x) \simeq {}^*Tf(x)$ for all standard x by the definition of ψ and the fact that ${}^*\psi$ is an extension of ψ. If $x \in {}^*[0,1]$ then

$$|{}^*Tf(x) - {}^*\psi(x)|$$
$$\le |{}^*Tf(x) - {}^*Tf(^\circ x)| + |{}^*Tf(^\circ x) - {}^*\psi(^\circ x)| + |{}^*\psi(^\circ x) - {}^*\psi(x)|,$$

and all terms on the right are infinitesimal by the preceding remarks and the continuity of ψ.

A word of caution here. The reader may think that the above proof is needlessly complicated since we could replace ψ by $\hat\psi(x) = \mathrm{st}(Tf(x))$ for all x in ${}^*[0,1]$ rather than $[0,1]$, in which case it would be obvious that $\hat\psi$ is near Tf. Unfortunately the $\hat\psi$ defined this way is usually external and thus not a standard element in ${}^*C([0,1])$. Notice also that an internal finite $f \in C([0,1])$ can be quite wild; e.g., $f(x) = \sin \omega x$, where ω is infinite.

The set of bounded linear operators can be made into a linear space in an obvious way. If $T, S \in B(X, Y)$ and $a \in R$ we define $(T + S)(x) = T(x) + S(x)$ and $(aT)(x) = aT(x)$. It is then not hard to see that the operator norm on $B(X, Y)$ makes $B(X, Y)$ into a normed vector space (Exercise 8).

4.14 Theorem Let X be a normed vector space and Y a Banach space. Then the normed vector space $B(X, Y)$ is complete and hence a Banach space. The set of compact operators in $B(X, Y)$ forms a closed linear subspace.

Proof: Let $\langle T_n \in B(X, Y) : n \in N \rangle$ be a Cauchy sequence. Then, for each $x \in X$, $T_n x$ is a Cauchy sequence and hence converges to an element y_x by completeness of Y. We define T by $Tx = \lim T_n x$. Then T is linear (check) and bounded since $\lim \|T_n\| = \|T\|$ (check). Finally, we show that T_n converges to T in norm. For given $\varepsilon > 0$ there is an N so that $\|T_n x - T_m x\| \le \|T_n - T_m\| \|x\| < \varepsilon \|x\|$ if $n, m \ge N$. Thus $\|T_n x - Tx\| \le \varepsilon \|x\|$ if $n \ge N$, and so $\|T_n - T\| \le \varepsilon$ for $n \ge N$, and we are through.

An easy exercise shows that the set of compact operators is a linear subspace of $B(X, Y)$. To show that it is closed, let $\langle T_n \rangle$ be a sequence of compact operators converging to an operator $T \in B(X, Y)$. If $y \in {}^*X$ is finite then it belongs to ${}^*B$, where $B = \{x \in X : \|x\| \le M\}$ for some standard real $M > 0$. Now note that $T_n x$ converges to Tx uniformly on the ball B, i.e., for any $\varepsilon > 0$ there is an $m(\varepsilon) \in N$ so that $\|T_n x - Tx\| < \varepsilon$ for all $n \ge m(\varepsilon)$ and all $x \in B$. Thus $\|{}^*T_{n_0} x - {}^*Tx\| < \varepsilon/2$ for $n_0 \ge m(\varepsilon/2)$ in N and all $x \in {}^*B$. Since T_{n_0} is compact, ${}^*T_{n_0} y$ is near a standard $z \in Y$ and so $\|{}^*Ty - z\| < \varepsilon$ by the triangle inequality. Since ε is arbitrary, ${}^*Ty$ is pre-near-standard. Since Y is complete, it follows from Proposition 3.14 that ${}^*Ty$ is near-standard. $\square$

The standard proof of the closedness of the set of compact operators usually involves the selection of infinite subsequences with certain desirable properties.

4.15 Corollary The space of bounded linear functionals on a normed vector space X is a Banach space.

The Banach space of this corollary is used sufficiently often for us to introduce some notation.

4.16 Definition The Banach space of bounded linear functionals on a normed linear space X is called the *dual space* of X and is denoted by X'. The dual of X' is denoted by X'' and is called the *second dual* of X. Similarly for X''', etc.

It is sometimes difficult to characterize the dual of a given Banach space, but the following example is an easy case.

4.17 Example: $l_1' = l_\infty$ Our aim is to define a mapping $T: l_\infty \to l_1'$ which is linear, 1–1, onto, and satisfies $\|Ty\| = \|y\|_\infty$ for $y \in l_\infty$. Let $y = \langle y_i : i \in N \rangle \in l_\infty$ and define $Ty: l_1 \to R$ by $Ty(x) = \sum_{i=1}^\infty x_i y_i$ for $x = \langle x_i \rangle \in l_1$. Then Ty is linear, and

$$|Ty(x)| \le \sup\{|y_i| : i \in N\} \sum_1^\infty |x_i| = \|y\|_\infty \|x\|_1,$$

so Ty is a bounded linear functional on l_1 with $\|Ty\| \le \|y\|_\infty$. We next show that $\|Ty\| \geq \|y\|_\infty$. We may assume $\|y\|_\infty > 0$. Given a positive $\varepsilon < \|y\|_\infty$, there is an n_0 so that $|y_{n_0}| > \|y\|_\infty - \varepsilon$. Now define $x = \langle x_i \rangle \in l_1$ by $x_i = 0$ for $i \neq n_0$ and $x_{n_0} = y_{n_0}/|y_{n_0}|$. Then $\|x\|_1 = 1$ and $|Ty(x)| = |y_{n_0}| > \|y\|_\infty - \varepsilon$, so $\|Ty\| \geq \|y\|_\infty$. We also see that T is 1–1, since if $Ty = \theta$ then $\|y\|_\infty = 0$ so $y = \theta$.

It only remains to show that T is onto. Let $f \in l_1'$. If $e^n \in l_1$ is defined by $e^n = \langle \delta_i^n \rangle$, where $\delta_i^n = 0$ if $i \neq n$ and $\delta_n^n = 1$, then $\|e^n\|_1 = 1$ for all $n \in N$. Put $f(e^n) = y_n \in R$. Then $|y_n| \leq \|f\|$, and so $y = \langle y_i \rangle \in l_\infty$. Now the functional Ty attached to y as in the first paragraph agrees with f on the elements e^n. A simple limiting argument (check) shows that $Ty = f$, and so $l_1' = l_\infty$.

In the case of a general normed vector space X, it is not at all obvious that X' contains any elements other than θ. The following result, which is basic to the study of duality, shows that X' always contains many elements.

4.18 Hahn–Banach Theorem Let X be a vector space and suppose that a given function $p: X \to R$ satisfies $p(x + y) \leq p(x) + p(y)$ and $p(ax) = ap(x)$ for each $a \geq 0 \in R$ and $x, y \in X$. Suppose that f is a linear functional defined on a subspace S of X with $f(x) \leq p(x)$ for all $x \in S$. Then there is a linear functional F on X which extends f [i.e., $F(x) = f(x)$ for all $x \in S$] and satisfies $F(x) \leq p(x)$ for $x \in X$.

Proof: Let g and h be linear functionals, each defined on a linear subspace of X. We say that g extends h and write $h \prec g$ if the domain of g contains the domain of h and $g = h$ on dom h. The relation $\prec$ partially orders the set of linear functionals.

Consider the set of all extensions g of f which satisfy $g(x) \leq p(x)$, for x in the domain of g. Applying Zorn's lemma (see the Appendix) to this set, partially ordered by $\prec$, we see that there is a maximal extension F. We need only show that the domain X_0 of F is all of X. Suppose this is not the case, i.e., there is a vector y in X but not in X_0. Then F may be extended to a functional g on the subspace $\hat{X} \supset X_0$ consisting of elements of the form $ay + x_0, x_0 \in X_0, a \in R$, by putting $g(ay + x_0) = ag(y) + F(x_0)$. Now g is specified uniquely by $g(y)$, and we need to show that $g(y)$ can be chosen so that $g(x) \leq p(x)$ for all $x \in \hat{X}$ in order to get a contradiction. For $x_1, x_2 \in X_0$ we have $F(x_2) - F(x_1) = F(x_2 - x_1) \leq p(x_2 - x_1) \leq p(x_2 + y) + p(-y - x_1)$, which yields $-p(-y - x_1) - F(x_1) \leq p(x_2 + y) - F(x_2)$. Since the left is independent of x_2 and the right is independent of x_1 there is a constant $c \in R$ so that

(i) $c \leq p(x_2 + y) - F(x_2)$,
(ii) $-p(-y - x_1) - F(x_1) \leq c$

for all $x_1, x_2 \in X_0$. We now put $g(y) = c$. Then for $x = ay + x_0 \in \hat{X}$ the inequality $g(x) = g(ay + x_0) = ac + F(x_0) \le p(ay + x_0)$ follows by replacing x_2 by x_0/a in (i) if $a > 0$ and x_1 by x_0/a in (ii) if $a < 0$. $\square$

4.19 Corollary If X is a normed vector space and $x \in X$, $x \ne \theta$, then there is an $x' \in X'$ so that $x'(x) = \|x\|$ and $\|x'\| = 1$.

Proof: Standard exercise. $\square$

We now show that X can be isometrically and isomorphically embedded in X''.

4.20 Theorem Let X be a normed vector space and define a map $T: X \to X''$ by $Tx(x') = x'(x)$ for all $x' \in X'$. Then T is a linear and norm-preserving embedding. If X is a Banach space then $T[X]$ is a closed linear subspace of X''.

Proof: The reader should check that T is linear. That Tx is bounded (as we have implied in the statement of the theorem) follows since $|Tx(x')| = |x'(x)| \le \|x\| \|x'\|$, and we see that $\|Tx\| \le \|x\|$. The result will be established when we show that $\|Tx\| \ge \|x\|$. This is trivial if $x = \theta$, so suppose $x \ne \theta$. From Corollary 4.19 there exists an $x' \in X'$ so that $\|x'\| = 1$ and $x'(x) = \|x\|$. Thus $\|x\| = |x'(x)| = |Tx(x')| \le \|Tx\| \|x'\| = \|Tx\|$. The rest is left to the reader. $\square$

Because of Theorem 4.20 we identify X with $T[X]$ and regard X as a subspace of X'' in the rest of this section without further explicit comment.

We end this section with a consideration of compactness properties in Banach spaces. We have seen in Example 3.22 that the closed unit ball in l_∞ is not norm-compact. This situation turns out to be typical of all infinite-dimensional spaces. In fact one can prove that a closed ball in a Banach space is norm-compact iff the space is finite-dimensional [14, Theorem IV.3.5]. It follows that no set in an infinite-dimensional Banach space X containing a closed ball can be norm-compact. Since this severely limits the sets which can be norm-compact we look for other topologies on a Banach space in which closed balls are compact.

4.21 Definition Let X be a normed vector space. The *weak topology* on X is the topology whose neighborhood system at a generic point $x \in X$ is generated by the subbase consisting of sets of the form $U(x; x', \varepsilon) = \{y \in X : |x'(y) - x'(x)| < \varepsilon\}$ for some $x' \in {}^*X$.

Let X' be the dual space of a normed vector space X. The *weak* topology* on X' is the topology whose neighborhood system at a generic point

$x' \in X'$ is generated by the subbase consisting of sets of the form $V(x'; x, \varepsilon) = \{y' \in X': |x(y') - x(x')| < \varepsilon\}$ for some $x \in X$ (regarded as embedded in X'').

Notice that in the definition of the subbase for the weak* topology we take only those $x \in X$ and not all $x'' \in X''$. This turns out to make a crucial difference. An easy exercise, which we leave to the reader, shows that the monads of points $x \in X$ and $x' \in X'$ in the weak and weak* topologies, respectively, are given by

$$m_w(x) = \{y \in {}^*X : {}^*x'(y) \simeq {}^*x'({}^*x) = x'(x) \text{ for all (standard) } x' \in X'\},$$
$$m_{w*}(x') = \{y' \in {}^*X' : {}^*x(y') \simeq {}^*x({}^*x') = x(x') \text{ for all (standard) } x \in X\}.$$

Using the Hahn–Banach theorem, we can show that the weak and weak* topologies are Hausdorff (exercise).

4.22 Alaoglu's Theorem The closed unit ball in X' is compact in the weak* topology.

Proof: Let B be the unit ball in X'. We must show that corresponding to every $y' \in {}^*B$ there is a point $x' \in B$ so that ${}^*x(y') \simeq x(x')$ for all $x \in X$. Fix $y' \in {}^*B$ and define a functional x' on X by $x'(x) = \mathrm{st}(y'({}^*x))$, $x \in X$. Then ${}^*x(y') \simeq x(x')$ for all standard $x \in X$. The linearity of x' is obvious, and, finally, $x' \in B$ since $|x'(x)| \leq {}^\circ(\|y'\| \|{}^*x\|) \leq \|x\|$ by transfer ($y' \in {}^*B$ so $\|y'\| \leq 1$). $\square$

The same result can be proved for a ball of any radius and also follows directly from Theorems 4.22 and 2.6. We obtain as a consequence the following corollary.

4.23 Corollary A norm-bounded and weak*-closed subset of X' is compact.

Proof: Use Theorems 4.22 and 2.4. $\square$

One might expect a similar result to be true for subsets of X in the weak topology. However, it turns out that the unit ball in X is weakly compact iff X is *reflexive*, which means that $X = X''$ [14, Theorem V.4.7].

Considering the importance of sequential compactness as emphasized in §III.3, we would like to know when the unit ball B in a Banach space X is weakly sequentially compact. A deep theorem due to Eberlein and Šmulian asserts that B is weakly sequentially compact iff B is weakly compact (iff X is reflexive by the above remark). A nonstandard proof of this result can be found in [47].

4.24 Example We will show that the unit sphere in l_∞ is not weak* sequentially compact even though it is weak* compact by Alaoglu's theorem. Consider the sequence $e^n \in l_1$ (regarded as embedded in l'_∞) defined by $e^n = \langle \delta_n^i : i \in N \rangle$. Then $\|e^n\|_1 = 1$. Suppose that $\langle e^n \rangle$ has a convergent subsequence $\langle e^{n_k} \rangle$. Define the element $x = \langle x_i : i \in N \rangle \in l_\infty$ by $x_i = 1$ if $i = n_k$ and k is even, and $x_i = 0$ otherwise. Then $e^{n_k}(x) = 1$ if k is even, and 0 if k is odd, so the sequence $\langle e^{n_k}(x) \rangle$ does not converge, i.e., $\langle e^{n_k} \rangle$ does not converge in the weak* topology. Note that by compactness (check) the sequence e^n has a weak* limit point y, but we cannot select a convergent subsequence since the neighborhood system at y does not have a countable base.

An extensive study of the structure of Banach spaces using nonstandard methods has been developed by Henson and Moore [16]. This study uses in an essential way the notion of the nonstandard hull of a Banach space. We present the definition of the nonstandard hull of a metric space in §III.6 to help the interested reader to understand these results.

Exercises III.4

1. Show that $d(x, y) \equiv \|x - y\|$ is a metric.
2. Show that $\| \ \|_1$ and $\| \ \|_\infty$ are norms on R^n.
3. Show that $B(S)$ with the sup norm $\| \ \|_\infty$ is a Banach space.
4. Show that $C(S)$ is a closed subspace of $B(S)$ if S is a topological space.
5. Show that for a normed space all monads are translates of the monad of zero.
6. Show that a linear operator is continuous if and only if it is continuous at θ.
7. Prove Corollary 4.8.
8. Show that the operator norm on $B(X, Y)$ makes $B(X, Y)$ into a normed vector space.
9. Show that the set of compact operators is a linear subspace of $B(X, Y)$.
10. Show that the weak and weak* topologies are Hausdorff.
11. Discuss the relationship between Alaoglu's theorem and the Tychonoff product theorem.
12. Two norms on a space X are equivalent if the corresponding metrics they define are equivalent.

    (a) Show that the norms $\|\cdot\|$ and $\|\|\cdot\|\|$ on X are equivalent iff there exist positive (nonzero) constants α and β in R so that $\alpha\|x\| \leq \|\|x\|\| \leq \beta\|x\|$ for all $x \in X$.
    (b) Show that any two norms on R^n are equivalent. (Hint: Show that any norm $\| \ \|$ is equivalent to $\|\cdot\|_\infty$. To do so you need only show that $\|\|x\|\|/\|x\|_\infty$ and $\|x\|_\infty/\|\|x\|\|$ are finite for all $x \in {}^*R^n$. Write $x = \sum_{i=1}^n x_i e_i$ and get estimates.)

13. Let X be a vector space with a topology $\mathcal{T}$. X is a *topological vector space* if both vector addition (as a map $X \times X \to X$) and scalar multiplication (as a map $R \times X \to X$) are continuous. Let $m(a)$ denote the monad of $a \in R$ and $\mu(x)$ denote the monad of $x \in X$. Show that if X is a topological vector space (of more than one dimension) then

    (a) $\mu(x) + \mu(y) = \mu(x) + y = \mu(x + y) = x + y + \mu(\theta)$,
    (b) $m(a)x \subset m(a)\mu(x) = a\mu(x) = \mu(ax)$,
    (c) $\mathcal{T}$ is Hausdorff iff $\mu(\theta) \cap X = \{\theta\}$,
    (d) if X is a topological vector space with topologies $\mathcal{T}_1$ and $\mathcal{T}_2$ having monads μ_1 and μ_2 then $\mathcal{T}_1 = \mathcal{T}_2$ iff $\mu_1(\theta) = \mu_2(\theta)$.

III.5 Inner-Product Spaces and Hilbert Spaces

In this section we consider those normed spaces and Banach spaces in which the norm is derived from an inner product. Most of the results and proofs of this section are standard. The canonical example of an inner product occurs in Euclidean space R^n where the scalar product of $x = \langle x_1, \ldots, x_n \rangle$ and $y = \langle y_1, \ldots, y_n \rangle$ is $(x, y) = \sum_1^n x_i y_i$. The angle θ between two nonzero vectors x and y is given by the familiar formula $\cos \theta = (x, y)/\|x\| \|y\|$. The scalar product is generalized to vector spaces as follows.

5.1 Definition Let H be a vector space. An *inner product* on H is a map $(,)$: $X \times X \to R$ which satisfies (for all x, y, z in X and $a, b \in R$)

    (i) $(x, y) = (y, x)$,
    (ii) $(ax + by, z) = a(x, z) + b(y, z)$,
    (iii) $(x, x) \geq 0$, and $(x, x) = 0$ iff $x = \theta$.

A vector space with an inner product is called an *inner product space*. A norm on H is obtained by setting $\|x\| = \sqrt{(x, x)}$ (exercise). If H is complete in this norm it is called a *Hilbert space*.

To prove that $\| \ \|$ is a norm on X one uses the following basic result.

5.2 Schwarz's Inequality For any x, y in an inner-product space H, $|(x, y)| \leq \|x\| \|y\|$.

    Proof: Let x and y be given. For any real λ we have $(x + \lambda y, x + \lambda y) = \|x\|^2 + 2\lambda(x, y) + \lambda^2 \|y\|^2 \geq 0$. Thus the quadratic expression in λ given by

$\|x\|^2 + 2\lambda(x, y) + \lambda^2\|y\|^2$ cannot have distinct real roots, and so the discriminant $|(x, y)|^2 - \|x\|^2\|y\|^2 \leq 0$.   □

5.3 Corollary (x, y) is continuous on $H \times H$ as a function of the variables $x, y \in H$.

Proof: Exercise.   □

5.4 Examples

1. In the linear space R^n we define the inner product of $x = \langle x_1, \ldots, x_n \rangle$ and $y = \langle y_1, \ldots, y_n \rangle$ by $(x, y) = \sum_{i=1}^n x_i y_i$. The reader should check that this defines an inner product on R^n. From Schwarz's inequality,

$$\left| \sum_{i=1}^n x_i y_i \right| \leq \left(\sum_{i=1}^n x_i^2 \right)^{1/2} \left(\sum_{i=1}^n y_i^2 \right)^{1/2}.$$

2. *The space l_2.* Let l_2 denote the space of all infinite sequences $x = \langle x_1, x_2, \ldots \rangle$ for which $\sum_{i=1}^\infty x_i^2 < \infty$. If $x = \langle x_1, x_2, \ldots \rangle$ and $y = \langle y_1, y_2, \ldots \rangle$ are two such sequences, we define $(x, y) = \sum_{i=1}^\infty x_i y_i$. To check that (x, y) is finite for $x, y \in l_2$ we have

$$\sum_{i=1}^n |x_i y_i| \leq \left(\sum_{i=1}^n x_i^2 \right)^{1/2} \left(\sum_{i=1}^n x_i^2 \right)^{1/2} \leq \left(\sum_{i=1}^\infty x_i^2 \right)^{1/2} \left(\sum_{i=1}^\infty y_i^2 \right)^{1/2},$$

and so $\sum_{i=1}^\infty |x_i y_i|$ converges. Using the fact that $(\sum_{i=1}^\infty x_i^2)^{1/2} = \|x\|$ is a norm, we can now easily check that l_2 is a linear space. We will see later that all separable Hilbert spaces are isomorphic to l_2.

Using the inner product, we can introduce a notion of orthogonality in an inner-product space.

5.5 Definition If H is an inner-product space then x and y in H are *orthogonal* if $(x, y) = 0$, in which case we write $x \perp y$. If $S \subseteq H$ then $S^\perp = \{x \in H : x \perp z$ for all $z \in S\}$.

5.6 Proposition For any $S \subseteq H$, $S^\perp$ is a closed linear subspace of H.

Proof: Let $x, y \in S^\perp$ and $a, b \in R$. Then, for any $z \in S$, $(ax + by, z) = a(x, z) + b(y, z) = 0$, so $S^\perp$ is a linear subspace. To show closure, let $x \in *S^\perp$

and $x \simeq y \in H$. Then $(y, z) \simeq (x, z) = 0$ for all $z \in S$ by the continuity of the inner product, and so $y \in S^\perp$. Thus $S^\perp$ is closed. □

Since the norm on an inner-product space H is derived from the inner product, we might expect that it has some special properties. It turns out that it is completely characterized by the following law.

5.7 Parallelogram Law A normed space $(H, \| \ \|)$ is an inner-product space iff for all $x, y \in H$

$$\|x - y\|^2 + \|x + y\|^2 = 2\|x\|^2 + 2\|y\|^2.$$

Proof: Suppose H is an inner-product space. Then

$$\|x - y\|^2 + \|x + y\|^2 = (x - y, x - y) + (x + y, x + y)$$
$$= \|x\|^2 - (y, x) - (x, y) + \|y\|^2 + \|x\|^2 + (y, x) + (x, y) + \|y\|^2$$
$$= 2\|x\|^2 + 2\|y\|^2.$$

The converse, which we omit, sets $(x, y) = \frac{1}{4}\{\|x + y\| - \|x - y\|\}$. □

Using this simple result, we now establish a sequence of results which are fundamental to all further analysis of Hilbert spaces.

5.8 Definition A subset K of a vector space H is *convex* if whenever $x, y \in K$ then $\alpha x + (1 - \alpha)y \in K$ for all real $\alpha \in [0, 1]$.

In the proof of the next result we use completeness in an essential way.

5.9 Theorem If K is a closed convex subset of a Hilbert space H, then there is a unique element $x_0 \in K$ so that $\|x_0\| \leq \|x\|$ for all $x \in K$, i.e., K has a unique element of smallest norm.

Proof: Let $d = \inf\{\|x\| : x \in K\}$. Then for each $\delta > 0$ there is an $x \in K$ so that $d \leq \|x\| < d + \delta$. By transfer, with δ infinitesimal, there is a $y \in {}^*K$ with $\|y\| \simeq d$. We now show that y is near-standard. Since K is complete by Corollary 3.15, it is enough to show that y is pre-near-standard (see Proposition 3.14). Fix $\varepsilon > 0$ in R. By transfer from the parallelogram law,

$$(5.1) \qquad \|x - y\|^2 + \|x + y\|^2 = 2\|x\|^2 + 2\|y\|^2$$

for any $x \in K$. If $x \in K$ then since $y \in {}^*K$, $(x + y)/2 \in {}^*K$, so $\|x + y\|^2 = 4\|(x + y)/2\|^2 \geq 4d^2$. It follows from (5.1) that $\|x - y\|^2 < 2\|x\|^2 + 2d^2 - 4d^2 + \eta = 2\|x\|^2 - 2d^2 + \eta$, where η is infinitesimal and $x \in K$. But we can find an $x \in K$ so that $\|x\|^2 < d^2 + \varepsilon/4$, and we get $\|x - y\|^2 < \varepsilon/2 + \eta < \varepsilon$.

Thus y is pre-near-standard, so y is near some $x_0 \in H$. The point $x_0 \in K$ since K is closed, and $\|x_0\| = d$ by the continuity of the norm. The uniqueness is another application of the parallelogram law (exercise). $\square$

5.10 Theorem Let E be a closed subspace of the Hilbert space H with $E \neq H$. There are unique linear operators $P: H \to E$, $Q: H \to E^{\perp}$ so that $x = Px + Qx$ for all $x \in H$. Further,

$$Px = x \quad \text{iff} \quad x \in E \quad \text{and} \quad Qx = x \quad \text{iff} \quad x \in E^{\perp}.$$

P and Q are called the *projections* of H onto E and $E^{\perp}$, respectively.

Proof: For $x \in H$ let $K = x + E = \{x + y : y \in E\}$. Then K is convex and closed (check). Let Qx be the unique element of smallest norm in K (existing by 5.9), and put $Px = x - Qx$. Then it is clear that $x = Px + Qx$ and $Px \in E$. To show that $(Qx, z) = 0$ for all $z \in E$, we put $Qx = y$. Assuming without loss of generality that $\|z\| = 1$, we have

$$\|y\|^2 \leq \|y - az\|^2 = (y - az, y - az) = \|y\|^2 - 2a(y, z) + |a|^2$$

for every $a \in R$, yielding $0 \leq -2a(y, z) + |a|^2$. If $a = (y, z)$ this gives $0 \leq -|(y, z)|^2$, and so $(y, z) = 0$. The uniqueness of P and Q follows from the fact that $E \cap E^{\perp} = \{\theta\}$. For if $x = x_1 + x_2$ with $x_1 \in E$, $x_2 \in E^{\perp}$, then $x_1 - Px = Qx - x_2$ and $x_1 - Px \in E$, $Qx - x_2 \in E^{\perp}$, so $x_1 = Px$ and $x_2 = Qx$. The rest of the proof is left to the reader. $\square$

The culmination of the preceding sequence of results is the following theorem, which probably has more applications than any other result on Hilbert spaces.

5.11 Riesz Representation Theorem To each bounded linear functional L on H there corresponds a unique element $y \in H$ so that $L(x) = (x, y)$ for each $x \in H$, and $\|L\| = \|y\|$.

Proof: We may assume that L is not identically zero (otherwise take $y = \theta$). Let $E = \{x \in H : Lx = 0\}$. Then E is a closed linear subspace (check) and $E^{\perp} \neq \{\theta\}$, so we may choose $z \neq \theta$ in $E^{\perp}$. Then, for any $x \in H$, $x - (Lx/Lz)z \in E$, so $(x, z) - (Lx/Lz)(z, z) = 0$. Thus $Lx = (x, [Lz/(z, z)]z)$, and we take $y = [Lz/(z, z)]z$. The rest is left as an exercise. $\square$

5.12 Corollary A Hilbert space H is self-dual; i.e., $H = H'$.

Next we investigate the generalization to Hilbert space of a familiar notion in R^n, that of an orthonormal basis. In R^n the vectors $e_1 = \langle 1, 0, 0, \ldots 0 \rangle$,

a given orthonormal sequence $S = \langle e_i : i \in N \rangle$. Then Bessel's inequality shows that this sequence is in l_2. Thus for a fixed countable orthonormal sequence $\langle e_i \rangle$ we obtain a mapping $T: H \to l_2$ defined by $Tx = \langle \hat{x}(i) : i \in N \rangle$. It is easy to check that T is a linear mapping. The next result, which requires that H be complete, shows that T maps H *onto* l_2.

5.18 Riesz–Fischer Theorem Let $\langle e_i : i \in N \rangle$ be a countable orthonormal sequence in the Hilbert space H. Then each element of l_2 is of the form $\hat{x}$ for some $x \in H$.

Proof: Let $\langle a_i : i \in N \rangle$ be a sequence in l_2 so that $\sum_{i=1}^{\infty} a_i^2 < \infty$. Then the sequence $x_n = \sum_{i=1}^{n} a_i e_i$ is a Cauchy sequence in H since $x_m - x_n = \sum_{i=n+1}^{m} a_i e_i$ if $m > n$, and so $\|x_m - x_n\|^2 = \sum_{i=n+1}^{m} a_i^2$. Since H is complete, there is an element x which is the (norm) limit of x_n. By the continuity of the inner product, $(x, e_i) = \lim(x_n, e_i) = a_i$ for any $i \in N$. $\square$

The element x which is obtained in Theorem 5.18 is often written $x = \sum_{i=1}^{\infty} a_i e_i$. One consequence of the next theorem is that if $\langle e_i : i \in N \rangle$ is a maximal orthonormal sequence, then the associated map T is one-to-one, and so an $x \in H$ can be written in only one way as $\sum_{i=1}^{\infty} a_i e_i$. For this reason a maximal orthonormal set (also called a *complete* orthonormal set) is sometimes called an *orthonormal basis*. It should be emphasized that this notion of basis must be understood in a limiting sense and not in the algebraic sense of vector space theory.

5.19 Theorem The orthonormal sequence $\langle e_i : i \in N \rangle$ in the Hilbert space H is maximal iff each $x \in H$ can be written uniquely as $x = \sum_{i=1}^{\infty} (x, e_i) e_i$.

Proof: Suppose $\langle e_i \rangle$ is maximal and $x \in H$. Then by the Riesz–Fischer theorem there is an element $y \in H$ so that $y = \sum_{i=1}^{\infty} (x, e_i) e_i$ and so $(y, e_i) = (x, e_i)$ for all $i \in N$. But then $(x - y, e_i) = 0$ for $i \in N$, so $x = y$ by the remark following Theorem 5.14.

Conversely, suppose each $x \in H$ can be written as $x = \sum_{i=1}^{\infty} (x, e_i) e_i$. If $\langle e_i \rangle$ is not maximal there is an $x \neq \theta$ in H so that $(x, e_i) = 0$ for $i \in N$. But then $x = \sum_{i=1}^{\infty} (x, e_i) e_i = \theta$ (contradiction). $\square$

5.20 Theorem   (Parseval's Identities)   If $\langle e_i : i \in N \rangle$ is a maximal orthonormal sequence in the Hilbert space H then

(i) $\sum_{i=1}^{\infty} |(x, e_i)|^2 = \|x\|^2$ for all $x \in H$;

(ii) $\sum_{i=1}^{\infty} (x, e_i)(y, e_i) = (x, y)$ for all $x, y \in H$.

Proof: We leave the proof of (ii) as an exercise. To prove (i) we see by Bessel's inequality that $\sum_{i=1}^{\infty} |(x, e_i)|^2 \leq \|x\|^2$. On the other hand, given $\varepsilon > 0$ there is a $z = \sum_{i=1}^{n} (x, e_i) e_i$ so that $\|x - z\| < \varepsilon$, whence $\|x\| < \|z\| + \varepsilon$. Thus

$$(\|x\| - \varepsilon)^2 \leq \|z\|^2 - \sum_{i=1}^{n} |(x, e_i)|^2 \leq \sum_{i=1}^{\infty} |(x, e_i)|^2,$$

and the result follows since ε is arbitrary. $\square$

The results above can now be used to show that l_2 is essentially the only separable infinite-dimensional Hilbert space.

5.21 Theorem Given a maximal orthonormal sequence $S = \langle e_i : i \in N \rangle$ in a separable Hilbert space H, the associated map $T: H \to l_2$ is one-to-one, onto, and satisfies $(x, y) = (Tx, Ty)$ for all $x, y \in H$, and so T is a Hilbert space isomorphism.

Proof: Use 5.18–5.20. $\square$

We end this section with an application of nonstandard analysis to prove a theorem concerning compact operators in Hilbert space. Much more can be done in this direction. In particular, Bernstein and Robinson [4] first proved that so-called polynomially compact operators have nontrivial invariant subspaces using refinements of the technique used here.

We are going to prove that every compact operator on a separable Hilbert space H can be approximated arbitrarily closely by an operator of "finite rank."

5.22 Definition An operator $Q: H \to H$ is of *finite rank* if there is a finite-dimensional subspace $E \subset H$ so that $Qx \in E$ for each $x \in H$.

Since every separable Hilbert space H is isomorphic to l_2 we will identify H and l_2 in the following discussion. Thus we will asume that an orthonormal sequence $\langle e_i \rangle$ is given and represent any $x \in H$ as either $x = \sum_{i=1}^{\infty} a_i e_i$ or $\langle a_1, a_2, \ldots \rangle$. First we need the following lemma.

5.23 Lemma If $x = \langle a_i : i \in {}^*N \rangle \in {}^*l_2$ is near-standard, then $\sum |a_i|^2 (i \in {}^*N, i \geq \omega)$ is infinitesimal for any $\omega \in {}^*N_\infty$.

Proof: If $y = \langle b_i : i \in N \rangle \in l_2$ then $\lim_{k \to \infty} \sum |b_i|^2 (i \in N, i \geq k) = 0$, so $\sum |{}^*b_i|^2 (i \in {}^*N, i \geq \omega) \simeq 0$ for any infinite ω. Now since $x \in {}^*l_2$ is near-standard there is a $y \in l_2$ with $\|x - {}^*y\|^2 = \sum |a_i - {}^*b_i|^2 (i \in {}^*N) \simeq 0$. By the transfer of the triangle inequality,

$$[\textstyle\sum |a_i|^2 (i \in {}^*N, i \geq \omega)]^{1/2}$$
$$\leq [\textstyle\sum |a_i - {}^*b_i|^2 (i \in {}^*N, i \geq \omega)]^{1/2} + [\textstyle\sum |{}^*b_i|^2 (i \in {}^*N, i \geq \omega)]^{1/2},$$

and both terms on the right are infinitesimal. $\square$

5.24 Theorem Let $T: H \to H$ be a compact linear operator. For each $\varepsilon > 0$ there is an operator Q of finite rank so that $\|T - Q\| < \varepsilon$.

Proof: For each $k \in {}^*N$ (finite or infinite) we define a projection operator $P_k: {}^*H \to {}^*H$ by $P_k x = \langle a_1, a_2, \ldots, a_k, 0, 0, \ldots \rangle$ when $x = \langle a_i : i \in {}^*N \rangle$. Then P_k is linear and $\|P_k x\| \leq \|x\|$ for any $x \in {}^*H$. Also, $\|(I - P_k)x\|^2 = \sum |a_i|^2 (i \in {}^*N, i \geq k + 1)$, and so, by Lemma 5.23, $\|(I - P_k)x\|$ is infinitesimal for k infinite and x near-standard. It follows that $\|{}^*T - P_k{}^*T\|$ is infinitesimal for all infinite k.

Now let $\varepsilon > 0$ in R be given. The internal set $A = \{n \in {}^*N : \|{}^*T - P_n{}^*T\| < \varepsilon\}$ contains all infinite natural numbers, and so contains a finite (standard) integer m by Corollary 7.2(ii) of Chapter II. Thus $\|{}^*T - P_m{}^*T\| < \varepsilon$. Transferring down shows that $\|T - P_m T\| < \varepsilon$. Finally, the operator $Q = P_m T$ is of finite rank since its range is contained in the subspace E generated by $\{e_1, \ldots, e_m\}$. $\square$

This result can be used as a starting point for the Fredholm theory of compact operators.

Exercises III.5

1. Show that if $(,)$ is an inner product on a vector space H then the map $\|\cdot\|: H \to R^+$ defined by $\|x\| = \sqrt{(x, x)}$ is a norm on H.
2. Prove Corollary 5.3
3. Show that the element x_0 of Theorem 5.9 is unique.
4. Complete the proof of Theorem 5.10.
5. Finish the proof of Theorem 5.11.
6. Show that S is a maximal orthonormal set iff $x \in H$ and $x \perp S \Rightarrow x = 0$.
7. Show that if any orthonormal set in an inner-product space H is either finite or countable, then H is separable.
8. Prove Theorem 5.17.
9. Prove Theorem 5.20(ii).
10. Establish the following converse to Lemma 5.23. If $x = \langle a_i : i \in {}^*N \rangle \in {}^*l_2$, $\|x\|^2 = \sum |a_i|^2 (i \in {}^*N)$ is finite, and $\sum |a_i|^2 (i \in {}^*N, i > \omega) \simeq 0$ for all infinite ω, then x is near-standard.

11. The Hilbert cube is the set of all $x = \langle x_i \rangle \in l_2$ such that $|x_i| \leq 1/i, i \in N$. Show that the Hilbert cube is compact.

12. Let H be a Hilbert space and let $B(H)$ denote the normed space of all bounded linear operators $A: H \to H$. A subbase for the weak operator topology on $B(H)$ is formed by the collection of all sets of the form $\{A: |((A - A_0)x, y)| < \delta\}$, $A_0 \in B(H)$, $x, y \in H$ and $\delta > 0$ in R. Show that the monad of A_0 in $B(H)$ in the weak topology is given by $\mu(A_0) = \{A \in {}^*B(H): (Ax, y) \simeq (A_0 x, y)$ for all standard $x, y \in H\}$.

13. (Standard) A bilinear form on H is a map $B: H \times H \to R$ such that $B(x, \cdot)$ is linear for each $x \in H$ and $B(\cdot, y)$ is linear for each $y \in H$. B is bounded if there exists $M \in R$ such that $|B(x, y)| \leq M\|x\|\|y\|$ for all $x, y \in H$. Show that if B is a bounded bilinear form, then there exists an operator $T \in B(H)$ such that $B(x, y) = (Tx, y)$ for all $x, y \in H$.

14. Use Exercises 5.12 and 5.13 to show that the unit ball in $B(H)$ is compact in the weak operator topology.

III.6 Nonstandard Hulls of Metric Spaces

In this short section we introduce the reader to the concept of the non-standard hull of a metric space. This notion was introduced by Luxemburg [36] and has proved to be a powerful tool in the nonstandard analysis of Banach spaces, as indicated by the survey paper of Henson and Moore [16]. The technique of nonstandard analysis, as applied to the theory of Banach spaces, is essentially equivalent to the use of Banach space ultrapowers, a technique which originated with Dacunha-Castelle and Krivine [10] and is now used extensively. Nonstandard methods, however, are more intuitive and usually easier to apply, especially when they involve concepts, such as the internal cardinality of a ∗-finite set, which are not easy to express in the ultraproduct setting.

In this section we will assume that the nonstandard analysis is carried out in a κ-saturated enlargement where $\kappa > \aleph_0$. Suppose that (X, d) is a metric space. Recall that the principal galaxy $G = \text{fin}(^*X)$ is the set of points in $^*X$ each of which is at a finite distance from a point in X (regarded as embedded in $^*X$). If $a, b \in {}^*X$ we say as usual that $a \simeq b$ if $^*d(a, b) \simeq 0$. Let $\hat{X}$ denote the equivalence classes of G under the equivalence relation $\simeq$. Alternatively, $\hat{X}$ is the set of monads, where each monad $m(a) = \{b \in G: {}^*d(a, b) \simeq 0\}$ for $a \in G$ (notice that if $a \in G$ and $b \simeq a$ then $b \in G$). Since $^*d(a, b)$ is finite for any $a, b \in G$, we can define

$$\hat{d}(x, y) = \text{st}(^*d(a, b))$$

when $x = m(a)$ and $y = m(b)$ in $\hat{X}$.

6.1 Proposition $(\hat{X}, \hat{d})$ is a metric space.

Proof: Exercise.   □

6.2 Definition $(\hat{X}, \hat{d})$ is called the *nonstandard hull* of (X, d).

We now use saturation to prove that $(\hat{X}, \hat{d})$ is complete [even if $(\hat{X}, \hat{d})$ is not]. Our construction is like that of Theorem 3.17, but here $\hat{X}$ consists of monads of finite points and not just pre-near-standard points.

6.3 Theorem Suppose that $*X$ lies in a κ-saturated superstructure with $\kappa > \aleph_0$. Then $(\hat{X}, \hat{d})$ is a complete metric space.

Proof: Let $\langle m(a_i) : i \in N \rangle$ be a Cauchy sequence in $(\hat{X}, \hat{d})$. Then for each $k \in N$ there is an $n(k) \in N$ so that $*d(a_i, a_j) < 1/k$ if i and j are both $> n(k)$; we can assume without loss of generality that $n(k) \to \infty$ as $k \to \infty$. Let $\phi(i) = a_i$. By Theorem 8.5 of Chapter II, the map $\phi: N \to *X$ can be extended to an internal map $\tilde{\phi}: \tilde{N} \to *X$, where $\tilde{N} \subseteq *N$ is internal and contains N and so contains some infinite integer. We would like to show that there is some infinite integer m' in $\tilde{N}$ so that $*d(a_i, a_{m'}) < 1/k$ for all $i \in N$ with $i > n(k)$, where $a_{m'} = \tilde{\phi}(m')$. For any $k \in N$ the set $E(k) = \{m \in \tilde{N} : *d(a_i, a_j) < 1/k$ for all $i, j \in \tilde{N}$ satisfying $n(k) < i \leq m, n(k) < j \leq m\}$ is internal and contains N. Therefore $E(k)$ also contains $\{m \in *N : m \leq m_k\}$ for some infinite integer m_k, and we may assume that $m_{k+1} \leq m_k$ for all $k \in N$. Again by Theorem 8.5 of Chapter II we may extend the sequence $\langle m_k \rangle$ to an internal decreasing mapping from an internal set $\hat{N} \subset *N$ into $*N$. Since $m_k > k$ for each finite $k \in N$, there is an infinite ω with $m_\omega \geq \omega$ and $m_\omega \in E(k)$ for all $k \in N$. Let $m' = m_\omega$. Then $*d(a_i, a_{m'}) < 1/k$ for all $i \in N$ with $i > n(k)$. It follows that $a_{m'}$ is finite and $\langle m(a_i) \rangle$ converges to $m(a_{m'})$.   □

If our metric space (X, d) is a normed vector space with norm $\| \ \|$, the nonstandard hull can be made into a normed vector space in an obvious way. For in this case G consists of all $x \in *X$ for which $\|x\|$ is finite, and so G is a vector space over the reals. We define addition and scalar multiplication of elements in $\hat{X}$ by

$$m(x) + m(y) = m(x + y), \qquad x, y \in G,$$

and

$$am(x) = m(ax).$$

Also we define a norm in $\hat{X}$ by $|||m(x)||| = \text{st}\|x\|$, $x \in G$. It is easy to check that $\hat{d}(m(x), m(y)) = |||m(x) - m(y)|||$. From Theorem 6.3 we see that $(\hat{X}, ||| \ |||)$ is a Banach space. The details are left to the reader.

Exercises III.6

1. Prove Proposition 6.1; in particular, show $\hat{d}$ is well defined.
2. Show that if $(X, \|\cdot\|)$ is a normed space, then $(\hat{X}, ||| \cdot |||)$ as a Banach space.
3. Show that there is an isometric embedding of a Banach space into its nonstandard hull.
4. Consider the sequence $\langle e^n \rangle$ which is 0 for $n \neq \omega \in {}^*N_\infty$ and 1 for $n = \omega$ to show that the mapping in Exercise 3 is not onto for l_2.
5. Consider the sequence $\langle x_n \rangle$, where $x_n = 1/\omega$ for $1 \leq n \leq \omega$, $\omega \in {}^*N_\infty$ and $x_n = 0$ for $n > \omega$, to show that the mapping in Exercise 3 is not onto for l_1.

*III.7 Compactifications

In this section we show how some Hausdorff spaces $(X, \mathscr{S})$ can be embedded as dense subsets of compact Hausdorff spaces $(Y, \mathscr{T})$. That is, there exists a 1–1 map $\psi: X \to \text{range } \psi \subseteq Y$ so that ψ is a homeomorphism and range ψ is dense in Y. In this case $(Y, \mathscr{T})$ is called a *compactification* of $(X, \mathscr{S})$. We usually identify X and range ψ, and so regard X as a subset of Y; we will denote Y by $\bar{X}$.

A given space X typically has many compactifications. For example, if one adjoins 0 and 1 to $(0, 1)$ one obtains the compact interval $[0, 1]$. Adjoining a single point to both ends of $(0, 1)$ gives a circle. Similarly the plane can be made into a sphere by adjoining a single point. We are interested here in compactifying a space X so that certain continuous functions on X have continuous extensions to $\bar{X}$. What, for example, should one adjoin to $(0, 1]$ to make $\sin(1/x)$ continuous on the resulting compact space?

7.1 Definition Let Q be a family of (perhaps not uniformly) bounded, continuous, real-valued functions on $(X, \mathscr{S})$. $(\bar{X}, \mathscr{T})$ is called a *Q-compactification* of $(X, \mathscr{S})$ if it is a compactification for which

(a) each $f \in Q$ has a continuous extension $\bar{f}$ to $(\bar{X}, \mathscr{T})$,

(b) if x and y are different points in $\bar{X} - X$ there is an $f \in Q$ whose extension $\bar{f}$ separates x and y, i.e., $\bar{f}(x) \neq \bar{f}(y)$. We sometimes write $\bar{X}^Q$ for $\bar{X}$.

In order to construct a Q-compactification we need to suppose that Q contains sufficiently many functions.

7.2 Definition A family Q of continuous functions *distinguishes points and closed sets* if, for each set $A \subset X$ and each $x \in X - A$, there is an $f \in Q$ so that $f(x) \notin \overline{f[A]}$.

It should be noted that not all Hausdorff spaces X admit sufficiently many continuous real-valued functions to distinguish points and closed sets. There are enough functions if X is completely regular [20].

The compactifications of this section will be constructed from $*X$. The original work on this construction was done by Gonshor [15], Luxemburg [36], Machover and Hirschfeld [37], and Robinson [43].

Let Q be a family of bounded, continuous, real-valued functions on $(X, \mathscr{S})$. Assuming that Q distinguishes points and closed sets, we construct $\bar{X}$ as follows. We call two points $y, z \in *X$ *equivalent*, and write $y \sim z$, if $*f(y) \simeq *f(z)$ for all $f \in Q$. It is easy to see (check) that $\sim$ is an equivalence relation. The equivalence class containing $x \in *X$ is denoted by $[x]$, and the set of all equivalence classes is $\bar{X}$.

Next we show that if $x \in X$, then $[x] = m(x)$, the monad of x. First note that if $y \in m(x)$, then $y \sim x$ since each $f \in Q$ is continuous. On the other hand, if U is an open set containing x, then there is an $f \in Q$ so that $f(x) \notin \overline{f[X - U]}$. Thus $f^{-1}[R - \overline{f[X - U]}]$ is an open set containing x and contained in U. We conclude that $[x] = m(x)$.

We extend each $f \in Q$ to a function on $\bar{X}$ (again denoted by f) by setting $f([y]) = \mathrm{st}(*f(y))$, $y \in *X$ (check that f is well defined). The set of extended functions is again denoted by Q. The topology $\mathscr{T}$ on $\bar{X}$ is the weak topology for the functions in Q. Thus U is open in $\bar{X}$ iff for each $[y] \in U$ there is a finite set $\{f_1, \ldots, f_n\} \subseteq Q$ and a positive number ε in R so that $\{[z] \in \bar{X}: |f_i([y]) - f_i([z])| < \varepsilon, 1 \le i \le n\} \subseteq U$.

In order that we may treat $\bar{X}$ as an element of the original superstructure $V(X)$ (which will be used in the proof of compactness), we may think of each point in $\bar{X}$ as a function on Q by the definition $[y](f) = f([y])$. Distinct points of $\bar{X}$ give distinct functions on Q. The standard construction of $\bar{X}$ is based on such a family of functions on Q. It is often helpful, however, to think of $\bar{X}$ as a quotient of $*X$ as we have done.

Let $\bar{X}^Q$ be constructed as above from a set Q of bounded, continuous, real-valued functions on $(X, \mathscr{S})$ which separate points and closed sets.

7.3 Theorem $(\bar{X}^Q, \mathscr{T})$ is a Q-compactification of $(X, \mathscr{S})$.

Proof: Let $\bar{X}^Q$ be denoted by $\bar{X}$. Define the map $\psi: X \to \bar{X}$ by $\psi(x) = [x]$, $x \in X$. Now $[x] = m(x)$ for $x \in X$, so the map ψ is 1–1 by 1.12(c) since X is Hausdorff.

To show ψ is a homeomorphism, we must show that ψ and ψ^{-1} are continuous. An easy exercise shows that ψ is continuous. To see that ψ^{-1} is continuous, we must show that if $x \in X$ and V is an open neighborhood of x in $\mathscr{S}$, then there is a $U \in \mathscr{T}_x$ so that $U \cap X \subseteq V$ (we regard X as contained in $\bar{X}$). Let $f \in Q$ be such that $f(x) \notin \overline{f[A]}$, where $A = X - V$. Then there is an $\varepsilon > 0$ in R so that $\{z \in X : |f(z) - f(x)| < \varepsilon\} \subseteq V$ (why?); we let $U = \{z \in \bar{X} : |f(z) - f(x)| < \varepsilon\}$.

To show $\psi[X]$ is dense in $\bar{X}$, let $[y] \in \bar{X} - \psi[X]$, and let $U \in \mathscr{T}$ be given by $U = \{[z] \in \bar{X} : |f_i([z]) - f_i([y])| < \varepsilon, 1 \le i \le n\}$. We must show that $[x] \in U$ for some $x \in X$. Let $\alpha_i = f_i([y])$, $1 \le i \le n$. Then the set $\{x \in {}^*X : |f_i(x) - \alpha_i| < \varepsilon, 1 \le i \le n\}$ is not empty (indeed it contains y). By downward transfer, the set $\{x \in X : |f_i(x) - \alpha_i| < \varepsilon, 1 \le i \le n\}$ is not empty, and we are through.

To show that $\bar{X}$ is compact we consider a mapping T on $\bar{X}$. For each $[y] \in \bar{X}$, $T([y])$ is the function from Q into R defined by setting $T([y])(f) = f([y])$ for each $f \in Q$. Let A be the range of T; then T is a 1–1 mapping from $\bar{X}$ onto A. We make T a homeomorphism by letting U be open in A iff $T^{-1}[U]$ is open in $\bar{X}$. Thus a typical neighborhood of an $a \in A$ is given by a finite set $\{f_1, \ldots, f_n\} \subset Q$ and an $\varepsilon > 0$ in R: it consists of those $b \in A$ with $|a(f_i) - b(f_i)| < \varepsilon, 1 \le i \le n$. Since X is dense in $\bar{X}$, each such neighborhood contains a $T([x])$ for some $x \in X$; i.e., $|a(f_i) - f_i(x)| < \varepsilon$ for $1 \le i \le n$. To show that $\bar{X}$ is compact, we need only show that A is compact. Fix $b \in {}^*A$. Let ε be a positive infinitesimal in ${}^*R$, and let Q_1 be a hyperfinite subset of Q such that ${}^*f \in Q_1$ for each $f \in Q$. By the transfer principle, there is an $x \in {}^*X$ such that $|b(f) - f(x)| < \varepsilon$ for each $f \in Q_1$. Let $c = T([x])$. For each $f \in Q$, $c(f) = T([x])(f) \simeq {}^*f(x) \simeq b({}^*f)$, so b is in the monad of the standard point $c \in A$. Thus A is compact.

Finally, by the construction, each member of Q has a continuous extension to $\bar{X}$, and the family of extensions separates the points of $\bar{X}$. $\square$

It is not hard to see that if Q_1 and Q_2 are two families as described above with $Q_1 \subseteq Q_2$, then there is a continuous map ζ from $\bar{X}^{Q_2}$ onto $\bar{X}^{Q_1}$ such that $\zeta(x) = x$ for all $x \in X$. In this case we write $\bar{X}^{Q_1} \le \bar{X}^{Q_2}$. It follows that a Q-compactification of X is unique up to a homeomorphism that leaves the points of X fixed (see, for example, [20, Theorem 22]).

Any compactification $\bar{X}$ of X is a Q-compactification; just let $Q = \{g_X : \text{the function } g: \bar{X} \to R \text{ continuous}\}$, where g_X denotes the restriction of g to X. It follows that if $\hat{Q}$ consists of all bounded, continuous, real-valued functions on X, then $\bar{X}^{\hat{Q}}$ is the largest compactification of X, i.e., $\bar{X}^{\hat{Q}} \ge \bar{X}$ for any other compactification $\bar{X}$ of X. $\bar{X}^{\hat{Q}}$ is called the *Stone–Čech compactification* and is often denoted by βX.

If $(X, \mathscr{T})$ is *locally compact* (i.e., each point $x \in X$ has at least one compact neighborhood), and Q_0 is any family of bounded, continuous, real-valued functions on X, then, following Constantinescu and Cornea [9], we may obtain Q by adjoining to Q_0 the family C_c of all continuous functions with compact support (i.e., vanishing off compact sets). If Q_0 is empty and so $Q = C_c$, then $\bar{X}^Q$ can be identified with $X \cup \{\infty\}$, where ∞ is a single point not in X. In this case we have a topology $\mathscr{S}$ whose members are all open sets in X, together with all sets U of $\bar{X}^Q$ such that $\bar{X}^Q - U$ is compact in X (check); the space $\bar{X}^Q$ is called the *one-point compactification* of X.

We end this section with a result concerning the Stone–Čech compactification $\bar{N}$ of the natural numbers N (with the discrete topology).

7.4 Theorem The points in $\bar{N} - N$ are in one-to-one correspondence with the free ultrafilters on N via the map $[\omega] \rightsquigarrow \mathscr{F}_{[\omega]}$, where $\mathscr{F}_{[\omega]} = \{A \subseteq N : \omega \in {}^*A\}$.

Proof: For each $A \subset N$, the characteristic function χ_A is in Q, the set of bounded continuous functions on N. It follows that for each equivalence class $[\omega] \in \bar{N} - N$ either $[\omega] \subset {}^*A$ or $[\omega] \subset {}^*N - {}^*A$. If $[\omega] \in \bar{N} - N$ then $\omega \in {}^*N_\infty$, and the family $\mathscr{F}_{[\omega]} = \{A \subseteq N : \omega \in {}^*A\}$ is a free ultrafilter (exercise). This is the same ultrafilter as $\{A \subseteq N : \chi_A([\omega]) = 1\}$, where χ_A has been extended to $\bar{N}$.

On the other hand, if $\mathscr{F}$ is a free ultrafilter on N, then the intersection monad $\mu(\mathscr{F}) = \bigcap {}^*F(F \in \mathscr{F})$ is a unique element $[\omega]$ in $\bar{N} - N$. To prove this, we assume it is false. Then there are at least two distinct equivalence classes $[\omega]$ and $[\gamma]$ in $\mu(\mathscr{F})$ and a bounded sequence $\langle s_n \rangle \in Q$ such that $a = {}^\circ s_\omega \neq {}^\circ s_\gamma = b$. We may assume that $a < b$ and choose $c \in R$ with $a < c < b$. Since either $\{n \in N : s_n \leq c\} \in \mathscr{F}$ or $\{n \in N : s_n > c\} \in \mathscr{F}$, we have a contradiction. $\square$

Exercises III.7

1. Let $(X, \mathscr{T})$ be locally compact, and let $\bar{X}$ denote the one-point compactification of X. Let A be an internal set of near-standard points in ${}^*X$. Use the fact that st$[A]$ is closed in X and a closed subset of $\bar{X}$ is compact to show that st$[A]$ is compact.

2. Show directly that the one-point compactification of a locally compact Hausdorff space is compact.

3. Show that, for $\omega \in {}^*N_\infty$, $\{A \subseteq N : \omega \in {}^*A\}$ is a free ultrafilter.

4. What is the Q-compactification of $(0, 1)$ when $Q = \{f(x) = x\}$?

5. What is the Q-compactification of $(0, 1)$ when $Q = \{f(x) = x, g(x) = \sin(1/x)\}$?

6. Show that X is open in a compactification $\bar{X}$ if and only if X is locally compact.

*III.8 Function Spaces

Let $(X, \mathcal{S})$ and $(Y, \mathcal{T})$ be Hausdorff topological spaces and F be a family of mappings from X into Y. This section will be concerned with two questions:

(a) For which topologies $\mathcal{F}$ on F is the map $\phi_A: F \times A \to Y$ defined by $\phi(f, x) = f(x)$ continuous for all subsets $A \subseteq X$ in a certain family $\mathcal{H}$? Such a topology $\mathcal{F}$ is said to be *jointly continuous* with respect to $\mathcal{H}$.

(b) *For which topologies on the space M of all mappings from X into Y* is the closure of F compact?

To answer these questions, we consider two important topologies, the topology of pointwise convergence and the compact-open topology. For a standard treatment the reader is referred to Kelley [20, Chapter 7]. Our treatment follows suggestions of Hirschfeld [18].

The nonstandard analysis will be done in an enlargement of a structure containing X and Y. Monads in $(X, \mathcal{S})$ and $(Y, \mathcal{T})$ will be denoted by $m_X(x)\,(x \in X)$ and $m_Y(y)\,(y \in Y)$, respectively, but we will denote nearness in both X and Y by $\simeq$ as in §III.1.

With each subset $A \subseteq X$ we associate an important pseudomonad $k_A(f)\,(f \in M)$ on the space M of all maps from X into Y by setting

(8.1) $k_A(f) = \{g \in {}^*M : g(x') \simeq f(x)$ for all $x \in A$ and $x' \in {}^*A$ with $x' \simeq x\}$.

The following result provides a nonstandard answer to question (a).

8.1 Proposition Let $\mathcal{F}$ be a topology for F with associated monads $m(f)$ $(f \in F)$. Then $\mathcal{F}$ is jointly continuous with respect to $\mathcal{H}$ iff $m(f) \subseteq \bigcap\{k_A(f):$ $A \in \mathcal{H}\}$ for all $f \in F$.

Proof: We need only show that, for each $A \in \mathcal{H}$, ϕ_A is continuous iff $m(f) \subseteq k_A(f)$ for all $f \in F$. But for $f \in F$ and $x \in A$, the monad of (f, x) in ${}^*F \times {}^*A$ is $m(f) \times m_A(x)$, where $m_A(x) = m_X(x) \cap {}^*A$. ϕ_A is continuous at each $(f, x) \in F \times A \Leftrightarrow {}^*\phi_A(m(f) \times m_A(x)) \subseteq m_Y(\phi_A(f, x))$ for each $f \in F$, $x \in A \Leftrightarrow$ if $f \in F$, $x \in A$, then whenever $g \in m(f)$ and $y \simeq x, y \in {}^*A$, we have $g(y) \simeq f(x) \Leftrightarrow m(f) \subseteq k_A(f)$ for each $f \in F$. $\square$

8.2 Definition

(a) The *topology of pointwise convergence* $\mathcal{P}$ on M is the weak topology for the family $\{\phi_x : x \in X\}$ of evaluation maps $\phi_x: M \to Y$ defined by $\phi_x(f) = f(x)$. The monads for $\mathcal{P}$ are denoted by $p(f)\,(f \in M)$.

(b) The *compact-open topology* $\mathcal{C}$ on M is generated by the subbase consisting of all sets of the form $W(K, U) = \{g \in M : g[K] \subset U\}$, where K is compact in $(X, \mathcal{S})$ and U is open in $(Y, \mathcal{T})$. We let $c(f)$ $(f \in M)$ denote the monads of $\mathcal{C}$.

From 1.18 we see that

(8.2) $p(f) = \{g \in {}^*M : g(x) \simeq f(x) \text{ for all standard points } x \in X\}$.

8.3 Proposition Let $\mathcal{K}$ be the family of compact subsets of $(X, \mathcal{T})$. Then, for each $f \in M$, $k_X(f) \subseteq \bigcap \{k_A(f) : A \in \mathcal{K}\} \subseteq c(f) \subseteq p(f)$.

Proof: (i) $k_X(f) \subseteq k_A(f)$ for any $A \subseteq X$, and the first containment follows.

(ii) Let K be compact in $(X, \mathcal{S})$ and U be an open set in $(Y, \mathcal{T})$ containing $f[K]$. If $g \in \bigcap \{k_A(f) : A \in \mathcal{K}\}$, then $g \in k_K(f)$, so $g(y) \simeq f(x)$ for all $x \in K$ and all $y \in {}^*K$ with $y \simeq x$. Since U is open, $g(y) \in {}^*U$ for all $y \in {}^*K$ with $y \simeq x \in K$. But this includes all $y \in {}^*K$ since K is compact, and so $g[{}^*K] \subseteq {}^*U$, i.e., $g \in {}^*W(K, U)$. Thus $\bigcap \{k_A(f) : A \in \mathcal{K}\} \subseteq {}^*W(K, U)$ for any K and U with $f[K] \subseteq U$, and the second containment follows.

(iii) A subbase for $\mathcal{P}$ consists of sets of the form $W(\{x\}, U)$, and so $\mathcal{P}$ is weaker than $\mathcal{C}$ and the third containment follows. $\quad\square$

8.4 Theorem Each topology which is jointly continuous with respect to the family of compact subsets of X is stronger than $\mathcal{C}$.

Proof: Immediate from 8.1 and 8.3. $\quad\square$

8.5 Theorem Assume $F \subset M$ is closed with respect to $\mathcal{P}$. Then F is compact in $(M, \mathcal{P})$ if for each x the set $\{f(x) : f \in F\}$ has compact closure in Y.

Proof: Our condition guarantees that, for any $x \in X$, every point in ${}^*\{f(x) : f \in F\} = \{g(x) : g \in {}^*F\}$ is near a standard point in Y. Given $g \in {}^*F$, let $f(x)$ be defined for each $x \in X$ by setting $f(x) = y_x$, where y_x is a point in Y with $y_x \simeq g(x)$ [such a point is unique since $(Y, \mathcal{T})$ is Hausdorff]. Then $f \in M$ and $f(x) \simeq g(x)$ for all $x \in X$, i.e., $g \in p(f)$. Since $g \in {}^*F$ and $F \subset M$ is closed, $f \in F$. Thus each $g \in {}^*F$ is near a standard $f \in F$. $\quad\square$

The fact that $\{f(x) : f \in F\}$ has compact closure for each $x \in X$ is an essential ingredient in obtaining a function $f \in F$ from a function $g \in {}^*F$. The argument of Theorem 8.5 does not work, however, for the compact-open topology since the condition $g(x) \simeq f(x)$ for all $x \in X$ is not sufficient to guarantee that $g \in c(f)$. If, however, $g(x') \simeq f(x)$ for all $x \in X$ and $x' \in X$ with $x' \simeq x$, then

$g \in k_X(f) \subseteq c(f)$ (by Proposition 8.3) and compactness follows. A standard condition guaranteeing that this holds is the following from Kelley [20].

8.6 Definition The family F is *evenly continuous* if for each $x \in X$, $y \in Y$ and each open neighborhood U of y, there are neighborhoods V of x and W of y so that for all $f \in F$ with $f(x) \in W$, we have $f[V] \subseteq U$.

8.7 Proposition The family F is evenly continuous iff the following condition holds: Given $x \in X$ and $y \in Y$, if $g \in {}^*F$ and $g(x) \simeq y$, then $g(x') \simeq y$ for all $x' \simeq x$ in ${}^*X$.

Proof: Assume first that F is evenly continuous. Fix a neighborhood $U \in \mathcal{T}_y$ and the corresponding sets $V \in \mathcal{S}_x$ and $W \in \mathcal{T}_y$ given by Definition 8.6. Since $g(x) \simeq y$, $g(x) \in {}^*W$, so by transfer $g[{}^*V] \subset {}^*U$. In particular, $g(x') \in {}^*U$ if $x' \simeq x$. This last statement is true for any $U \in \mathcal{T}_y$, and so $g(x') \simeq y$ if $x' \simeq x$.

To prove the converse, fix $U \in \mathcal{T}_y$ and let V and W be $*$-open sets in ${}^*\mathcal{S}_x$ and ${}^*\mathcal{T}_y$, respectively, with $V \subseteq m_X(x)$ and $W \subseteq m_Y(y)$. Now if $g \in {}^*F$ and $g(x) \in W$, then $g(x) \simeq y$. By assumption, for all $x' \in V$, $g(x') \in m_Y(y) \subseteq {}^*U$. The rest follows by downward transfer. $\square$

As a corollary we get a generalized Ascoli theorem due to Kelley [20].

8.8 Ascoli Theorem If $F \subset M$ is closed in $\mathscr{C}$ and evenly continuous, and $\{f(x): f \in F\}$ has compact closure for each $x \in X$, then F is compact in $(M, \mathscr{C})$.

Proof: Immediate from the discussion preceding Definition 8.6. $\square$

For the rest of this section we assume that $(Y, \mathcal{T})$ is a metric space with metric d. In this context, a notion which is closely related to even continuity is the notion of equicontinuity, which has already been presented in the real-variable case in Definition I.13.6.

8.9 Definition A family $F \subset M$ is called *equicontinuous* on X if, for each $x \in X$ and each $\varepsilon > 0$ in R, there is a $V \in \mathcal{S}_x$ such that, for *any* $f \in F$, if $x' \in V$, then $d(f(x'), f(x)) < \varepsilon$.

8.10 Proposition The family $F \subset M$ is equicontinuous on X iff, for any $x \in X$ and any $g \in {}^*F$, $g(x') \simeq g(x)$ whenever $x' \simeq x$.

Proof: Exercise. $\square$

If F is the family $\{n + nx : n \in N\}$ then F is evenly continuous but not equi-continuous on $[0, 1]$. By Propositions 8.7 and 8.10, any equicontinuous family $F \subset M$ is evenly continuous.

If $F \subset M$ is a family of continuous functions, then the compact-open topology in F is the same as the *topology of uniform convergence on compact sets*, or the *topology of compact convergence*. For the latter topology, a typical basic open neighborhood of $f \in F$ is of the form $\{g \in F : d(f(x), g(x)) < \varepsilon$ for all $x \in K\}$ for some compact $K \subseteq X$ and $\varepsilon > 0$ in R (see [20, p. 229]). It follows from Theorem 8.8 that if F is an equicontinuous family in M (whence each $f \in F$ is continuous), and F is closed in M with respect to the topology of uniform convergence on compact sets with $\{f(x) : f \in F\}$ having compact closure in Y for each $x \in X$, then F is compact with respect to the topology of uniform convergence on compact sets. Moreover, for an equicontinuous family F, the topology of pointwise convergence is jointly continuous on compact sets (exercise), and hence coincides with the topology of uniform convergence on compact sets.

Exercises III.8

1. Use Theorem 8.5 to prove Alaoglu's theorem, 4.22.
2. Prove Proposition 8.10.
3. (a) Show that the set of real-valued continuous functions on R (with the usual topology) is closed with respect to the topology of uniform convergence on compact sets.

   (b) Show that part (a) is no longer true if we replace the usual topology on R with a topology $\mathscr{S}$ such that $\{r\} \in \mathscr{S}$ for each $r \neq 0$ in R, and U is an open neighborhood of 0 if $0 \in U$ and $R - U$ is countable. [Hint: what are the compact sets? Is g continuous if $g(0) = 1$ and $g(r) = -1$ for $r \neq 0$?]
4. Show that if $(Y, \mathscr{T})$ is a metric space and F is an equicontinuous family, then the topology of pointwise convergence is jointly continuous on compact sets and hence coincides with the topology of uniform convergence on compact sets.
5. Let C denote the set of real-valued continuous functions on $I = [0, 1]$. Then the map $d : C \times C \to R^+$ defined by $d(f, g) = \max\{|f(x) - g(x)| : x \in I\}$ is a metric on C. Show that the compact-open topology on C coincides with the metric topology.
6. Show that the space $C(X, Y)$ of continuous mappings from $(X, \mathscr{S})$ to $(Y, \mathscr{T})$ with the compact-open topology is Hausdorff if $(Y, \mathscr{T})$ is Hausdorff.

CHAPTER IV

Nonstandard Integration Theory

In trying to apply the theory of the Riemann integral we are faced with the following technical problem. Suppose we are given a converging infinite series $\sum_{n=1}^{\infty} f_n(x) = f(x)$ of functions on $[a, b]$ and are asked to calculate $\int_b^a f(x) \, dx$. The answer is often simple if we can write

$$\int_a^b f(x) \, dx = \sum_{n=1}^{\infty} \int_a^b f_n(x) \, dx.$$

Thus we need to find conditions under which integration and infinite summation can be interchanged. Equivalently [letting $g_n(x) = \sum_{i=1}^{n} f_i(x)$] we need conditions under which, if $g(x) = \lim_{n \to \infty} g_n(x)$, then

$$\lim_{n \to \infty} \int_a^b g_n(x) \, dx = \int_a^b g(x) \, dx$$

for a sequence $\{g_n(x)\}$ of Riemann-integrable functions on $[a, b]$. It turns out that we can reduce the discussion to sequences $\{g_n(x)\}$ which are monotone increasing, i.e., $g_{n+1}(x) \geq g_n(x)$ for all $n \in N$ [this is the case if $f_n(x) \geq 0$ for all $n \in N$]. Thus, assuming that $\{g_n(x)\}$ is a monotone increasing sequence of integrable functions and $g_n(x)$ converges to $g(x)$ on $[a, b]$, we need conditions which insure that $g(x)$ is integrable and the above equation holds. A result of this type is known as a monotone convergence theorem.

Unfortunately, the conditions under which a monotone convergence theorem holds for Riemann integration are quite restrictive (for example, it holds if the sequence $\{g_n\}$ converges *uniformly* on $[a, b]$). This fact led Lebesgue [26] and others to generalize the process of integration in such a way that the conditions for a monotone convergence theorem were *considerably* relaxed. The procedure was to generalize the concept of the length of an interval so that one could measure the "length" of a very general subset of $[a, b]$ called a measurable set. The theory of integration then developed systematically from this "measure theory."

An alternative approach was developed by P. Daniel [11]. He began with the general notions of a lattice L of functions on a set X and an integral I on L. As indicated in Definition 1.2, a lattice of functions is a linear space which is also closed under the operation of taking absolute valves, and an integral I on L is a linear functional which is also positive [i.e., $f \geq 0$ implies $I(f) \geq 0$]. Daniel showed that if I satisfied the additional continuity condition "If $\{f_n\}$ decreases to 0 then $I(f_n)$ decreases to 0," (L, I) could be enlarged to a structure $(\hat{L}, \hat{I})$ which satisfied the monotone convergence theorem.

Our nonstandard approach to integration follows the Daniel approach except that we begin with an "internal" integration structure (L, I) on an internal set X in some enlargement. We show that, without any continuity assumption, we can construct from (L, I) a standard integration structure $(\hat{L}, \hat{I})$ *on the same internal set* X, and that structure satisfies the monotone convergence theorem. In §IV.2 we show that the usual measure-theoretic approach can be recovered from any structure $(\hat{L}, \hat{I})$ satisfying the monotone convergence theorem. The usual Lebesgue theory on R^n is developed in §IV.3 by using the standard part map to carry results on *R^n down to R^n. Some important convergence theorems which hold in any structure for which the monotone convergence theorem is valid are developed in §IV.4. A nonstandard approach to the Fubini theorem, which is an analogue of the iterated integration procedure for the Riemann integral, is developed in §IV.5. Finally, in §IV.6 we apply the nonstandard integration theory developed in the previous sections to study several important stochastic processes, including the Poisson process and Brownian motion. These processes are represented as processes on a *-finite probability space and indicate the usefulness of an integration theory on nonstandard sets.

References to the original work on nonstandard integration theory will be given in the body of this chapter, with the exception, as noted in the Preface, of [27, 29, 32, 33] by the second author.

IV.1 Standardizations of Internal Integration Structures

The Riemann integral for continuous functions on an interval $[a, b]$ (see §I.12) has the properties

(1.1) $$\int_a^b [\alpha f(x) + \beta g(x)] \, dx = \alpha \int_a^b f(x) \, dx + \beta \int_a^b g(x) \, dx,$$

(1.2) $$\int_a^b f(x) \, dx \geq 0 \qquad \text{if} \quad f(x) \geq 0 \text{ on } [a, b].$$

Implicit in (1.1) is the fact that a linear combination of continuous functions is continuous. It is also true that $|f|$ is continuous if f is continuous. A general theory of integration should specify (A) a class L of "integrable" functions on a space X corresponding to the continuous functions on $[a, b]$ in the above example, and (B) a real-valued function I on L whose value at $f \in L$ we denote by If (a numerical-valved function on a set of functions is usually called a functional). Here If corresponds to the Riemann integral of f. In general, the analogues of the properties above should be satisfied. We abstract these properties in the notion of an *integration structure*. It consists of a lattice of functions and a positive linear functional on this lattice as in Definition 1.2 below. This definition incorporates the standard (real) and nonstandard (hyperreal) notions of integration structures since we want to consider internal analogues of integration structures when the functions are internal and hyperreal-valued.

Our main objective in this section is to show how, beginning with an internal integration structure (L, I) on an internal set X, we can construct a real integration structure $(\hat{L}, \hat{I})$ on the same internal set X by a process called *standardization*. The important fact is that the real integration structures so obtained satisfy a closure property called the monotone convergence theorem. This theorem states roughly that a monotone increasing sequence $\langle f_n \rangle$ of functions in $\hat{L}$, whose integrals $\hat{I}f_n$ are uniformly bounded, converges to a function $f \in \hat{L}$, and $\hat{I}f$ is the limit of $\langle \hat{I}f_n \rangle$. It is the basic tool in all further developments of integration theory.

We begin with a definition summarizing standard notation.

1.1 Definition Let X be a set and $E \subseteq X$. The functions χ_E, 1, and 0 on X are defined by

$$\chi_E = \begin{cases} 1, & x \in E, \\ 0, & x \notin E, \end{cases}$$

$1 = \chi_X$, and $0 = \chi_\varnothing$, where $\varnothing$ is the empty set.

If f and g are functions on X, we write $f \leq g$ if $f(x) \leq g(x)$ for all $x \in X$; we define $\alpha f, f + g, fg, f/g$ (if g does not vanish at any point in X), and $|f|$ as usual by assigning the values $\alpha f(x), f(x) + g(x), f(x)g(x), f(x)/g(x)$, and $|f(x)|$ at $x \in X$.

1.2 Definition A set L of real- or hyperreal-valued functions on a set X is a real (hyperreal) *lattice* if

(a) $f, g \in L$ implies $\alpha f + \beta g \in L$ for all real (hyperreal) α, β,
(b) $f \in L$ implies $|f| \in L$.

A real- or hyperreal-valued function I on L is called a real (hyperreal) *positive linear functional* (p.l.f.) if

(c) $I(\alpha f + \beta g) = \alpha If + \beta Ig$ for all $f, g \in L$ and real (hyperreal) α, β,
(d) $If \geq 0$ if $f \geq 0$.

The pair (L, I) then forms a real (hyperreal) *integration structure* on X. The integration structure $(\tilde{L}, \tilde{I})$ on X is an extension of the integration structure (L, I) if $L \subseteq \tilde{L}$ and $\tilde{I}f = If$ when $f \in L$.

If the sets X and L (and hence all $f \in L$) and the functional I are internal in some enlargement $V(^*S)$ of a superstructure $V(S)$, then we say that (L, I) is an *internal* integration structure.

A lattice L always contains 0 (check), and is also closed under the operations of taking maxima and minima, defined as follows.

1.3 Definition If f and g are (real- or hyperreal-valued) functions defined on X, we define the *maximum* and *minimum* of f and g by

$$\max(f, g) = f \vee g = (f + g + |f - g|)/2,$$
$$\min(f, g) = f \wedge g = (f + g - |f - g|)/2$$

and the *positive* and *negative* parts of f by $f^+ = f \vee 0, f^- = (-f) \vee 0$.

Clearly, if L is a lattice and $f, g \in L$ then $f \vee g, f \wedge g \in L$. Conversely, if L is a set of functions on X which is closed under linear combinations and for which $f, g \in L$ implies $f \vee g$ and $f \wedge g \in L$, then L is a lattice (Exercise 1). Notice that if $f, g \in L$ and $f \geq g$, then the inequality $If \geq Ig$ follows from 1.2(d). This fact will be used frequently in the development.

The following are examples of real integration structures of real-valued functions.

1.4 Examples

1. Let $C[a, b]$ denote the set of all continuous real-valued functions on the finite interval $[a, b] \subset R$. Define the linear functional $\int_a^b$ on $C[a, b]$ by $\int_a^b f = \int_a^b f(x)\, dx$ (Riemann integral). Then $(C[a, b], \int_a^b)$ is a real integration structure on $[a, b]$ (exercise). Note that $1 \in C[a, b]$.

2. Let $C_c(R)$ denote the set of all continuous real-valued functions f on R with compact support, where the *support* of f is the set

$$\operatorname{supp} f = \overline{\{x : f(x) \neq 0\}}.$$

(a) Let $\int$ denote the functional on $C_c(R)$ defined by $\int f = \int_a^b f(x)\, dx$ if $\operatorname{supp} f \subseteq [a, b]$. (The definition of $\int$ is independent of the choice of a and

b satisfying this condition.) Then $(C_c(R), \int)$ is a real integration structure (exercise). Note that $1 \notin C_c(R)$.

(b) Let $\{\ldots, x_{-2}, x_{-1}, x_0, x_1, \ldots\}$ be a countable set of points in R with no limit point. For each $f \in C_c(R)$ let $\sum f = \sum_{i=-\infty}^{\infty} f(x_i)$. Then $(C_c(R), \sum)$ is a real integration structure on R (exercise).

3. A *step function* on R is a function f of the form $f = \sum_{i=1}^{n} c_i \chi_{E_i}$, where the sets E_i are disjoint finite intervals (open, closed, or semiopen; this includes the case where the end points are equal and E_i is thus a single point). Let $S(R)$ denote the set of step functions on R. Define the functional $\int$ on $S(R)$ by $\int f = \sum_{i=1}^{n} c_i(b_i - a_i)$ if $f = \sum_{i=1}^{n} c_i \chi_{E_i}$ and E_i has the end points a_i and b_i, $a_i \le b_i$. Then $(S(R), \int)$ is a real integration structure on R (exercise).

4. With $Y = \{x_1, \ldots, x_n\}$ a finite set, let $B(Y)$ denote the set of all real-valued functions on Y. If $a_1, \ldots, a_n$ are fixed real numbers with $a_i > 0$, $1 \le i \le n$, define the functional $\sum$ on $B(X)$ by $\sum f = \sum_{i=1}^{n} a_i f(x_i)$. Then $(B(Y), \sum)$ is a real integration structure on Y (exercise).

5. With Y any nonempty set, let $B_0(Y)$ denote the set of all real-valued functions on Y, each of which is zero except for finitely many $x \in Y$. If a is a positive real-valued function on Y, let $\sum_0$ denote the functional on $B_0(Y)$ defined by $\sum_0 f = \sum_{i=1}^{n} a(x_i) f(x_i)$, where $\operatorname{supp} f = \{x_1, \ldots, x_n\}$. Then $(B_0(Y), \sum_0)$ is a real integration structure on Y (exercise). If Y is a finite set, this example degenerates to Example 1.4.4.

The next proposition, easily proved using the transfer principle, shows that each standard real integration structure on a set Y (in particular, each of Examples 1.4) gives rise to an internal integration structure on $^*Y$ by transfer. We now fix an enlargement of a structure containing Y, with the associated monomorphism $*$.

1.5 Proposition If (L, I) is a real integration structure on a set Y, then $(^*L, ^*I)$ is an internal integration structure on $X = ^*Y$.

Proof: Exercise. $\square$

There are internal integration structures which cannot be obtained from a real integration structure by using Proposition 1.5, as the following example shows.

1.6 Hyperfinite Integration Structures Let X be an internal $*$-finite set $\{x_1, \ldots, x_\omega\}$ in an enlargement $V(^*S)$ of some superstructure $V(S)$. Let $B_\omega(X)$ denote the set of all hyperreal-valued internal functions on X. With $\{a_1, \ldots, a_\omega\}$ a fixed set of hyperreal nonnegative numbers of the same internal cardinality as X, let $\sum_\omega$ denote the hyperreal functional on $B_\omega(X)$ defined by $\sum_\omega f = \sum_{i=1}^{\omega} a_i f(x_i)$, where the summation is the extension of finite summation. Then $(B_\omega(X), \sum_\omega)$ is a hyperreal integration structure on X (Exercise 5). Such "hyperfinite" integration structures have recently been used as the starting point in an extensive nonstandard treatment of Brownian motion and other stochastic processes. An introduction to this theory is presented in §IV.6.

Now let (L, I) be an internal hyperreal integration structure on an internal set X in an enlargement $V(*S)$ of a superstructure $V(S)$ containing the reals. Our main objective in this section is to construct a *real* integration structure $(\hat{L}, \hat{I})$ on the same internal set X so that the monotone convergence theorem is valid. $(\hat{L}, \hat{I})$ will be called the *standardization* of (L, I). To prove the convergence theorem and other results we need to assume that $V(*S)$ is $\aleph_1$-saturated. Thus we assume from now on *without further explicit comment* that any internal structure (L, I) being standardized lies in an $\aleph_1$-saturated enlargement $V(*S)$ of a superstructure $V(S)$. $\hat{L}$ is now defined as follows.

1.7 Definition Let (L, I) be an internal integration structure on an internal set X. We define the set L_0 of *null* functions to be the set of hyperreal-valued (possibly external) functions g on X such that, for each $\varepsilon > 0$ in R, there is a $\psi \in L$ with $|g| \leq \psi$ and $^\circ I \psi < \varepsilon$. Further we define $\hat{L}$ to be the set of real-valued functions f on X such that $f = \phi + g$, where $\phi \in L$, $^\circ I |\phi| < \infty$, and $g \in L_0$.

1.8 Lemma

   (a) If $f = \phi + g \in \hat{L}$ with $\phi \in L$, $^\circ I|\phi| < \infty$, $g \in L_0$, and we also have $f = \tilde{\phi} + \tilde{g}$ with $\tilde{\phi} \in L$, $\tilde{g} \in L_0$, then $^\circ I|\tilde{\phi}| < \infty$ and $\phi - \tilde{\phi} \in L_0$, so $I\phi - I\tilde{\phi} = I(\phi - \tilde{\phi}) \simeq 0$.
   (b) If $f_i \in \hat{L}$ with $f_i = \phi_i + g_i$, $\phi_i \in L$, $g_i \in L_0$ $(i = 1, 2)$, then $(f_1 \vee f_2) - (\phi_1 \vee \phi_2)$ and $(f_1 \wedge f_2) - (\phi_1 \wedge \phi_2)$ are in L_0.

   Proof: (a) Since $\tilde{\phi} - \phi = g - \tilde{g} \in L_0$, we have $|^\circ I(\tilde{\phi} - \phi)| \leq {}^\circ I|\tilde{\phi} - \phi| = 0$ and $|^\circ I|\tilde{\phi}| - {}^\circ I|\phi|| \leq {}^\circ I|\tilde{\phi} - \phi|$ (Exercise 6). It follows that $^\circ I|\tilde{\phi}| < \infty$ and $|^\circ I\tilde{\phi} - {}^\circ I\phi| = 0$.
   (b) Given $\varepsilon > 0$ in R, there is a $\psi \in L$ with $|g_i| < \psi$ $(i = 1, 2)$ and $^\circ I\psi < \varepsilon$ (why?). From the inequalities

$$(\phi_1 \vee \phi_2) - \psi = (\phi_1 - \psi) \vee (\phi_2 - \psi)$$
$$\leq (\phi_1 + g_1) \vee (\phi_2 + g_2) = f_1 \vee f_2$$
$$\leq (\phi_1 \vee \phi_2) + \psi,$$

it follows that $(f_1 \vee f_2) - (\phi_1 \vee \phi_2) \in L_0$. Similarly $(f_1 \wedge f_2) - (\phi_1 \wedge \phi_2) \in L_0$. $\square$

1.9 Theorem The sets L_0 and $\hat{L}$ are real lattices.

Proof: We show only that L_0 is a real lattice. The proof that $\hat{L}$ is a real lattice is left as an exercise [use Lemma 1.8(b)].

Let $g_1, g_2 \in L_0$ and $\alpha, \beta \in R$ with $\alpha^2 + \beta^2 > 0$. Given $\varepsilon > 0$ in R there exists a function $\psi \in L$ so that $|g_i| \leq \psi (i = 1, 2)$ and $I\psi < \varepsilon/2(\max(|\alpha|, |\beta|))$. Then $|\alpha g_1 + \beta g_2| \leq \tilde{\psi}$, where $\tilde{\psi} = 2\psi \max(|\alpha|, |\beta|) \in L$ and $I\tilde{\psi} < \varepsilon$. Property 1.2(b) for L_0 is obvious. $\square$

If $f \in \hat{L}$ has two representations $f = \phi + g = \tilde{\phi} + \tilde{g}$ as in Lemma 1.8(a) then $^\circ I\phi = {^\circ I\tilde{\phi}} < \infty$, so we may unambiguously make the following definition.

1.10 Definition For each $f = \phi + g \in \hat{L}$, where $\phi \in L$ and $g \in L_0$, we set $\hat{I}f = {^\circ I\phi}$. The real number $\hat{I}f$ is called the *integral* of f [with respect to the hyperreal integration structure (L, I)].

1.11 Theorem The functional $\hat{I}$ is a real p.l.f. on $\hat{L}$.

Proof: The linearity of $\hat{I}$ follows from that of I. If $f \geq 0$ then $f = \phi + g$ where $\phi \in L$, $g \in L_0$, and we may take $\phi \geq 0$ by Lemma 1.8(b). Thus $\hat{I}f \geq 0$. $\square$

1.12 Corollary The pair $(\hat{L}, \hat{I})$ is a real integration structure on X.

1.13 Definition The structure $(\hat{L}, \hat{I})$ constructed from (L, I) is called the *standardization* of (L, I).

The next result gives another useful characterization of functions in $\hat{L}$. In its proof we use saturation.

1.14 Theorem A real-valued function f on X is in $\hat{L}$ iff for each $\varepsilon > 0$ in R there exist functions ψ_1 and ψ_2 in L with $\psi_1 \leq f \leq \psi_2, {^\circ I(|\psi_1|)} < \infty$, and $I(\psi_2 - \psi_1) < \varepsilon$, in which case $^\circ I\psi_1 \leq \hat{I}f \leq {^\circ I\psi_1} + \varepsilon$.

Proof: First assume that $f = \phi + g \in \hat{L}$ with $\phi \in L$, $g \in L_0$, and $^\circ I|\phi| < \infty$. For each $n \in N$ choose $\phi_n \in L$ with $|g| \leq \phi_n$ and $I\phi_n < \varepsilon/n$ for some fixed $\varepsilon > 0$ in R. Setting $\psi_1 = \phi - \phi_2$, and $\psi_2 = \phi + \phi_2$, we have $\psi_1 \leq f \leq \psi_2$, $^\circ I|\psi_1| < \infty$, and $I(\psi_2 - \psi_1) < \varepsilon$. If ψ_1 and ψ_2 are any elements of L satisfying

$\psi_1 \leq f \leq \psi_2$ and $I(\psi_2 - \psi_1) < \varepsilon$, then $\psi_1 - \phi_n \leq \phi \leq \psi_2 + \phi_n$, and so

$$^\circ I\psi_1 - \varepsilon/n \leq {}^\circ I\phi = \hat{I}f \leq {}^\circ I\psi_2 + \varepsilon/n \leq {}^\circ I\psi_1 + \varepsilon + \varepsilon/n$$

for each $n \in N$. It follows that $^\circ I\psi_1 \leq \hat{I}f \leq {}^\circ I\psi_1 + \varepsilon$.

To prove the converse we use saturation. Assume that f is an arbitrary real-valued function on X for which the conditions of the theorem hold. Then there exists an increasing sequence $\{\psi_n : n \in N\}$ (i.e., $\psi_{n+1} \geq \psi_n$ for all n) and a decreasing sequence $\{\psi'_n : n \in N\}$ in L with $\psi_n \leq f \leq \psi'_n$, $^\circ I\psi_n < \infty$, and $I(\psi'_n - \psi_n) < 1/n$ for each $n \in N$. We now apply Theorem II.8.5 with $C = N$, $D = L$, and $\phi : N \to L$ and $\phi' : N \to L$ defined by $\phi(n) = \psi_n$ and $\phi'(n) = \psi'_n$. Then there are internal extensions $\tilde{\phi} : {}^*N \to L$ and $\tilde{\phi}' : {}^*N \to L$. By the permanence principle, Theorem II.7.1(i), we may find a $k \in {}^*N_\infty$ so that ψ_n and ψ'_n form increasing and decreasing sequences and $\psi_n \leq \psi'_n$ for $n \leq k$. Thus, for some infinite $\omega \leq k$, $\psi_n \leq \psi_\omega \leq \psi'_\omega \leq \psi'_n$, and so $\psi_n - \psi'_n \leq f - \psi_\omega \leq \psi'_n - \psi_n$ for all $n \in N$. It follows that $f - \psi_\omega \in L_0$ and $f \in L$. $\square$

We now come to a result, called the monotone covergence theorem, which is central to the further development of the subject, both practically and theoretically. The result says, roughly speaking, that $\hat{L}$ is closed under monotone limits if the integrals are uniformly bounded. We will later generalize the result to a larger class of functions.

1.15 Monotone Convergence Theorem for $(\hat{L}, \hat{I})$ Suppose that $\langle f_n \in \hat{L} : n \in N \rangle$ is a monotone increasing (i.e., $f_{n+1} \geq f_n$ for all $n \in N$) sequence of functions in $\hat{L}$ for which

(a) $\lim_{n \to \infty} f_n(x) = f(x)$ exists for all $x \in X$,
(b) $\sup\{\hat{I}f_n : n \in N\} = \lim_{n \to \infty} \hat{I}f_n < \infty$.

Then $f \in \hat{L}$ and $\hat{I}f = \lim_{n \to \infty} \hat{I}f_n$.

Proof: We may assume without loss of generality that $f_n \geq 0$ (otherwise consider $f_n - f_1$). By 1.8(b) we may find representations $f_n = \phi_n + g_n$ with $\phi_n \in L$, $g_n \in L_0$, and $0 \leq \phi_n \leq \phi_{n+1}$ (check). Let $B = \lim_{n \to \infty} \hat{I}f_n$. Then given $\varepsilon > 0$ in R we may find an $m \in N$ so that, for $n \geq m$ in N, $B - \varepsilon < \hat{I}f_n \leq B$, and hence $B - \varepsilon < I\phi_n < B + \varepsilon$ for any $\varepsilon > 0$ in R.

We now use saturation again. As in the proof of Theorem 1.14 we can extend the sequence $\langle \phi_n \in L : n \in N \rangle$ to $\langle \phi_n \in L : n \in {}^*N \rangle$ so that it is still increasing (if necessary repeat some $\phi \in L$ for all $n \geq$ some k in ${}^*N_\infty$). Thus, for some infinite ω, $\phi_\omega \geq \phi_n$ for each $n \in N$ and $^\circ I\phi_\omega = \sup\{^\circ I\phi_n : n \in N\}$ (Exercise 8). We need only show that $f - \phi_\omega \in L_0$.

Fix $\varepsilon > 0$ in R, and for each $n \in N$ choose a $\psi_n \in L$ with $|g_n| \leq \psi_n$ and $I\psi_n < \varepsilon/2^n$. Again by $\aleph_1$-saturation we may extend the sequence $\langle \psi_n : n \in N \rangle$

to $\langle \psi_n : n \in {}^*N \rangle$ so that, for some infinite $k \in {}^*N$, $\psi_n \geq 0$ and $I\psi_n < \varepsilon/2^n$ for each $n \leq k$. Let $\psi = \sum_{n=1}^{k} \psi_n$. Then $I\psi < \varepsilon$ and

(1.3) $\qquad \phi_n - \psi \leq \phi_n - \psi_n \leq \phi_n + g_n \leq f \leq (1 + \varepsilon)(\phi_\omega + \psi)$

for each $n \in N$, so that

$$(\phi_n - \phi_\omega) - \psi \leq f - \phi_\omega \leq \varepsilon\phi_\omega + (1 + \varepsilon)\psi.$$

We may choose $n \in N$ so large that

$$-2\varepsilon < I(\phi_n - \phi_\omega) - I\psi.$$

Also,

$$I(\varepsilon\phi_\omega + (1 + \varepsilon)\psi) < \varepsilon I\phi_\omega + \varepsilon + \varepsilon^2.$$

Since ε is arbitrary, it follows that $f - \phi_\omega \in L_0$ (check). $\square$

Our next theorem is a result which is useful in many applications. It gives conditions under which the *standard part* ${}^\circ\phi$ of a function $\phi \in L$ is in $\hat{L}$ and $\hat{I}({}^\circ\phi) = {}^\circ I\phi$. In general we define ${}^\circ\phi$ by

$$
{}^\circ\phi(x) = \begin{cases} \mathrm{st}(\phi(x)), & \phi(x) \text{ finite}, \\ \infty, & \phi(x) \in {}^*R_\infty^+, \\ -\infty, & \phi(x) \in {}^*R_\infty^-. \end{cases}
$$

1.16 Theorem If $\phi \in L$ takes only finite values, and for some $\psi \geq 0$ in L with ${}^\circ I\psi < \infty$ we have $\{x \in X : \phi(x) \neq 0\} \subseteq \{x \in X : \psi(x) \geq 1\}$, then $\phi - {}^\circ\phi \in L_0$, ${}^\circ\phi \in \hat{L}$, and $\hat{I}({}^\circ\phi) = {}^\circ I\phi$.

Proof: For each $\varepsilon > 0$ in R, $|\phi - {}^\circ\phi| \leq \varepsilon\psi$, and so $\phi - {}^\circ\phi \in L_0$. Since $|\phi(x)| \leq n\psi(x)$ for all infinite n and all $x \in X$, an easy argument using the permanence principle shows that $|\phi| \leq n\psi$ for some finite n, and so ${}^\circ I\phi < \infty$. Thus ${}^\circ\phi = \phi + ({}^\circ\phi - \phi) \in \hat{L}$ and $\hat{I}({}^\circ\phi) = {}^\circ I\phi$. $\square$

It is important to note that if $1 \in L$ and ${}^\circ I1 < \infty$ then the conclusion of Theorem 1.16 holds under the sole assumption that ϕ is finite-valued, for then we may take $\psi = 1$. In this case we now show that if f is a real-valued function on X and $f = \phi + g$, $\phi \in L$, $g \in L_0$, then $I|\phi|$ is automatically finite.

1.17 Theorem Assume the function $1 \in L$ and ${}^\circ I1 < \infty$. Then $1 \in \hat{L}$. Moreover, if $f = \phi + g$ is a real-valued function on X with $\phi \in L$ and $g \in L_0$ then ${}^\circ I|\phi| < \infty$, i.e., $f \in \hat{L}$.

Proof: If $1 \in L$ and ${}^\circ I1 < \infty$ then, by Theorem 1.16, $1 = {}^\circ 1 \in \hat{L}$. Moreover, given $f = \phi + g$ as above, we fix $\psi \in L$ with $|g| \leq \psi$ and $I\psi < 1$. Then

$\phi - \psi \leq f \leq \phi + \psi$. Since f is real-valued and both $\phi - \psi$ and $\phi + \psi$ are internal, there is an $n \in N$ with $\phi - \psi \leq n$ and $-n \leq \phi + \psi$ (permanence principle). Thus $^\circ I|\phi| < \infty$. $\square$

1.18 Examples

1. Let $(L, I) = (*C_c(R), *\int)$ be the internal integration structure on $X = *R$ constructed from Example 1.4.2 using Proposition 1.5.

(a) We first show that if g vanishes off the bounded interval $[a, b] \subset X$ and takes only infinitesimal values, then $g \in L_0$. We may assume $b - a \geq 1$. For each $\varepsilon > 0$ in R, $|g(x)| \leq \phi$, where ϕ is the simplest piecewise linear function which is $\varepsilon/2(b - a)$ on $[a, b]$ and positive inside and zero outside $(a - \frac{1}{2}, b + \frac{1}{2})$. Then $\phi \in L$, and we see by transfer that $I\phi \leq \varepsilon$. Thus $g \in L_0$.

(b) Suppose now that $f \in C_c(R)$ and thus vanishes outside an interval $[a, b] \subset R$. We will show that $^\circ(*f) \in \hat{L}$, and $\hat{I}(^\circ(*f)) = {}^\circ I*f$. Since f is continuous we have $|f| \leq M$ for some positive $M \in R$, so $*f$ is everywhere finite. The result now follows from Theorem 1.16 if we choose ψ to be piecewise linear, 1 on $*[a, b]$, and 0 outside $[a - \frac{1}{2}, b + \frac{1}{2}]$. Note also that $I(*f) = \int f \, dx$ by transfer.

(c) Lastly we show that $\hat{L}$ contains the characteristic function of each nonstandard interval of finite length in X. Only closed intervals will be considered; the other cases are similar. Suppose that $a, b \in *R$ with $a < b$ and $|a - b|$ finite; we want to show that $\chi_{[a,b]} \in \hat{L}$. By Exercise 10 we may assume that $b \not\approx a$ (check). For each $a, b \in R$ and $n \in N$, the simplest function $f_{n,a,b}$ which is piecewise linear, continuous, and 1 if $a \leq x \leq b$, 0 if $x \leq a - 1/n$ or $x \geq b + 1/n$, is in $C_c(R)$, and $\int f_{n,a,b}(x) \, dx \leq b - a + 2$. By transfer, for each $a, b \in *R$ with $a < b$ and $|a - b|$ finite but not infinitesimal, and for each $n \in *N$, the function $F_{n,a,b}$ on $*R$ defined by

$$F_{n,a,b}(x) = \begin{cases} 1, & a \leq x \leq b, \\ 0, & x \leq a - 1/n, \quad x \geq b + 1/n, \\ n(x - a) + 1, & a - 1/n \leq x \leq a, \\ 1 - n(x - b), & b \leq x \leq b + 1/n, \end{cases}$$

is in L and $^\circ I F_{n,a,b} \leq {}^\circ(b - a) + 2 < \infty$. Now let $\omega \in *N_\infty$, and consider $g = \chi_{[a,b]} - F_{\omega,a,b}$. Then for each $n \in N$ with $2/n \leq b - a$, $|g| \leq \phi_n$, where $\phi_n \in L$ is defined by

$$\phi_n(x) = \begin{cases} n(x - a + 1/n), & a - 1/n \leq x \leq a, \\ n(a + 1/n - x), & a \leq x \leq a + 1/n, \\ n(x - b + 1/n), & b - 1/n \leq x \leq b, \\ n(b + 1/n - x), & b \leq x \leq b + 1/n, \\ 0 & \text{otherwise.} \end{cases}$$

By transfer $I\phi_n = 2/n$, and we conclude that $\chi_{[a,b]} \in \hat{L}$. An easy calculation shows that $\hat{I}\chi_{[a,b]} = {}^\circ|b - a|$.

2. Let $(L, I) = (*B(Y), *\sum)$ be the internal structure on $X = *Y = Y = \{x_1, \dots, x_n\}$ constructed from Example 1.4.4 using Proposition 1.5. We first show that L_0 consists exactly of the functions g which take infinitesimal values. Suppose that $g(x_i) \simeq 0$ for all $1 \leq i \leq n$. Then $|g| \leq \phi$, where $\phi = \varepsilon/2 \sum_1^n a_i$, and ${}^\circ I\phi < \varepsilon$ by transfer. Conversely, if $g(x_i) = r \not\simeq 0$ for some i, then when $|g| \leq \phi$, $\phi \geq |r|\chi_{\{x_i\}}$, so $I\phi \geq a_i|r| \not\simeq 0$. Thus $g \notin L_0$. It is now easy to show that $\hat{L}$, consists of all real-valued functions on X, and that $(\hat{L}, \hat{I}) = (B(Y), \sum)$.

3. Let X be any internal set and let $x_0 \in X$. Let L consist of all hyperreal-valued functions which vanish except at x_0. Put $If = f(x_0)$ for $f \in L$. Then (L, I) is an internal integration structure. It is easy to check that L_0 consists of all functions which vanish except at x_0, where they are infinitesimal, and that $\hat{L}$ consists of all real-valued functions f which vanish except at x_0, where $f(x_0)$ is finite, and $\hat{I}(f) = f(x_0)$.

Exercises IV.1

1. Show that if L is closed under linear combinations and $f, g \in L \Rightarrow f \vee g \in L$ and $f \wedge g \in L$, then $f \in L \Rightarrow |f| \in L$.
2. (Standard) Show that the structures in Examples 1.4 are real integration structures.
3. Let $X = l_2$ and $y = \langle y_i \rangle \in l_2$; assume that $y_i \geq 0$ for all $i \in N$. Show that (X, I) is an integration structure if we define $Ix = (x, y)$ for all $x \in X$.
4. Prove Proposition 1.5.
5. Show that the structure in 1.6 is a hyperreal integration structure.
6. (a) Show that, for functions ϕ and $\tilde{\phi}$ in L, $|{}^\circ I|\tilde{\phi}| - {}^\circ I|\phi|| \leq {}^\circ I|\tilde{\phi} - \phi|$.
   (b) Show that if $\phi \in L \cap L_0$, then ${}^\circ I|\phi| = 0$
7. Prove that $\hat{L}$ is a real lattice.
8. In the proof of Theorem 1.15, show that for some infinite $\omega \in *N$, $\phi_\omega \geq \phi_n$ for all $n \in N$, and ${}^\circ I\phi_\omega = \sup\{{}^\circ I\phi_n : n \in N\}$.
9. Show that one cannot in general replace $(1 + \varepsilon)(\phi_\omega + \psi)$ with $(\phi_\omega + \psi)$ in the right-hand side of Eq. (1.3) in the proof of Theorem 1.15.
10. Let (L, I) be an internal integration structure on the internal set X and suppose that the function f is real-valued and nonnegative. Show that $f \in L_0$ iff $f \in \hat{L}$ on X and $\hat{I}f = 0$.
11. (Comparison Theorem) Let (L, I) and (L', I') be two internal integration structures on the internal set X. Suppose that $L_0 \subseteq L_0'$ and that for each $\phi \in L$ there exists a $\psi \in L'$ so that $I\phi \simeq I'\psi$ and $\phi - \psi \in L_0'$. Show that $\hat{L} \subseteq \hat{L}'$ and $\hat{I}f = \hat{I}'f$ for all $f \in \hat{L}$.

12. Use Exercise 11 to show that if (L, I) and (L', I') are the $*$-transfers $(*C_0(R), *\int)$ and $(*S(R), *\int)$ of the structures in Examples 1.4.2(a) and 1.4.3 respectively, then $(\hat{L}, \hat{I}) = (\hat{L}', \hat{I}')$.

13. In the standardization of Example 1.18.1 give an example of a function $g \in L_0$ which takes infinitely large values.

14. (Standard) A collection of subsets of a set X is a *ring* if $A, B \in \mathscr{S}$ implies that $A \cup B$ and $A - B \in \mathscr{S}$. A function $v : \mathscr{S} \to R^+$ is a *finitely additive measure on S* if $v(A \cup B) = v(A) + v(B)$ for $A, B \in \mathscr{S}$ with $A \cap B = \varnothing$.

  (a) Show that if $A, B \in \mathscr{S}$ then $A \cap B$ and $A \triangle B = (A - B) \cup (B - A) \in \mathscr{S}$.

  (b) Show that if $\mathscr{E}$ is any collection of subsets of X then there is a unique ring $\mathscr{S}$ containing $\mathscr{E}$. (Hint: $\mathscr{S}$ is the intersection of all rings containing $\mathscr{E}$.)

  (c) Show that the set L of all linear combinations of characteristic functions of disjoint sets in $\mathscr{S}$ is a lattice.

  (d) Show that if $\phi = \sum_{i=1}^n a_i \chi_{A_i} \in L$, we may unambiguously define $I\phi = \sum_{i=1}^n a_i v(A_i)$ and that (L, I) is an integration structure.

15. Develop the internal analogues of the notions in Exercise 14.

IV.2 Measure Theory for Complete Integration Structures

In the last section we showed that the monotone convergence theorem holds for the integration structure $(\hat{L}, \hat{I})$ obtained from an internal structure (L, I) by standardization. In this section we develop a measure theory for *any* integration structure $(\hat{L}, \hat{I})$ for which the monotone convergence theorem is valid. Such structures will be called complete.

2.1 Definition A real integration structure $(\hat{L}, \hat{I})$ on a set X is *complete* if whenever $\langle f_n \in \hat{L} : n \in N \rangle$ is a monotone increasing sequence for which

  (a) $\lim_{n \to \infty} f_n(x) = f(x)$ exists for all $x \in X$,
  (b) $\sup\{\hat{I}f_n : n \in N\} = \lim_{n \to \infty} \hat{I}f_n < \infty$,

then $f \in \hat{L}$ and $\hat{I}f = \lim_{n \to \infty} \hat{I}f_n$.

Throughout this section $(\hat{L}, \hat{I})$ will denote a *complete* integration structure. Our first objective is to introduce a set $\hat{M}$ of functions which includes the set $\hat{L}$. The functions in $\hat{M}$ are called *measurable* functions. Roughly speaking, measurable functions will have the same regularity as functions in $\hat{L}$ but may not have finite integrals. We will find that products of measurable functions are measurable, a useful fact that is not in general true for functions in $\hat{L}$. We then extend the functional $\hat{I}$ to a subset $\hat{L}_1$ of M, and obtain a real integration structure which is an extension of $(\hat{L}, \hat{I})$.

We will also study the basic properties of those sets, called measurable, whose characteristic functions are in $\hat{M}$. This leads to a discussion of measure theory which is often taken as the starting point for a standard development of integration theory and is important in many areas of analysis; in particular, it is basic to probability theory. We will show that the two approaches are equivalent. Most of the proofs are standard except at the end of the section where we establish connections with §IV.1.

The functions in $\hat{M}$ will be extended real-valued functions; that is, they may take the values $+\infty$ and $-\infty$. Thus we make the following definition.

2.2 Definition The *extended real number system* is the set $\bar{R} = R \cup \{-\infty, +\infty\}$. By convention $-\infty < x$, and $x < +\infty$ for all $x \in R$. The rules of arithmetic for R are supplemented by the following rules: If $x \in R$ then

$$(\pm\infty) + (\pm\infty) = x + (\pm\infty) = (\pm\infty) + x = \pm\infty,$$

$$(\pm\infty)(\pm\infty) = +\infty,$$

$$(\pm\infty)(\mp\infty) = -\infty,$$

$$x(\pm\infty) = (\pm\infty)x = \begin{cases} \pm\infty & \text{if } x > 0 \\ 0 & \text{if } x = 0 \\ \mp\infty & \text{if } x < 0 \end{cases}$$

$$x/(\pm\infty) = 0 \qquad \text{for all } x \in R.$$

If a set $A \subseteq R$ is not bounded above we define $\sup A = +\infty$, and if A is not bounded below we define $\inf A = -\infty$, with a similar convention for lim sup and lim inf. As usual, we often denote $+\infty$ by ∞.

Notice that we have not defined $(\pm\infty) + (\mp\infty)$, $(\pm\infty)/(\pm\infty)$, or $(\pm\infty)/(\mp\infty)$.

2.3 Definition $\hat{L}^+$ denotes the set of nonnegative functions in $\hat{L}$. We denote by $\hat{M}^+$ the set of nonnegative $\bar{R}$-valued functions h on X such that $h \wedge f \in \hat{L}$ for each $f \in \hat{L}$. If $h \in \hat{M}^+$ we define

$$\hat{J}h = \sup\{\hat{I}(h \wedge f) : f \in \hat{L}\}.$$

$\hat{J}$ is an $\bar{R}$-valued function on $\hat{M}^+$.

We denote by $\hat{M}$ the set of $\bar{R}$-valued function h on X whose positive and negative parts $h^+ = h \vee 0$ and $h^- = -h \vee 0$ are both in $\hat{M}^+$. If $h \in \hat{M}$ and either $\hat{J}h^+$ or $\hat{J}h^-$ is finite, we define

$$\hat{J}h = \hat{J}h^+ - \hat{J}h^-.$$

2.4 Remarks

1. Since $\hat{L}$ is a lattice we see that $\hat{M} \supset \hat{L}$, and it is easy to check that if $h \in \hat{L}$ then $\hat{J}h = \hat{I}h$.

2. In defining $\hat{M}^+$ and $\hat{J}h$ for $h \in \hat{M}^+$, we may assume that $f \in \hat{L}^+$, where $\hat{L}^+$ is the set of nonnegative functions in $\hat{L}$. That is, fix $h \geq 0$ and suppose that $h \wedge f \in \hat{L}$ for all $f \in \hat{L}^+$. Then if $f = f^+ - f^- \in \hat{L}$, we have $h \wedge (f^+ - f^-) = (h \wedge f^+) - f^- \in \hat{L}$. Similarly, $\hat{J}h = \sup\{\hat{I}(h \wedge f) : f \in \hat{L}^+\}$ for $h \in \hat{M}^+$.

3. An easy calculation shows that $\hat{J}h = \sup\{\hat{I}f : 0 \leq f \leq h, f \in \hat{L}\}$ for $h \in \hat{M}^+$. This formula will be used later without explicit comment.

4. Suppose that $(\hat{L}, \hat{I})$ is obtained by standardization from (L, I). For $h \in \hat{M}^+$, $\hat{J}h$ may be less than the supremum of the integrals $^\circ I\phi$ for $\phi \in L$, $0 \leq \phi \leq h$. For example, let $X = \{x, y\}$, and let L be the internal set of $*R$-valued functions on X. For $\phi \in L$ define $I\phi = \phi(x) + \omega\phi(y)$, where $\omega \in *N_\infty$. Then each $f \in \hat{L}$ vanishes at y, $1 \in \hat{M}$, $\hat{J}1 = 1$, but $\sup\{^\circ I\phi : \phi \in L, 0 \leq \phi \leq 1\} = \infty$.

2.5 Proposition If $h_1, h_2 \in \hat{M}^+$ and $\alpha \in R^+$, then $h_1 + h_2$, αh_1, $h_1 \wedge h_2$, and $h_1 \vee h_2$ are in $\hat{M}^+$. Also $\hat{J}(h_1 + h_2) = \hat{J}h_1 + \hat{J}h_2$, $\hat{J}(\alpha h_1) = \alpha\hat{J}h_1$ for $\alpha \in R$, and $\hat{J}h_1 \leq \hat{J}(h_2)$ if $h_1 \leq h_2$.

Proof: Let $f \in \hat{L}^+$. Then

$$(h_1 + h_2) \wedge f = [(h_1 \wedge f) + (h_2 \wedge f)] \wedge f \in \hat{L}.$$

For $\alpha > 0$, $(\alpha h_1 \wedge f) = \alpha(h_1 \wedge (1/\alpha)f) \in \hat{L}$. Similarly $h_1 \wedge h_2$ and $h_1 \vee h_2 \in \hat{M}^+$.

For any $f \in \hat{L}^+$, the reader should check that $(h_1 + h_2) \wedge f \leq (h_1 \wedge f) + (h_2 \wedge f)$. Thus

$$\hat{I}((h_1 + h_2) \wedge f) \leq \hat{I}(h_1 \wedge f) + \hat{I}(h_2 \wedge f) \leq \hat{J}h_1 + \hat{J}h_2.$$

Taking the supremum on the left-hand side, we obtain

$$\hat{J}(h_1 + h_2) \leq \hat{J}h_1 + \hat{J}h_2.$$

On the other hand, suppose $f_1, f_2 \in \hat{L}$ and $f_1 \leq h_1$, $f_2 \leq h_2$. Then $f_1 + f_2 \leq h_1 + h_2$, so

$$\hat{I}f_1 + \hat{I}f_2 = \hat{I}(f_1 + f_2) \leq \hat{J}(h_1 + h_2),$$

and hence $\hat{J}h_1 + \hat{J}h_2 \leq \hat{J}(h_1 + h_2)$. Thus $\hat{J}(h_1 + h_2) = \hat{J}h_1 + \hat{J}h_2$. The rest is left as an exercise. $\square$

Our next result extends the monotone convergence theorem to $(\hat{M}, \hat{J})$. In considering its meaning remember that $\hat{J}$ is an extended real-valued function and takes on the value $+\infty$ for many functions in $\hat{M}^+$.

2.6 Monotone Convergence Theorem for $(\hat{M}^+, \hat{J})$ If $\langle h_n \in \hat{M}^+ : n \in N \rangle$ is an increasing sequence in $\hat{M}^+$, then $h = \sup h_n \in M^+$ and $\hat{J}h = \sup\{\hat{J}h_n : n \in N\} = \lim_{n \to \infty} \hat{J}h_n$.

Proof: Let $f \in \hat{L}^+$. Then $h_n \wedge f \in \hat{L}$ for each n, the sequence $\langle h_n \wedge f : n \in N \rangle$ increases to $h \wedge f$, and $\sup\{\hat{I}(h_n \wedge f) : n \in N\} \leq \hat{I}f) < \infty$. By completeness, $h \wedge f \in \hat{L}$ and $\hat{I}(h \wedge f) = \lim \hat{I}(h_n \wedge f)$. Thus $h \in \hat{M}^+$ and

$$\hat{J}h = \sup\{\hat{I}(h \wedge f) : f \in \hat{L}\}$$
$$= \sup\{\sup\{\hat{I}(h_n \wedge f) : f \in \hat{L}\} : n \in N\}$$
$$= \sup\{\hat{J}h_n : n \in N\}. \quad \square$$

It is now natural to restrict our attention to those functions in $\hat{M}$ whose integrals are finite.

2.7 Definition We define $\hat{L}_1$ to be the set of $\bar{R}$-valued functions $h \in \hat{M}$ for which $\hat{J}h$ is finite and $\hat{L}_1^+$ to be the set of nonnegative functions in $\hat{L}_1$.

The functions in $\hat{L}_1$ are extended real-valued functions. For this reason they cannot, in general, be added without encountering difficulties with expressions of the form $\infty - \infty$. We can, however, restrict ourselves to the real-valued functions in $\hat{L}_1$ and obtain an integration structure. Later we will show that with any function $f \in \hat{L}_1$ is associated a real-valued function $\tilde{f} \in \hat{L}_1$ (which equals f almost everywhere; see §IV.4) such that $\hat{J}f = \hat{J}\tilde{f}$.

2.8 Proposition The set of real-valued functions in $\hat{L}_1$ together with $\hat{J}$ forms a complete integration structure on X, $\hat{L}_1 \supseteq \hat{L}$, and $\hat{J}f = \hat{I}f$ if $f \in \hat{L}$.

Proof: Exercise. $\square$

2.9 Remarks

1. To show that a given function h is in $\hat{L}_1$ it suffices to show that $h \in \hat{M}$ and $|h| \leq g$ for some $g \in \hat{L}_1^+$ (exercise).
2. If $1 \in \hat{L}$, then every real-valued function in $\hat{L}_1$ is in $\hat{L}$ (exercise).
3. In general, $\hat{L}_1$ properly contains $\hat{L}$. In Example I.18.3, $\hat{L}$ consists of all real-valued functions which vanish except perhaps at x_0, while $\hat{L}_1$ consists

of all $\bar{R}$-valued functions f which are finite at x_0, and $\hat{J}f = f(x_0)$. In particular, $1 \in \hat{L}_1 - \hat{L}$.

To proceed we need to make a further assumption on L due, in the standard development of the subject, to Marshall Stone.

2.10 Definition A lattice L (real or hyperreal) is *Stonian* if $\phi \in L$ implies $\phi \wedge 1 \in L$. An integration structure (L, I) is Stonian if L is Stonian.

2.11 Remarks

1. If $1 \in L$, then L is Stonian.
2. If $\hat{L}$ is Stonian, then $1 \in \hat{M}^+$.
3. Each of the real lattices in Examples 1.4 is a Stonian lattice.
4. If L is a real Stonian lattice on a standard set Y, then $*L$ is an internal Stonian lattice on $X = *Y$.
5. If L is Stonian, then $\phi \wedge \alpha \in L$ for any $\alpha > 0$ since $\phi \wedge \alpha = \alpha((1/\alpha)\phi \wedge 1)$.

2.12 Proposition If (L, I) is an internal Stonian integration structure on the internal set X, then the standardization $(\hat{L}, \hat{I})$ is a Stonian integration structure.

Proof: Let $\varepsilon > 0$ in R and $f \in \hat{L}$ be given. By Theorem 1.14 there are functions $\psi_1, \psi_2 \in L$ so that $\psi_1 \leq f \leq \psi_2$, $°I\psi_1 < \infty$, and $°I(\psi_2 - \psi_1) < \varepsilon$. Then $\psi_1 \wedge 1 \leq f \wedge 1 \leq \psi_2 \wedge 1$ and $°I(\psi_2 \wedge 1 - \psi_1 \wedge 1) \leq °I(\psi_2 - \psi_1) < \varepsilon$, so $f \wedge 1 \in \hat{L}$ by Theorem 1.14. $\square$

The above results show that all of the examples of integration structures encountered so far have been Stonian. In the rest of this chapter we will assume *without further explicit comment* that all integration structures are Stonian.

To lead into our discussion of measurable sets we give an alternative characterization of measurable function in terms "good" sets which are defined as follows.

2.13 Definition We let $\hat{\mathscr{L}}$ denote the collection of all sets $A \subseteq X$ for which $\chi_A \in \hat{L}^+$.

2.14 Proposition If $A = \{x \in X : f(x) > \alpha\}$, where $f \in \hat{L}^+$ and $\alpha > 0$, then $A \in \hat{\mathscr{L}}$.

Proof: By considering $(1/\alpha)f$ we may assume $\alpha = 1$. Then $\tilde{f} = f - f \wedge 1 \in \hat{L}$, and if $B = \{x \in X : \tilde{f}(x) > 0\}$ then $A = B$. Also $1 \wedge n\tilde{f} \in \hat{L}$, $\hat{I}(1 \wedge n\tilde{f}) \le \hat{I}(1 \wedge f) \le \hat{I}f$ for all $n \in N$, and so $\chi_B = \lim(1 \wedge n\tilde{f}) \in \hat{L}$ by completeness. $\quad\square$

2.15 Proposition $\hat{M}^+$ consists of all nonnegative extended real-valued functions h such that $h \wedge n\chi_A \in \hat{L}$ for each $n \in N$ and $A \in \hat{\mathscr{L}}$. Given $h \in \hat{M}^+$, $\hat{J}h = \sup\{\hat{I}(h \wedge n\chi_A) : n \in N, A \in \hat{\mathscr{L}}\}$.

Proof: Given $f \ge 0$ in $\hat{L}$, let $A_n = \{x \in X : f(x) > 1/n\}$, $n \in N$. Then $\chi_{A_n} \in \hat{L}$ by 2.14, and the result follows from completeness and the fact that $h \wedge f = \lim_{n \to \infty} [h \wedge n\chi_{A_n} \wedge f]$ and $h \wedge n\chi_{A_n} \wedge f \le h \wedge n\chi_{A_n} \le h$. $\quad\square$

We are now ready to consider the notions of measurable set and measure. These notions were the starting point of the integration theory developed by Lebesgue. He proposed attaching a real number $\mu(A)$, called the measure of A, to a subset A of a set X. The measure of a subset can be thought of as a generalization of the length of an interval on the real line, or the area of a rectangle in the plane. Thus it is natural to require that the measure of a disjoint union of sets is the sum of the measures of the sets, at least for finite unions. Unfortunately it is usually impossible to define μ on all subsets of a given set X. The best we can expect is that the subsets, called measurable, on which μ is defined are closed under countable unions and complements, and that the measure is "countably additive". The general definitions of measurable sets and measure as presented by Lebesgue are as follows.

2.16 Definition A collection $\mathscr{M}$ of subsets of a set X is called a *σ-algebra* if

(a) $X \in \mathscr{M}$,
(b) $A \in \mathscr{M}$ implies that the complement A' of A is in $\mathscr{M}$,
(c) $\{A_i \in \mathscr{M} : i \in N\}$ implies $\bigcup A_i \, (i \in N) \in \mathscr{M}$.

Each set in $\mathscr{M}$ is called *measurable*, and $(X, \mathscr{M})$ is called a *measurable space*. A nonnegative function $\mu : \mathscr{M} \to \bar{R}^+$ is called a *measure on* $\mathscr{M}$ if $\mu(\varnothing) = 0$ and

(d) for each collection $\{A_i \in \mathscr{M} : i \in N\}$ which is disjoint (i.e., $A_i \cap A_j = \varnothing$ if $i \ne j$) we have

$$\mu(\bigcup A_i \, (i \in N)) = \sum \mu(A_i) \, (i \in N).$$

This property is called *countable additivity*. A measure μ on $\mathscr{M}$ is *complete* if

(e) whenever $A \in \mathscr{M}$ with $\mu(A) = 0$ and $B \subset A$, then $B \in \mathscr{M}$ (and thus $\mu(B) = 0$ since $\mu(B) \le \mu(A - B) + \mu(B) = \mu(A)$).

The triple $(X, \mathscr{M}, \mu)$ is called a *measure space*.

2.17 Remarks

1. $\emptyset = X'$.
2. If $\{A_i : i \in N\} \subset \mathcal{M}$ then, by De Morgan's law, $\bigcap A_i \, (i \in N) = (\bigcup A_i' \, (i \in N))' \in \mathcal{M}$.
3. Finite unions and intersections of sets in $\mathcal{M}$ are again in $\mathcal{M}$.
4. If $A, B \in \mathcal{M}$ then $A - B = A \cap B'$ and the symmetric difference $A \triangle B = (A - B) \cup (B - A)$ are in $\mathcal{M}$.
5. If μ is a measure on $(X, \mathcal{M})$, then for any collection $\{A_n \in \mathcal{M} : n \in N\}$ we have $\mu(\bigcup_1^\infty A_n) \leq \sum_1^\infty \mu(A_n)$. If $A_1 \subseteq A_2$ then $\mu(A_1) \leq \mu(A_2)$. (Exercise).
6. The term "complete" for measures is not related to completeness for integration structures.

Now we will show how to use a complete integration structure $(\hat{L}, \hat{I})$ on X to introduce a measure theory on X.

2.18 Definition A set $A \subseteq X$ is *measurable* with respect to $(\hat{L}, \hat{I})$ if $\chi_A \in \hat{M}^+$. The collection of these measurable sets is denoted by $\hat{\mathcal{M}}$. For each $A \in \hat{\mathcal{M}}$ define $\hat{\mu}(A) = \hat{J}\chi_A$.

Note that $\hat{\mathscr{L}} \subseteq \{A \in \hat{\mathcal{M}} : \mu(A) < \infty\}$.

2.19 Theorem $\hat{\mathcal{M}}$ is a σ-algebra on X and $\hat{\mu}$ is a measure on $\hat{\mathcal{M}}$.

Proof: (a) By Remark 2.11.2, $1 = \chi_X \in \hat{M}^+$.
(b) If $A \in \hat{\mathcal{M}}$ then $\chi_A \in \hat{M}^+$ and so $\chi_{A'} = 1 - \chi_A \in \hat{M}^+$.
(c) Suppose $A_i \in \mathcal{M} \, (i \in N)$ and put $A = \bigcup_{i=1}^\infty A_i$ and $B_n = \bigcup_{i=1}^n A_i$. Then $\langle \chi_{B_n} : n \in N \rangle$ is an increasing sequence of functions in $\hat{M}^+$. Since $\chi_A = \lim_{n \to \infty} \chi_{B_n}$, $\chi_A \in \hat{M}^+$ by the monotone convergence theorem, 2.6, and hence $A \in \mathcal{M}$.
(d) In the notation of (c) we have

$$\hat{\mu}(A) = \hat{J}\chi_A = \lim_{n \to \infty} \hat{J}\chi_{B_n} \qquad \text{by monotone convergence}$$

$$= \lim_{n \to \infty} \sum_{i=1}^n \hat{J}\chi_{A_i} \qquad \text{since the } \{A_i\} \text{ are disjoint}$$

$$= \sum_{i=1}^\infty \hat{\mu}(A_i). \qquad \square$$

2.20 Examples

1. Let $(\hat{L}, \hat{I})$ be the standardization of $(L, I) = ({}^*C_c(R), {}^*\!\int)$ on $X = {}^*R$ (see Example 1.18.1).

(a) $\hat{\mathscr{L}}$ contains all intervals of finite length, including intervals of infinitesimal length and (the degenerate case) single points [see Example 1.18.1(c)].
   (b) $\hat{\mathscr{M}}$ contains each interval on ${}^*R$ (exercise).
   (c) The set G of finite numbers in ${}^*R$ is in $\hat{\mathscr{M}}$ (exercise).
   (d) The set of numbers infinitesimally close to any $a \in R$ is in $\hat{\mathscr{L}}$ (exercise).

2. In Example 1.18.3, $\hat{\mathscr{L}}$ consists of $\{x_0\}$, and $\hat{\mathscr{M}}$ consists of all sets.

In the standard developments of integration, one begins with a measure on a σ-algebra $\mathscr{M}$. Using $\mathscr{M}$, one then defines the notions of measurable function and associated integral. We now present this development. Our eventual aim is to show that if we begin with the $\hat{\mathscr{M}}$ and $\hat{\mu}$ obtained from $(\hat{L}, \hat{I})$ then the measurable functions and integrals obtained from the standard development coincide with those obtained from $(\hat{L}, \hat{I})$.

In the next few results μ will be a measure on an arbitrary σ-algebra $\mathscr{M}$.

2.21 Definition An extended real-valued function h on X is *measurable with respect to* $\mathscr{M}$ if $A_\alpha = \{x \in X : f(x) > \alpha\} \in \mathscr{M}$ for each $\alpha \in R$. The set of functions f which are measurable with respect to $\mathscr{M}$ is denoted by M.

We will see presently that $M = \hat{M}$ in our situation, but a few results must first be established. We want to show that each $h \in M$ is the limit of a sequence of functions in M, each of which takes only finitely many values.

2.22 Definition A function $v \in M$ is *simple* if it takes only finitely many distinct real values $a_1, \ldots, a_n$, and the sets $A_i = \{x \in X : v(x) = a_i\} \in \mathscr{M}$ ($i = 1, \ldots, n$). The representation $v(x) = \sum_{i=1}^n a_i \chi_{A_i}$ is called the *reduced representation* of v.

2.23 Proposition Each nonnegative function $h \in M$ is the limit of a monotonically increasing sequence $\langle v_n \in M : n \in N \rangle$ of nonnegative simple functions.

Proof: Define

$$v_n(x) = \begin{cases} (k-1)/2^n, & \text{if } (k-1)/2^n \le h(x) < k/2^n, \quad 1 \le k \le n\,2^n, \\ n, & \text{if } h(x) \ge n, \end{cases}$$

(drawing a picture helps here). Then $0 \le h(x) - v_n(x) \le 1/2^n$ if $h(x) \le n$, and $v_n = n$ if $h(x) > n$. Also v_n increases monotonically to h.   □

In the standard development of integration that we are following, the integral of a nonnegative function $h \in M$ is defined as follows.

2.24 Definition Let the measure μ on $\mathcal{M}$ be given. If $v = \sum_{i=1}^{n} a_i \chi_{A_i}$ is a simple function with each $a_i \ge 0$, we define the *integral* of v by $\int v \, d\mu = \sum_{i=1}^{n} a_i \mu(A_i)$. One can show that the integral is well defined (Exercise 7). If $h \in M$ is nonnegative we define the integral of h by

$$\int h \, d\mu = \sup \left\{ \int v \, d\mu : v \text{ simple}, 0 \le v \le h \right\}.$$

If $h \in M$ and $h = h^+ - h^-$ we define $\int h \, d\mu = \int h^+ \, d\mu - \int h^- \, d\mu$ if one of the integrals is finite.

We now show that our development of integration coincides with this standard development.

2.25 Theorem Let $(\hat{L}, \hat{I})$ be a complete integration structure with measurable functions $\hat{M}$, and let M be the functions measurable with respect to the σ-algebra $\hat{\mathcal{M}}$ obtained from $(\hat{L}, \hat{I})$. Then an $\bar{R}$-valued function h is in $\hat{M}^+$ iff it is in M^+, and $\hat{J}h = \int h \, d\hat{\mu}$, where $\hat{\mu}$ is the measure obtained from $(\hat{L}, \hat{I})$.

Proof: Assume that $h \in \hat{M}^+$. For $\alpha > 0$ let $A = \{x \in X : h(x) > \alpha\}$; fix $m > \alpha$ in N and $C \in \hat{\mathcal{L}}$. For any $n \in N$, $\chi_A \wedge n\chi_C = \chi_{A \cap C}$, and

$$A \cap C = \{x \in X : h \wedge m\chi_C > \alpha\} \in \hat{\mathcal{L}}$$

by 2.14 and 2.15, so $A \in \hat{\mathcal{M}}$. Moreover,

$$\{x \in X : h(x) > 0\} = \bigcup_{n \in N} \{x \in X : h(x) > 1/n\} \in \hat{\mathcal{M}},$$

and so $h \in M^+$.

Now assume that $h \in M^+$, and fix $C \in \hat{\mathcal{L}}$ and $n \in N$. Then $h \wedge n\chi_C$ is the limit of an increasing sequence of simple functions from $\hat{L}$ by 2.23. Thus $h \wedge n\chi_C \in \hat{L}$ by completeness, so $h \in \hat{M}^+$.

To show that $\hat{J}h = \int h \, d\hat{\mu}$, note that $\int v \, d\hat{\mu} = \hat{J}v$ for nonnegative simple functions and that

$$\int h \, d\hat{\mu} = \sup\{\hat{J}v : v \text{ simple}, 0 \le v \le h\}$$
$$\le \sup\{\hat{J}f : f \in \hat{M}, 0 \le f \le h\}$$
$$= \hat{J}h.$$

But if $f \in \hat{L}$ and $0 \leq f \leq h$, then there exists an increasing sequence $\langle v_n :$ $n \in N \rangle$ of simple functions with $0 \leq v_n \leq f$ and $\lim_{n \to \infty} v_n = f$, so that

$$\hat{I}f = \lim_{n \to \infty} \hat{I}v_n \leq \int h \, d\hat{\mu}.$$

Hence

$$\hat{J}h \leq \int h \, d\hat{\mu}. \quad \square$$

2.26 Corollary $\hat{M} = M$ and $\hat{J}h = \int h \, d\hat{\mu}$ for all $h \in \hat{M}$ for which $\hat{J}$ is defined.

Proof: To show that $M \subseteq \hat{M}$ let $h = h^+ - h^- \in M$. By Theorem 2.25, $h^+ \in M^+ = \hat{M}^+$ and $h^- \in M^+ = \hat{M}^+$, and so $h \in \hat{M}$.

To show that $\hat{M} \subseteq M$ we proceed in the same way, using the fact that if $f, g \in \hat{M}^+$ and $f \cdot g = 0$, then $f - g \in M$. To prove this we have

$$\{x \in X : f(x) - g(x) > \alpha\} = \begin{cases} \{x \in X : f(x) > \alpha\} & \text{if } \alpha \geq 0 \\ \{x \in X : g(x) < -\alpha\} & \text{if } \alpha < 0. \end{cases}$$

Now $\{x \in X : f(x) > \alpha\} \in \hat{\mathcal{M}}$. Also

$$\{x \in X : g(x) < -\alpha\} = \{x \in X : g(x) \geq -\alpha\}'$$
$$= (\bigcap \{x \in X : g(x) > -\alpha - 1/n\}(n \in N))'$$

is in $\hat{\mathcal{M}}$ by Theorem 2.19. $\quad \square$

2.27 Notation Let $(\hat{L}, \hat{I})$ be a complete integration structure with associated sets and measure $\hat{M}$, $\hat{\mathcal{M}}$, and $\hat{\mu}$. With Corollary 2.26 in mind we will denote the value of $\hat{J}$ at $h \in \hat{M}$ by the standard notation $\int h \, d\hat{\mu}$.

We can now show that the set of measurable functions is closed under many limiting and algebraic operations.

2.28 Proposition If $\langle h_n \in \hat{M} : n \in N \rangle$ is a sequence of functions in $\hat{M}$, then the functions $h, H, \tilde{h}, \tilde{H}$ defined by

$$h(x) = \inf\{h_n(x) : n \in N\}, \qquad H(x) = \sup\{h_n(x) : n \in N\},$$
$$\tilde{h}(x) = \liminf h_n(x), \qquad \tilde{H}(x) = \limsup h_n(x)$$

are in $\hat{M}$.

Proof: Since $\{x \in X : H(x) > \alpha\} = \bigcup_{n=1}^{\infty} \{x \in X : h_n(x) > \alpha\}$ we see that $H \in \hat{M}$ by 2.26. Then $h \in \hat{M}$ since $\inf\{h_n\} = -\sup\{-h_n\} \in \hat{M}$. Finally $\tilde{h} = \sup\{\inf\{h_m : m \geq n\}\} \in \hat{M}$ and similar $\tilde{H} \in \hat{M}$. $\quad \square$

2.29 Proposition If $f, g \in \hat{M}$ and H is a continuous function on the plane R^2, then the function h defined by $h(x) = H(f(x), g(x))$ is in $\hat{M}$. In particular, $f + g$ and $fg \in \hat{M}$.

Proof: Since H is continuous, the sets $U_\alpha = \{\langle u, v \rangle : H(u, v) > \alpha\}$ are open, and so each can be written as a union of open boxes:

$$U_\alpha = \bigcup_{n=1}^{\infty} \{\langle u, v \rangle : \langle u, v \rangle \in (a_n, b_n) \times (c_n, d_n)\}.$$

Therefore

$$\{x : h(x) > \alpha\} = \bigcup_{n=1}^{\infty} \left(\{x : f(x) \in (a_n, b_n)\} \cap \{x : g(x) \in (c_n, d_n)\} \right)$$

is measurable (why?), and so h is measurable. $\quad\square$

The preceding two propositions can be used to show that most functions commonly encountered in analysis are measurable.

2.30 Notation If $f \in \hat{M}$ and $\int f \, d\hat{\mu}$ is defined, then $f\chi_A \in \hat{M}$ for any $A \in \hat{\mathcal{M}}$ by Proposition 2.29. We put $\int_A f \, d\hat{\mu} = \int f\chi_A \, d\hat{\mu}$.

It follows from Proposition 2.29 that if f, $g \in \hat{M}$ and $E \in \hat{\mathcal{M}}$ then the function h defined by

$$h(x) = \begin{cases} f(x), & x \in E, \\ g(x), & x \in E', \end{cases}$$

is in $\hat{M}$ (exercise). This fact will be used later without explicit reference.

We end this section with several results which hold when the complete integration structure $(\hat{L}, \hat{I})$ is the standardization of an internal structure (L, I). We begin by showing that $^\circ\phi \in \hat{M}$ for any $\phi \in L$.

2.31 Proposition If $\phi \in L$ then $^\circ\phi \in \hat{M}$.

Proof: We need only show that $^\circ\phi \wedge \chi_C \in \hat{L}$ if $\phi \in L^+$, and $C \in \hat{\mathcal{L}}$. The rest follows by considering ϕ^+ and ϕ^- and rescaling. Given $\varepsilon > 0$, choose ψ_1 and ψ_2 in L with $0 \le \psi_1 \le \chi_C \le \psi_2 \le 1$, $^\circ I \psi_2 < \infty$, and $I(\psi_2 - \psi_1) < \varepsilon$. Then

$$-\varepsilon\psi_2 + (\phi \wedge \psi_1) \le {}^\circ\phi \wedge \chi_C \le (\phi \wedge \psi_2) + \varepsilon\psi_2$$

and

$$I((\phi \wedge \psi_2) - (\phi \wedge \psi_1) + 2\varepsilon\psi_2) \le I(\psi_2 - \psi_1) + 2\varepsilon I\psi_2 \le \varepsilon + 2\varepsilon I\psi_2.$$

Since ε is arbitrary, the result follows from 1.14. $\quad\square$

The following result shows that if $h \in \hat{M}^+$, we may often be able to find a function $\phi \in L$ which is "close" to h in an appropriate sense.

2.32 Proposition Assume that $1 \in \hat{L}$. For each $h \in \hat{M}^+$ there is a function $\phi \in L$ so that $|(h \wedge n) - (\phi \wedge n)| \in L_0$ for each $n \in N$, and thus

$$\hat{J}h = \sup\{\hat{J}(h \wedge n):n \in N\} = \sup\{°I(\phi \wedge n):n \in N\} = °I(\phi \wedge \omega)$$

for some $\omega \in {}^*N_\infty$.

Proof: By Theorem 1.14 we may choose sequences $\langle \phi_n:n \in N \rangle$ and $\langle \psi_n:n \in N \rangle$ in L such that $\phi_n \leq h \wedge n \leq \psi_n$, $\phi_n \leq \phi_{n+1}$, and $I(\psi_n - \phi_n) < 1/n$ for each $n \in N$. Given $k \geq m \geq n$ in N, we obtain $\psi_m \wedge n \geq h \wedge n \geq \phi_k \wedge n \geq \phi_m \wedge n$ and $I((\psi_m \wedge n) - (\phi_m \wedge n)) \leq I(\psi_m - \phi_m) < 1/m$. By $\aleph_1$-saturation we may find $\phi \in L$ such that $\psi_m \wedge n \geq \phi \wedge n \geq \phi_m \wedge n$ for every $m, n \in N$ with $m \geq n$. Clearly $|(h \wedge n) - (\phi \wedge n)| \in L_0$. $\square$

The function ϕ is sometimes called a "lifting" of h.

Given a ${}^*R$-valued function $\phi \in L$, where (L, I) is an internal integration structure, it is important to know whether $°\phi \in \hat{L}$ or $°\phi \in \hat{L}_1$ and whether $\hat{J}(°\phi) = °I\phi$. Note that if $\phi(x) \simeq 0$ for all x, and $I\phi \not\simeq 0$, then $°\phi = 0$ and $\hat{J}(°\phi) \neq °I\phi$. If $\phi \geq 0$ then we always have $\hat{J}(°\phi) \leq °I\phi$ (Exercise 17).

2.33 Proposition Let ϕ be a finite-valued element of L. Then $°\phi \in \hat{L}$ iff $\sup \{°I(|\phi| - (1/n \wedge |\phi|)):n \in N\} < \infty$.

Proof: We may assume that $\phi \geq 0$. For each $n \in N$,

$$\{x \in X:\phi - (1/n \wedge \phi) > 0\} = \{x \in X:\phi > 1/n\}$$
$$= \{x \in X:\phi - (1/2n \wedge \phi) > 1/2n\}$$
$$\subseteq \{x \in X:2n[\phi - (1/2n \wedge \phi)] \geq 1\}.$$

Moreover, $°\phi = \lim_{n \to \infty} °[\phi - (1/n \wedge \phi)]$. If $\sup \{°I[\phi - (1/n \wedge \phi)]:n \in N\} < \infty$ then, for each $n \in N$, $°[\phi - (1/n \wedge \phi)] \in \hat{L}$ by Theorem 1.16, and so $°\phi \in \hat{L}$ by Theorem 1.15. The converse follows from the fact that if $\phi \geq 0$ is in L and $0 \leq °\phi \leq \psi \in L$ with $°I\psi < \infty$, then, for each $n \in N$, $\phi - (1/n \wedge \phi) \leq \psi$.
  $\square$

In the following treatment "S-integrability", we have replaced Anderson's original definition [2] by a condition which is a direct consequence of the definition of the general integral, and is often easier to apply.

2.34 Definition A function $\phi \in L$ is *S-integrable* if $°\phi \in \hat{L}_1$ and $\hat{J}(°\phi) = °I\phi$.

2.35 Proposition A function $\phi \in L$ is S-integrable iff $I(|\phi| - (|\phi| \wedge \omega)) \simeq 0$ and $I(|\phi| \wedge 1/\omega) \simeq 0$ for each $\omega \in {}^*N_\infty$.

Proof: We may assume that $\phi \geq 0$. For each $n \in {}^*N$, $\phi = (\phi - (\phi \wedge n)) + ((\phi \wedge n) - (\phi \wedge 1/n)) + (\phi \wedge 1/n)$. Assume that $I(\phi - (\phi \wedge \omega)) \simeq 0$ and $I(\phi \wedge 1/\omega) \simeq 0$ for each $\omega \in {}^*N_\infty$. Then by the permanence principle $I(\phi - (\phi \wedge n))$ and $I(\phi \wedge 1/n)$ are finite for some $m \in N$ and all $n \geq m$ in N. Fix $n \geq m$ in N. Then $\phi \wedge n \leq n^2(\phi \wedge 1/n)$, so $I((\phi \wedge n) - (\phi \wedge 1/n))$ is also finite, whence $I\phi$ is finite. Moreover, $(\phi \wedge n) - (\phi \wedge 1/n)$ is finite-valued and $\{x \in X : (\phi \wedge n) - (\phi \wedge 1/n) > 0\} = \{x \in X : \phi > 1/n\} \subseteq \{x \in X : n(\phi \wedge 1/n) \geq 1\}$, and so $({}^\circ \phi \wedge n) - ({}^\circ \phi \wedge 1/n) \in \hat{L}$ and ${}^\circ I((\phi \wedge n) - (\phi \wedge 1/n)) = \hat{J}((\phi \wedge n) - ({}^\circ \phi \wedge 1/n))$ by Theorem 1.16. Now by our assumption and Theorem 2.6,

$$
\begin{aligned}
{}^\circ I\phi &= \lim_{n \to \infty} {}^\circ I((\phi \wedge n) - (\phi \wedge 1/n)) \\
&= \lim_{n \to \infty} \hat{J}(({}^\circ \phi \wedge n) - ({}^\circ \phi \wedge 1/n)) \\
&= \hat{J}({}^\circ \phi).
\end{aligned}
$$

If, on the other hand, $I\phi$ is finite, then the second and third in this string of equalities hold as before. If we also have ${}^\circ I\phi = \hat{J}({}^\circ \phi)$, then $I\phi \simeq I((\phi \wedge \omega) - (\phi \wedge 1/\omega))$, and so $I(\phi - (\phi \wedge \omega)) \simeq 0$ and $I(\phi \wedge 1/\omega) \simeq 0$ for each $\omega \in {}^*N_\infty$. $\square$

Note that the condition $I(|\phi| \wedge 1/\omega) \simeq 0$ is automatically satisfied for any $\phi \in L$ and $\omega \in {}^*N_\infty$ if $1 \in L$ and ${}^\circ I(1) < \infty$.

Exercises IV.2

1. (Standard) Finish the proof of Proposition 2.5.
2. (Standard) Prove Proposition 2.8.
3. (Standard) Show that if $h \in \hat{M}$ and $|h| \leq g$ for some $g \in \hat{L}_1^+$ then $h \in \hat{L}_1$.
4. (Standard) Show that if $1 \in \hat{L}$, then every real-valued function in $\hat{L}_1$ is in $\hat{L}$.
5. (Standard) Show that if $(X, \mathcal{M}, \mu)$ is a measure space and $A_n \in \mathcal{M}, n \in N$, then $\mu(\bigcup_1^\infty A_n) \leq \sum_1^\infty \mu(A_n)$, and if $A_1 \subseteq A_2$ then $\mu(A_1) \leq \mu(A_2)$
6. Verify the statements in (b)–(d) of Example 2.20.1.
7. (Standard) Show that, for a simple function v, $\int v \, d\mu$ is well defined in Definition 2.24. That is, if $v = \sum a_i \chi_{A_i} = \sum b_j \chi_{B_j}$, $a_i \geq 0$, $b_j \geq 0$, show that $\sum a_i \mu(A_i) = \sum b_j \mu(B_j)$.
8. Show that the measure $\hat{\mu}$ obtained from the standardization $(\hat{L}, \hat{I})$ of an internal integration structure (L, I) is complete (Definition 2.16(e)). [Hint: Use Exercise IV 1.10]

9. (Standard)   Show that if $f, g \in \hat{M}$ and $E \in \hat{\mathcal{M}}$, then the function h defined by

$$h(x) = \begin{cases} f(x), & x \in E, \\ g(x), & x \in E', \end{cases}$$

is in $\hat{M}$.

10. (Standard)   Given a sequence $\langle f_n \rangle$ of measurable functions, show that the set E of points where $\lim_{n \to \infty} f_n(x)$ exists is measurable. [Hint: Consider $\limsup f_n$ and $\liminf f_n$].

11. Prove that if (L, I) is an internal integration structure with standardization $(\hat{L}, \hat{I})$, then for each $\varepsilon > 0$ and $A \in \hat{\mathcal{L}}$ there is a $\phi \in L$ with $0 \le \phi \le \chi_A$ and $\hat{\mu}(A) - {}^\circ I(\phi) < \varepsilon$. In particular, if $\hat{\mu}(A) > 0$ there is a $\phi \in L$ with $0 \le \phi \le \chi_A$ and $I(\phi) > \hat{\mu}(A)/2$.

12. (Standard)   Show that $\hat{\mathcal{M}}$ consists of those sets C such that $C \cap A \in \hat{\mathcal{L}}$ for each $A \in \hat{\mathcal{L}}$.

13. Let S be an internal hyperfinite subset of an internal set X. If $\mathcal{A}$ is the set of internal subsets of X, define the function $v: \mathcal{A} \to {}^*R^+$ by $v(A) = |A \cap S|/|S|$, where $|\cdot|$ denotes internal cardinality.

   (a) Show that v is finitely additive, i.e., $v(A \cup B) = v(A) + v(B)$ for $A, B \in \mathcal{A}$ and $A \cap B = \varnothing$.

   (b) Show how you may use the theory of §IV.1 to define a measure μ on a σ-algebra $\mathcal{M}$ of subsets of X (see 1.6 in particular) so that $\mathcal{M} \supset \mathcal{A}$ and $\mu(A) = {}^\circ v(A)$ for $A \in \mathcal{A}$. Note that $0 \le \mu(A) \le 1$ for all $A \in \mathcal{M}$.

14. (Nonmeasurable sets)   Consider Exercise 13 where $X = \{n \in {}^*N : 0 \le n < \omega, \omega \in {}^*N_\infty\}$. Define an operation $\oplus$ on X by $n \oplus m = n + m$ if $n + m < \omega$, and $n \oplus m = n + m - \omega$ if $n + m \ge \omega$. Call n and m in X equivalent if there is a standard $k \in N$ with either $n \oplus k = m$ or $m \oplus k = n$ (this is an equivalence relation). Using the axiom of choice, choose one point from each equivalence class to form a set B. Show that $B \notin \mathcal{M}$. (Hint: Show that $X = \bigcup[(B \oplus n) \cup (B \oplus (\omega - n)](n \in N))$.

15. Let (L, I) be an internal lattice. Give an example of a function $\phi \in L$ for which $I\phi$ is finite but ϕ is not S-integrable.

16. Let (L, I) be an internal lattice.

   (a) Show that if $f, g \in L$, g is S-integrable, and $|f| \le |g|$, then f is S-integrable.

   (b) Show that if $f \in L$ is S-integrable and $g \in L$ satisfies $|g| \le n$ for some $n \in N$, then fg is S-integrable.

   (c) Show that if f, g are S-integrable and $a, b \in {}^*R$ are finite, then $af + bg$ is S-integrable.

17. Modify the proof of Proposition 2.33 to show that for $\phi \geq 0$ in L, $\hat{J}(^\circ\phi) \leq {^\circ}I\phi$. (Hint: we may assume $^\circ I\phi < \infty$).

18. Use Theorem 1.14 and 1.17 to show that if (L, I) is an internal Stonian integration structure, then the function 1 is in $\hat{L}$ iff $1 \in L$ and $^\circ I(1) < \infty$.

19. State and prove Proposition 2.35 with the additional simplifying assumption that the function $1 \in L$ and $^\circ I(1) < \infty$.

20. Let (L, I) be the hyperfinite integration structure of Example 1.6, and let $(\hat{L}, \hat{I})$ be the standardization of (L, I), with associated $\hat{\mathscr{L}}$, $\hat{\mathscr{M}}$, $\hat{\mu}$, etc. Assume that $\sum a_i$ $(i \in I)$ is finite.

    (a) Show that $A \in \hat{\mathscr{L}}$ iff for every $\varepsilon > 0$ in R there exist internal subsets B and C of X such that $B \subseteq A \subseteq C$ and $\sum a_i$ $(i \in C - B) < \varepsilon$.

    (b) Show that $A \in \hat{\mathscr{L}}$ iff there is an internal set B such that $\hat{\mu}((A - B) \cup (B - A)) = 0$. (Hint: use $\aleph_1$-saturation and the permanence principle.)

IV.3 Integration on R^n; the Riesz Representation Theorem

Let X be any open or closed subset of R^n and suppose that I_0 is a positive linear functional (p.l.f.) on the lattice $C_c(X)$ of continuous functions with compact support on X (of course $C_c(X) = C(X)$ if X is compact). For example, $I_0(f)$ could denote the Reimann integral of $f \in C_c(X)$ or, more generally, the Riemann–Stieltjes integral of f with respect to an increasing integrator. In particular, $I_0(f)$ could be evaluation of f at some point $x_0 \in X$. We want to use the theory developed in the previous sections to define a measure space $(X, \mathscr{M}_X, \mu_X)$ and a corresponding *complete* integration structure (L_X, I_X) on X which is an extension of the structure $(C_c(X), I_0)$. Most of these results are easy to prove and are left as exercises. The measure μ_X will be shown to satisfy an additional condition known as regularity. This and other associated results are more technical, and can be skipped if desired. All of the above results taken together yield the Riesz representation theorem. With minor modifications except in one place, the results and proofs of this section carry over to the case that X is any locally compact Hausdorff space. One essential difficulty arises in the proof of Lemma 3.8, which, for the general case, requires Usysohn's lemma [20]. Also, if X is not compact a "countability" condition is needed for the general case to show "outer regularity."

Without further explicit comment, the nonstandard analysis in this section will be carried out in a κ-saturated enlargement $V(^*R)$ of $V(R)$. We assume that $\kappa \geq \aleph_1$. For a general space X we would need $\varkappa > \text{card } \mathscr{T}$, where $\mathscr{T}$ is the collection of open sets in X.

Let (L, I) be the internal integration structure $(*C_c(X), *I_0)$ on $*X$, with $(\hat{M}, \hat{J})$, $(\hat{L}_1, \hat{J})$, $\hat{\mathcal{M}}$, $\hat{\mathcal{L}}$, $\hat{\mu}$ denoting the objects constructed from (L, I) by the procedures of §§IV.1 and IV.2.

Recall that if G denotes the near-standard elements in $*X$ then the standard part map st: $G \to X$ maps G onto X. The basic idea of this section is to use the standard part map to lift functions from X to $*X$ as follows.

3.1 Definition For each $\bar{R}$-valued function f on X we define the function $\tilde{f}$ on $*X$ by

$$\tilde{f}(x) = \begin{cases} f(\text{st}(x)), & x \in G, \\ 0, & x \notin G, \end{cases}$$

and for each $A \subseteq X$ we define $\tilde{A} = \text{st}^{-1}(A) \cap *X$.

3.2 Remarks

1. $\tilde{f}$ is constant on the monads of standard points in $*X$, and zero at all points which are remote (i.e., not near-standard). In particular, $\tilde{f}(x) = 0$ if $x \in *X$ and the norm of x is infinite.

2. $\widetilde{af} = a\tilde{f}$, $\widetilde{f + g} = \tilde{f} + \tilde{g}$, $\widetilde{f \vee g} = \tilde{f} \vee \tilde{g}$, $\widetilde{f \wedge g} = \tilde{f} \wedge \tilde{g}$ (exercise).

3. $\tilde{\chi}_A = \chi_{\tilde{A}}$ (exercise).

We now obtain measure-theoretic structures on X with the following definition.

3.3 Definition We let $M_X = \{f : \tilde{f} \in \hat{M}\}$ and define J_X by putting $J_X(f) = \hat{J}(\tilde{f})$ when $\hat{J}(\tilde{f})$ is defined. For each set $A \subseteq X$ with $\tilde{A} \in \hat{\mathcal{M}}$, i.e., $\chi_A \in M_X$, we set $\mu_X(A) = \hat{\mu}(\tilde{A})$; the set $\mathcal{M}_X = \{A \subseteq X : \tilde{A} \in \hat{\mathcal{M}}\}$. We let L_X denote the *real-valued* functions f in M_X for which $J_X f$ is defined and finite.

3.4 Proposition (L_X, J_X) is a complete integration structure which extends $(C_c(X), I_0)$. Moreover, $(X, \mathcal{M}_X, \mu_X)$ is a measure space such that $f \in M_X$ iff f is $\mathcal{M}_X$-measurable, and $\int f \, d\mu_X = J_X f$ when $J_X f$ is defined.

Proof: That (L_X, J_X) is an integration structure is left as an exercise.

To show that (L_X, J_X) extends $(C_c(X), I_0)$, let $f \in C_c(X)$. By the uniform continuity of f, $*f(y) \simeq *f(x)$ if $y \simeq x$ and $*f$ is zero at any remote point since f has compact support. Thus $\tilde{f} = °(*f)$. By the obvious extension of Example 1.18.1(b), $\tilde{f} \in \hat{L}$ and $\hat{J}\tilde{f} = \hat{I}\tilde{f} = °I*f = I_0 f$.

To show that (L_X, J_X) is complete, let $\langle f_n \rangle$ be a monotone increasing sequence of functions in L_X for which $\lim_{n \to \infty} f_n(x) = f(x)$ exists for all $x \in X$ and $\sup\{J_X f_n : n \in N\} < \infty$. Then $\langle \tilde{f}_n \rangle$ is a monotone increasing sequence of functions in $\hat{L}_1$ and $\sup\{\hat{J}\tilde{f}_n : n \in N\} < \infty$. Also $\lim_{n \to \infty} \tilde{f}_n(z) = \tilde{f}(z)$ for all $z \in {}^*X$ (check), so $\tilde{f} \in \hat{L}_1$ and $\hat{J}\tilde{f} = \lim \hat{J}\tilde{f}_n$ by the monotone convergence theorem for $(\hat{L}_1, \hat{J})$. Therefore $f \in L_X$ and $J_X f = \lim_{n \to \infty} J_X f_n$. The rest is left to the reader (Exercise 2); the equality $\int f \, d\mu_X = J_X f$ follows from the corresponding fact for simple functions. $\square$

When we start with I_0 being the p.l.f. given by ordinary Riemann integration, then $\mathcal{M}_X$ is called the class of *Lebesgue-measurable sets* and μ_X is called *Lebesgue measure*. In that case we write $\int f \, d\mu_X$ as $\int f \, dx$.

3.5 Examples In the following examples we consider the case in which $X = R$ and I_0 is given by Riemann integration.

1. The characteristic function of any bounded interval in X is in L_X (i.e., these intervals are in $\mathcal{M}_X$). This follows from Example 1.18.1(c). The corresponding result for bounded rectangles holds if $X = R^n$.

2. Next we show that L_X contains the function

$$f(x) = \begin{cases} 1/\sqrt{x}, & 0 < x \le 1, \\ 0 & \text{otherwise}, \end{cases}$$

and hence contains unbounded functions. If $A = (0, 1]$ then χ_A and hence $n\chi_A$ are in L_X by Example 1. Thus $f_n = n\chi_A \wedge 1/\sqrt{x} \in L_X$ by the lattice property. Now the sequence $\langle f_n \rangle$ is monotone increasing and converges to f. An easy calculation shows that $J_X f_n \le 2$, so the result follows from completeness.

3. If $E \in \mathcal{M}_X$ is bounded then $\mu_X(E) < \infty$ (Exercise 3). This again generalizes to $X = R^n$.

The following results give more detailed information about $\mathcal{M}_X$ and μ_X and center about the notions of regularity, which is defined as follows.

*3.6 Notation Let $\mathcal{K}$ and $\mathcal{T}$ be the collections of subsets of X that are *compact* and *open* in X, respectively. Recall that, for $X \subseteq R^n$, $V \subseteq X$ is open in X if $V = X \cap W$ for some open $W \subseteq R^n$. A set K is compact in X iff it is compact in R^n. We write $K \prec f$ if $K \in \mathcal{K}, f \in C_c(X), 0 \le f \le 1$, and $f(x) = 1$ for all $x \in K$. We write $f \prec V$ if $V \in \mathcal{T}, f \in C_c(X), 0 \le f \le 1$, and $\operatorname{supp} f \subseteq V$. The notation $K \prec f \prec V$ means that $K \prec f$ and $f \prec V$.

*3.7 Definition A measure μ on a σ-algebra $\mathcal{M} \supseteq \mathcal{K} \cup \mathcal{T}$ of subsets of a metric space X is *inner regular* if

(a) $\mu(A) = \sup\{\mu(K):K \subseteq A,\, K \in \mathcal{K}\},\, A \in \mathcal{M}$,

outer regular if

(b) $\mu(A) = \inf\{\mu(V):A \subseteq V,\, V \in \mathcal{T}\},\, A \in \mathcal{M}$,

and *regular* if it is both inner and outer regular.

We first show that $\mathcal{M}_X \supseteq \mathcal{K} \cup \mathcal{T}$. To do so we need the following fact about continuous functions.

*3.8 Lemma Suppose $K \in \mathcal{K}$, $V \in \mathcal{T}$, and $K \subset V$. Then there exists a function $f \in C_c(X)$ so that $K \prec f \prec V$.

Proof: Let U be an open set with compact closure $\bar{U}$ such that $K \subseteq U \subseteq \bar{U} \subseteq Y$. For any set $A \subset X$ and $x \in X$, let $\rho(x, A)$ be the distance from x to A, i.e., $\rho(x, A) = \inf\{|y - x|: y \in A\}$, where $|\cdot|$ is the norm in X. Then $\rho(x, A)$ is continuous as a function of x and $\rho(x, A) = 0$ if $x \in A$. Now define f by $f(x) = \rho(x, U')/[\rho(x, U') + \rho(x, K)]$. □

*3.9 Proposition If $V \in \mathcal{T}$, then $V \in \mathcal{M}_X$ and $\mu_X(V) = \sup\{I_0 f : f \prec V\}$.

Proof: Let $A \in \hat{\mathcal{L}}$ and $\varepsilon > 0$ in R be given. We may choose $\psi_1, \psi_2 \in L$ with $0 \le \psi_1 \le \chi_A \le \psi_2 \le 1$ and $I(\psi_2 - \psi_1) < \varepsilon/3$ by Theorem 1.14 (the inequality $\psi_2 \le 1$ uses the fact that L is Stonian). Let $\mathcal{K}_0 = \{K \in \mathcal{K} : K \subset V\}$. For each $K \in \mathcal{K}_0$ let

$$\alpha_K = \inf\{{}^\circ I(\psi_1 \wedge {}^*f):K \prec f\},$$

$$\beta_K = \inf\{{}^\circ I(\psi_2 \wedge {}^*f):K \prec f\},$$

$$\alpha = \sup\{\alpha_K:K \in \mathcal{K}_0\}, \qquad \beta = \sup\{\beta_K:K \in \mathcal{K}_0\}.$$

For each $K \in \mathcal{K}_0$, $\beta_K - \alpha_K \le \varepsilon/3$, so $\beta - \alpha \le \varepsilon/3$. By definition of α, we may choose a standard $f \in C_c(X)$ with $f \prec V$ such that $I(\psi_1 \wedge {}^*f) > \alpha - \varepsilon/3$. By κ-saturation we may choose a $K' \in {}^*\mathcal{K}_0$ and a $\phi \in L$ so that $K' \supset {}^*K$ for each $K \in \mathcal{K}_0$, $0 \le \phi \le 1$, $\phi|K' \equiv 1$, and $I(\psi_2 \wedge \phi) < \beta + \varepsilon/3$ (check). It follows that $\psi_1 \wedge {}^*f \le \chi_A \wedge \chi_{\tilde{V}} \le \psi_2 \wedge \phi$ and $I(\psi_2 \wedge \phi) - I(\psi_1 \wedge {}^*f) < (\beta - \alpha) + 2\varepsilon/3 \le \varepsilon$, and hence $\chi_A \wedge \chi_{\tilde{V}} \in \hat{L}$ for each $A \in \hat{\mathcal{L}}$. We conclude that $\tilde{V} \in \hat{\mathcal{M}}$ and $V \in \mathcal{M}_X$.

We assume $\mu_X(V) < \infty$, and leave the case $\mu_X(V) = \infty$ to the reader. Given $\varepsilon > 0$ there exists an $A \in \hat{\mathcal{L}}$ so that $\hat{J}(\chi_{\tilde{V}}) \le \hat{\mu}(A \cap \tilde{V}) + \varepsilon$ since $\hat{J}\chi_{\tilde{V}} = \sup\{\hat{J}(\chi_A \wedge \chi_{\tilde{V}}): A \in \hat{\mathcal{L}}\}$. With the ψ_1 and f obtained for this ε and A as in the first paragraph, we have ${}^\circ I(\psi_1 \wedge {}^*f) \le \hat{\mu}(A \cap \tilde{V}) \le {}^\circ I(\psi_1 \wedge {}^*f) + \varepsilon$, and so

$^{\circ}I(\psi_1 \wedge {}^*f) \leq \hat{J}\chi_{\tilde{V}} = \mu_X(V) \leq {}^{\circ}I(\psi_1 \wedge {}^*f) + 2\varepsilon$. Also, by Theorem 1.16, $^{\circ}({}^*f) = \tilde{f} \in \hat{L}$, and $^{\circ}I(\psi_1 \wedge {}^*f) \leq {}^{\circ}I^*f = I_0f = \hat{J}\tilde{f} \leq \hat{J}\chi_{\tilde{V}} = \mu_X(V)$ since $\tilde{f} \leq \chi_{\tilde{V}}$. We conclude that $\mu_X(V) = \sup\{I_0f : f \prec V\}$. $\square$

*3.10   Proposition   If $K \in \mathscr{K}$ then $\tilde{K} \in \hat{\mathscr{L}}$, so $K \in \mathscr{M}_X$, and $\mu_X(K) = \inf\{I_0f : K \prec f\}$.

Proof: Let $\alpha = \inf\{I_0f : K \prec f\}$. There is a $\phi \in L$ with $0 \leq \phi \leq 1, \phi|^*K \equiv 1$, and $\phi|^*X - \tilde{K} = 0$ such that $^{\circ}I\phi = \alpha$ (check). Given $f \in C_c(X)$ with $K \prec f$ and $\varepsilon > 0$ in R, we have

$$(1 + \varepsilon)^*f \geq \chi_{\tilde{K}} \geq \phi,$$

whence $\chi_{\tilde{K}} - \phi \in L_0$, $\chi_{\tilde{K}} \in \hat{L}$, and $\hat{\mu}(\tilde{K}) = \hat{J}\chi_{\tilde{K}} = {}^{\circ}I\phi = \alpha$. $\square$

*3.11   Corollary   If $K \in \mathscr{K}$ then $\mu_X(K) = \inf\{\mu_X(V) : V \in \mathscr{T}, V \supseteq K\}$.

Proof: Exercise. $\square$

*3.12   Theorem   The measure μ_X on $\mathscr{M}_X$ is regular.

Proof: (a) We first show that μ_X is inner regular. Let $A \in \mathscr{M}_X$. For any $\varepsilon > 0$ in R and $n \in N$, choose $h \in \hat{L}^+$ so that if $\hat{J}\chi_{\tilde{A}} < \infty$ we have $\hat{J}(h \wedge \chi_{\tilde{A}}) > \hat{J}\chi_{\tilde{A}} - \varepsilon$ and if $\hat{J}\chi_{\tilde{A}} = \infty$ we have $\hat{J}(h \wedge \chi_{\tilde{A}}) > n$. Now choose $\psi \in L$ so that $0 \leq \psi \leq h \wedge \chi_{\tilde{A}}$ and $^{\circ}I\psi \geq \hat{J}(h \wedge \chi_{\tilde{A}}) - \varepsilon$. Let $K = \mathrm{st}\{y \in {}^*X : \psi(y) > 0\}$. Then K is the standard part of an internal set which is near-standard (i.e., contained in G) since $0 \leq \psi \leq \chi_{\tilde{A}}$, and $K \subseteq A$. Thus K is compact by Exercise III.3.11. Finally, if $\hat{J}\chi_{\tilde{A}} < \infty$ we have

$$\hat{J}\chi_{\tilde{A}} \geq \hat{J}\chi_{\tilde{K}} \geq {}^{\circ}I\psi \geq \hat{J}\chi_{\tilde{A}} - 2\varepsilon,$$

and $\varepsilon > 0$ is arbitrary, so (a) is established in this case. A similar argument works if $\hat{J}\chi_{\tilde{A}} = \infty$.

(b) Now we show that μ_X is outer regular. Let $A \in \mathscr{M}_X$. The result is trivial if $\mu_X(A) = \infty$, so suppose that $\mu_X(A) < \infty$. First assume that W is open in X and that $A \subseteq W \subseteq \bar{W} \subseteq X$ and $\bar{W}$ is compact. Given $\varepsilon > 0$ in R we may use (a) to find a compact $K \subseteq W - A$ so that $\mu_X[(W - A) - K] < \varepsilon$. Then the open set $V = W - K \supseteq A$ and $\mu_X(V) - \mu_X(A) < \varepsilon$. In general there exits an increasing sequence $\langle W_n \rangle$ of sets open in X with $X = \bigcup W_n (n \in N)$, and $\bar{W}_n$ compact and contained in X for each n (exercise). Let $A_k = A \cap W_k \in \mathscr{M}_X$, and put $B_1 = A_1$, $B_k = A_k - A_{k-1}$, $k \geq 2$, so that the B_k are disjoint and $\bigcup_{k=1}^{\infty} B_k = A$. For each k we may find an open set $V_k \supseteq B_k$ with

$\mu_X(V_k) < \mu_X(B_k) + \varepsilon/2^k$. Then $V = \bigcup_{k=1}^{\infty} V_k$ is open and $\mu_X(V) \le \sum_{k=1}^{\infty} \mu_X(V_k) \le \sum_{k=1}^{\infty} \mu_X(B_k) + \varepsilon = \mu_X(A) + \varepsilon.$ $\square$

The following result summarizes this section. In its proof we use the notation $\int f \, d\mu$ for integration based on a measure μ on $\mathcal{M}_X$.

*3.13 Riesz Representation Theorem Let T be a p.l.f. on $C_c(X)$. Then there exists a σ-algebra $\mathcal{M}_X$ on X which contains all open and compact subsets of X and a unique *complete* regular measure μ_X on $\mathcal{M}_X$ so that $T(f) = \int f \, d\mu_X$ for all $f \in C_c(X)$.

Proof: From the previous results, all that remains is to show the uniqueness and completeness of μ_X. To show uniqueness, let μ be any other regular measure on $\mathcal{M}_X$ so that $Tf = \int f \, d\mu$ for all $f \in C_c(X)$. It suffices to show that $\mu(K) = \mu_X(K)$ for all $K \in \mathcal{K}$ by regularity. Let $K \in \mathcal{K}$ and $\varepsilon > 0$ in R be fixed. By regularity there is a $V \supseteq K$ with $\mu(V) < \mu(K) + \varepsilon$. Let f satisfy $K \prec f \prec V$. Then

$$\mu_X(K) \le \int f \, d\mu_X = Tf = \int f \, d\mu \le \int \chi_V \, d\mu = \mu(V) < \mu(K) + \varepsilon.$$

This is true for any $\varepsilon > 0$, so that $\mu_X(K) \le \mu(K)$. Similarly $\mu(K) \le \mu_X(K)$, and the uniqueness follows. The completeness of μ_X follows easily from the completeness of $\hat{\mu}$ (see Exercise IV.2.8) and is left as an exercise. $\square$

Exercises IV.3

1. Prove the validity of Remarks 3.2.2 and 3.2.3.
2. Show that (L_X, J_X) as defined in Definition 3.3 is an integration structure, and finish the proof of Proposition 3.4.
3. Show that if $E \in \mathcal{M}_X$ is bounded, then $\mu_X(E) < \infty$.
4. Show that if X is an open or closed subset of R^n and $K \subset X$ is compact, then there is an open set V in X (i.e., $V = X \cap W$ for some open $W \subset R^n$) such that $K \subseteq V$ and the closure of V is both compact and contained in X.
5. Finish the proof of Proposition 3.9 by showing that if $\mu_X(V) = \infty$, then $\mu_X(V) = \sup\{I_0 f : f \prec V\}$.
6. Assume that X is compact in R^n, and deduce Proposition 3.9 from Proposition 3.10.
7. Prove Corollary 3.11.
8. Show that if X is open or closed in R^n, then there is an increasing sequence $\langle W_n \rangle$ of sets open in X with $X = \bigcup W_n (n \in N)$ and each $\bar{W}_n$ compact and contained in X.
9. Prove that μ_X is a complete measure in Theorem 3.13.

10. Show that in the case of Lebesgue integration the function f on R defined by

$$f(x) = \begin{cases} 1/x, & x \in (0, 1), \\ 0 & \text{otherwise} \end{cases}$$

is Lebesgue-measurable but not Lebesgue-integrable.

11. Replace $*I_0$ with any internal positive linear functional I on $*C_c(X)$ such that $I(*f) < \infty$ for each $f \in C_c(X)$. Prove that if we define (L_X, J_X) on X as in Definition 3.3, then (L_X, J_X) is a complete integration structure with $J_X f = {}^\circ If$ for each $f \in C_c(X)$.

12. Let $\Delta x = 1/n!$ with $n \in *N_\infty$ be a fixed infinitesimal and let $T = \{u \in *R: u = n\Delta x, n \in *Z\}$. Let $L = *C_c(R)$ and for $f \in L$ put $I(f) = \sum f(x)\Delta x$ $(x \in T)$ (note that for any $f \in L$ the sum is equal to the $*$-finite transfer of finite summation).

    (a) Show that (L, I) is an internal integration structure.
    (b) Show that if I' is the $*$-transfer of Riemann integration in $C_c(R)$, then there are internal functions f for which $If \neq I'f$.
    (c) If $(\hat{L}, \hat{I})$ is the standardization of (L, I), modify the procedure in Exercise IV.2.14 to produce a subset E of $*[0, 1]$ which is not in $\mathcal{M}$.

13. Let (L_X, J_X), $X = R$, denote the integration structure on X obtained from the (L, I) of Exercise 12 by the procedure of Exercise 11.

    (a) Show that the associated $\mathcal{M}_X$ contains all compact and open sets.
    (b) Show that the associated μ_X is regular.
    (c) Prove, hence, that (L_X, I_X) coincides with the Lebesgue integration structure. (Hint: Use Theorem 3.13, especially uniqueness.)

14. (Standard)   Let D be the unit disk $\{z: |z| < 1\}$ in the complex plane, and let C be its boundary $\{z: |z| = 1\}$. It is well known that for each continuous function f on C, there is a unique continuous function h_f on $\bar{D} = D \cup C$ such that $h_f|C = f$ and $h_f|D$ is harmonic, that is, $(\partial^2 h_f/\partial x^2) + (\partial^2 h_f/\partial y^2) = 0$. Moreover, $h \geq 0$ if $f \geq 0$. Use Theorem 3.13 to show that for each $x \in D$ there is a measure μ_x on C such that $h_f(x) = \int_C f \, d\mu_x$ for all continuous functions f on C.

IV.4 Basic Convergence Theorems

In this section we will present several convergence theorems which complement those which have been presented in §IV.2. Our first concern is to establish analogues of the monotone convergence theorem in which we deal with sequences of integrable functions which are not necessarily monotone. The basic results here are Fatou's lemma and the dominated convergence theorem. Next we present several results concerning various types of convergence for sequences of measurable functions, including almost uniform convergence and convergence in measure. The proofs are standard; we include these results to fill out the standard theory.

Throughout the section we will be dealing with classes M and L_1 of $\bar{R}$-valued measurable and integrable functions on a measure space $(X, \mathcal{M}, \mu)$. Integrals of functions f in M and L_1 will be denoted by $\int f \, d\mu$.

Before embarking on a presentation of the convergence theorems, we consider the role played by sets of measure zero in the discussion. These occur frequently enough for us to make the following definition.

4.1 Definition A proposition $P(x)$, which depends on $x \in X$, holds μ–*almost everywhere* (*a.e.*) if there is a set E of measure zero so that $P(x)$ is true for all $x \in E'$ (the complement of E in X). When the measure μ is understood we write a.e. instead of μ-a.e.

For example, a function f is bounded a.e. if there is a constant $B > 0$ so that $\mu(\{x : |f(x)| > B\}) = 0$. Similarly, we say that $f = g$ a.e. if there is a set $A \subseteq X$ with $\mu(A) = 0$ and $\{x : f(x) \neq g(x)\} \subseteq A$. If μ is a complete measure or f and g are measurable, we need only specify that $\mu(\{x : f(x) \neq g(x)\}) = 0$. The relation of equality a.e. is easily seen to be an equivalence relation (Exercise 1).

The basic fact is that sets of measure zero can be ignored as far as integration is concerned, as indicated by the following results.

4.2 Theorem

(a) If $f \in M$ is zero a.e. then $\int f \, d\mu = 0$.
(b) If $f \in M^+$ and $\int f \, d\mu = 0$ then $f = 0$ a.e.

Proof: Let $E = \{x : f(x) \neq 0\}$; then $E \in \mathcal{M}$.

(a) Suppose first that $f \in M^+$ and $\mu(E) = 0$. Letting $v_n = n\chi_E$, we have $v_n \in M^+$ and $\int v_n \, d\mu = n\mu(A) = 0$. With $h = \lim v_n$ it follows from Theorem 2.6 that $h \in M^+$ and $\int h \, d\mu = \sup\{\int v_n \, d\mu : n \in N\} = 0$. Finally $f \leq h$, and hence $0 \leq \int f \, d\mu \leq \int h \, d\mu = 0$, so that $\int f \, d\mu = 0$. For general f we write $f = f^+ - f^-$. If $f = 0$ a.e. then f^+ and f^- are both zero a.e., and the result follows by linearity of the integral.

(b) The sets $E_n = \{x : f(x) \geq 1/n\}$ are in $\mathcal{M}$ and $E = \bigcup E_n (n \in N)$. Since $f \geq (1/n)\chi_{E_n}$, we have $0 = \int f \, d\mu \geq (1/n)\mu(E_n) \geq 0$, so $\mu(E_n) = 0$. Hence $\mu(E) = 0$ by countable additivity. $\square$

4.3 Corollary If $f, g \in M$ and $f = g$ a.e. then $\int f\,d\mu = \int g\,d\mu$.

Proof: If $E = \{x : f(x) = g(x)\}$, then $\int f\chi_{X-E}\,d\mu = \int g\chi_{X-E}\,d\mu = 0$ by 4.2(a), $\int f\,d\mu = \int f\chi_E\,d\mu = \int g\chi_E\,d\mu = \int g\,d\mu$.

4.4 Theorem If $f \in M$ and $\int |f|\,d\mu < \infty$, then f is finite a.e.

Proof: Let $E = \{x : |f(x)| = \infty\}$. Then $E \in \mathcal{M}$ (check) and $n\chi_E \leq |f|$, and so $n\mu(E) \leq \int |f|\,d\mu < \infty$ for any $n \in N$. We conclude that $\mu(E) = 0$. □

Most of the results in §IV.2 can be improved by replacing assumptions which hold everywhere by corrresponding assumptions holding almost everywhere. We illustrate this by proving a final version of the monotone convergence theorem.

4.5 Lebesgue's Monotone Convergence Theorem Let $f_n\ (n \in N)$ and g belong to M. If $f_n \geq g$ a.e. where $\int g\,d\mu > -\infty$, and $f_n \leq f_{n+1}$ a.e. for all $n \in N$, then f_n converges a.e. to a function $f \in M$ and $\lim_{n \to \infty} \int f_n\,d\mu = \int f\,d\mu$.

Proof: By combining the countably many sets (where $f_n < g, f_n > f_{n+1}$) into one set E of measure zero, we may set each f_n and g equal to 0 on E without changing the integrals. We may also assume that $0 \geq g(x) > -\infty$ for all x (check), so $-\infty < \int g\,d\mu \leq 0$. The result now follows from the monotone convergence theorem applied to $f_n - g$. □

4.6 Fatou's Lemma If $\langle f_n \rangle$ is a sequence of nonnegative measurable functions, then $\int (\liminf f_n)\,d\mu \leq \liminf \int f_n\,d\mu$.

Proof: If $g_n = \inf f_i\ (i \geq n)$, then $g_n \in M^+$ and $\langle g_n : n \in N \rangle$ is an increasing sequence which converges to $\liminf f_n$. Also, if $n \leq m$, then $g_n \leq f_m$, so $\int g_n\,d\mu \leq \int f_m\,d\mu$; hence $\int g_n\,d\mu \leq \liminf \int f_n\,d\mu$. Therefore $\int (\liminf f_n)\,d\mu = \lim_{n \to \infty} \int g_n\,d\mu \leq \liminf \int f_n\,d\mu$ by the monotone convergence theorem. □

4.7 Lebesgue's Dominated Convergence Theorem Suppose that $\langle f_n \rangle$ is a sequence of measurable functions which converges a.e. to a measurable function f. If there is nonnegative function $g \in L_1$ so that $|f_n| \leq g$ a.e. for each $n \in N$, then $f \in L_1$ and $\int f\,d\mu = \lim_{n \to \infty} \int f_n\,d\mu$.

Proof: Fix a set $E \in \mathcal{M}$ with $\mu(E) = 0$ so that $\langle f_n \rangle$ converges to f except possibly on the set E, and $|f_n| \leq g$ except possibly on the set E. If $\tilde{f}_n =$

$f_n \chi_{X-E}$, $\tilde{f} = f\chi_{X-E}$, and $\tilde{g} = g\chi_{X-E}$ then the sequence $\langle \tilde{f}_n \rangle$ of measurable functions converges everywhere to $\tilde{f}$, $|\tilde{f}_n| \leq \tilde{g}$ on X, and finally $\int \tilde{f} \, d\mu = \int f \, d\mu$ and $\int \tilde{f}_n \, d\mu = \int f_n \, d\mu$ by Corollary 4.3.

Since $|\tilde{f}| \leq \tilde{g}$ and $\tilde{f} \in M$, $\tilde{f} \in L_1$, as is each of the functions $\tilde{f}_n$. Now $\tilde{g} + \tilde{f}_n \geq 0$, and so by Fatou's Lemma

$$\int \tilde{g} \, d\mu + \int \tilde{f} \, d\mu = \int (\tilde{g} + \tilde{f}) \, d\mu \leq \liminf \int (\tilde{g} + \tilde{f}_n) \, d\mu$$

$$= \liminf \left[\int \tilde{g} \, d\mu + \int \tilde{f}_n \, d\mu \right] = \int \tilde{g} \, d\mu + \liminf \int \tilde{f}_n \, d\mu.$$

Hence $\int \tilde{f} \, d\mu \leq \liminf \int \tilde{f}_n \, d\mu$. Similarly, applying Fatou's lemma to $\tilde{g} - \tilde{f} \geq 0$, we obtain

$$\int \tilde{g} \, d\mu - \int \tilde{f} \, d\mu = \int (\tilde{g} - \tilde{f}) \, d\mu \leq \liminf \int (\tilde{g} - \tilde{f}) \, d\mu$$

$$= \int \tilde{g} \, d\mu - \limsup \int \tilde{f}_n \, d\mu.$$

Thus $\limsup \int \tilde{f}_n \, d\mu \leq \int \tilde{f} \, d\mu$, and the result follows. □

The rest of this section will center on various convergence properties of sequences of measurable functions without special concern for the convergence of their integrals. The first of these is the famous result of Egoroff which states that a.e. convergence "almost" implies uniform convergence. To be specific we introduce the following definition.

4.8 Definition A sequence $\langle f_n \rangle$ converges *almost uniformly* if for each $\varepsilon > 0$ there exists a set $E \in \mathcal{M}$ with $\mu(E) < \varepsilon$ so that $\langle f_n \rangle$ converges uniformly on E'.

4.9 Egoroff's Theorem If $\mu(X)$ is finite and $\langle f_n \rangle$ converges a.e. to f on X then $\langle f_n \rangle$ converges almost uniformly to f.

Proof: For each k and n define the set $E_{kn} \in \mathcal{M}$ by $E_{kn} = \bigcap_{m=n}^{\infty} \{x : |f_m(x) - f(x)| < 1/k\}$. Notice that if E is the set on which $\langle f_n \rangle$ converges then for each k we have $\bigcup E_{kn}(n \in N) \supseteq E$. For fixed k we have $E_{kn} \subseteq E_{km}$ if $n \leq m$, and so $\lim_{n \to \infty} \mu(E_{kn}) = \mu(\bigcup E_{kn}(n \in N)) \geq \mu(E) = \mu(X)$. Thus, for a given $\varepsilon > 0$, we see that with each $k \in N$ is associated an $n_k \in N$ so that $\mu(E'_{kn_k}) < \varepsilon/2^k$. If $F = \bigcap E_{kn_k}(k \in N)$ then $\mu(F') \leq \sum_{k=1}^{\infty} \mu(E'_{kn_k}) < \sum_{k=1}^{\infty} \varepsilon/2^k = \varepsilon$. Finally we show that $\langle f_n \rangle$ converges uniformly on F. Let $\varepsilon > 0$ be given and find a k so that $1/k < \varepsilon$. Then $|f_m(x) - f(x)| < \varepsilon$ for all $m \geq n_k$ if $x \in E_{kn_k}$. Since $F \subseteq E_{kn_k}$ we have uniform convergence on F. □

Another type of convergence which is important in probability theory is that of convergence in measure.

4.10 Definition A sequence $\langle f_n \rangle$ of measurable real-valued functions on X *converges in measure* to a real-valued function f if for every real $\varepsilon > 0$ we have $\lim_{n \to \infty} \mu(\{x : |f_n - f| \geq \varepsilon\}) = 0$. Similarly $\langle f_n \rangle$ is *Cauchy in measure* if for each $\varepsilon > 0$ we have $\lim_{n,m \to \infty} \mu(\{x : |f_n(x) - f_m(x)| \geq \varepsilon\}) = 0$.

It is easy to see that if $\langle f_n \rangle$ is convergent in measure to f then it is Cauchy in measure.

Recall that Egoroff's theorem has been established only for sets of finite measure (see Exercise 2). The following result shows that, in general, almost uniform convergence is stronger than both convergence a.e. and convergence in measure.

4.11 Theorem If a sequence $\langle f_n \rangle$ converges to f almost uniformly then it converges a.e. and in measure.

Proof: For each $k \in N$ let $\langle f_n \rangle$ converge uniformly to f on F_k where $\mu(F'_k) < 1/k$. Then $\langle f_n \rangle$ converges on F where $F = \bigcup F_k (1 \leq k < \infty)$ and $\mu(F') \leq \mu(F'_k) < 1/k$ for each $k \in N$, so that $\mu(F') = 0$. Thus $\langle f_n \rangle$ converges a.e.

To prove convergence in measure let $\varepsilon > 0$ be given and choose k with $1/k < \varepsilon$. Since f_n converges uniformly on F_k, there is an m such that $\{x : |f_n(x) - f(x)| \geq \varepsilon\} \subseteq F'_k$ for all $n \geq$ some m depending on k. Thus $\mu(\{x : |f_n(x) - f(x)| \geq \varepsilon\}) < 1/k < \varepsilon$ for all $n \geq m$, and the result follows. $\square$

The following example shows that a sequence can converge in measure but fail to converge at any point.

4.12 Example Represent each $n \in N$ as $n = k + 2^m$, $m \geq 1$, $0 \leq k < 2^m$, and define $f_n(x)$ on $[0,1]$ to be $\chi_{[k2^{-m},(k+1)2^{-m}]}$ (the reader should draw some pictures). Then for any $x \in [0,1]$ and any n_0 there is an $m_1 \geq n_0$ and an $m_2 \geq n_0$ so that $f_{m_1}(x) = 0$ and $f_{m_2}(x) = 1$. Thus f_n does not converge at any point. On the other hand, given $\varepsilon > 0$, the Lebesgue measure of $\{x : |f_n(x)| > \varepsilon\} \leq 2/n$, so that $f_n \to 0$ in measure.

In this example it is possible to select a subsequence of $\langle f_n \rangle$ which converges a.e. This is true in general, as we now show.

4.13 Theorem If $\langle f_n \rangle$ converges in measure to f, then there is a subsequence $\langle f_{n_k} \rangle$ which converges almost uniformly and hence a.e. to f.

Proof: Given k we can find an n_k so that $\mu(\{x : |f_n(x) - f(x)| \geq 2^{-k}\}) < 2^{-k}$ for $n \geq n_k$. We may assume that $n_{k+1} > n_k$. Now let $E_k = \{x : |f_{n_k}(x) - f(x)| \geq 2^{-k}\}$. Given ε, let m be chosen so that $2^{-m+1} < \varepsilon$. If $x \notin \bigcup_{k=m}^{\infty} E_k = A$ then $|f_{n_k}(x) - f(x)| < 2^{-k}$ for $k \geq m$, so $f_{n_k}(x)$ converges uniformly to $f(x)$ on A'. But $\mu(A) \leq \sum_{k=m}^{\infty} \mu(E_k) \leq \sum_{k=m}^{\infty} 2^{-k} = 2^{-m+1} < \varepsilon$, and the result follows. $\square$

Exercises IV.4

1. (Standard)   Show that the relation $\equiv$ on the set of functions on a measure space $(X, \mathcal{M}, \mu)$ defined by $f \equiv g$ if $f = g$ a.e. is an equivalence relation.
2. (Standard)   Show that Egoroff's theorem does not hold for Lebesgue measure on all of R.
3. (Standard)   Show that if for each $n \in N$, $f_n \in L_1$ and $\sum_{n=1}^{\infty} \int |f_n| \, d\mu < \infty$, then the series $\sum_{n=1}^{\infty} f_n$ converges absolutely and almost everywhere to an integrable function f and $\int f \, d\mu = \sum_{n=1}^{\infty} \int f_n \, d\mu$.
4. (Standard)   Show that if $\lim_{n \to \infty} \int |f - f_n| \, d\mu = 0$ then f_n converges to f in measure.

In the following problems, (L, I) will be an internal integration structure and $(\hat{L}, \hat{I})$ the complete integration structure of §IV.1 with associated measurable structure of §IV.2.

5. Show that if $g \in L_0$ then $g \simeq 0$ $\hat{\mu}$-a.e. (Hint: Assuming $g \geq 0$, for any $\varepsilon > 0$, there is a $\psi \in L$ with $0 \leq g \leq \psi$ and $I\psi < \varepsilon$. Use Proposition 2.33, Exercise 2.17, and the fact that $\{x : g \leq 1/n\} \subseteq \{x : \psi \geq 1/n\} \subseteq \{x : \psi \geq 1/2n\}$)
6. (Lifting of Measurable Functions)   Assume that $1 \in \hat{L}$. A function f is in $\hat{M}$ iff there exists a $\phi \in L$ such that $^\circ\phi = f$ $\hat{\mu}$-a.e. If f is bounded then ϕ can be obtained with the same bound and $\int f \, d\hat{\mu} = {}^\circ I\phi$. (Hint: Use Proposition 2.32 and Exercise 5.) Any function $\phi \in L$ satisfying these conditions is called a *lifting* of f.
7. (Lifting of Integrable Functions)   Assume that $1 \in \hat{L}$. Show that $f \in \hat{L}_1$ iff f has an S-integrable lifting ϕ, in which case $\int f \, d\hat{\mu} = {}^\circ I\phi$.

IV.5 The Fubini Theorem

A familiar process in the theory of Riemann integration for functions of several variables is that of iterated integration. If, for example, $f(x, y)$ is a continuous function on the set $[a, b] \times [c, d]$ in $R \times R$ then we have the equality

$$\int_a^b \int_c^d f(x, y) \, dx \, dy = \int_a^b \left(\int_c^d f(x, y) \, dy \right) dx = \int_c^d \left(\int_a^b f(x, y) \, dx \right) dy.$$

The purpose of this section is to establish a nonstandard version of this equality in the contexts of the earlier sections of this chapter. The general result is known as the Fubini theorem, after its originator, G. Fubini. The nonstandard version is then applied to establish a Fubini theorem for integration structures on Euclidean spaces.

First some notation. We will be dealing with integration structures (internal or standard) on product spaces $U \times V$ (internal or standard). These structures will typically be denoted by $(L_{U \times V}, I_{U \times V})$. We will also be given integration structures (L_U, I_U) and (L_V, I_V) on U and V, respectively. Given a function $f \in L_{U \times V}$ we may find that $f(u, \cdot) \in L_V$ for $u \in U$, in which case $I_V f$ is a function of u. If $g = I_V f$ is also in L_U then we denote its integral $I_U g$ by $I_U I_V f$ (a slight abuse of notation since we are suppressing variables).

5.1 Definition Let (L_U, I_U), (L_V, I_V), and (L_W, I_W) be integration structures on U, V, and $W = U \times V$, respectively. If the integration structures are standard, we say that a function $f \in L_W$ has the *strong Fubini property* with respect to I_U, I_V, and I_W if

(i) $f(u, \cdot) \in L_V$ for all $u \in U$ and $f(\cdot, v) \in L_U$ for all $v \in V$,
(ii) $I_V f$ is in L_U and $I_U f$ is in L_V,
(iii) $I_W f = I_U I_V f = I_V I_U f$.

If "all" in (i) is replaced by "almost all" (i.e., the conditions hold a.e.), and (ii) and (iii) hold if $I_U f$ and $I_V f$ are set equal to zero when not otherwise defined, then we say that f has the *Fubini property*. If the integration structures are internal and (i), (ii) and (iii) hold without exception, we say that f has the *internal strong Fubini property*.

To begin we need the following basic result.

5.2 Lemma Let (L_U, I_U), (L_V, I_V), and (L_W, I_W) with $W = U \times V$ be real complete integration structures on U, V, and W, respectively. Suppose that each function $f_n \in L_W$ in the sequence $\{f_n : n \in N\}$ has the Fubini property with respect to I_U, I_V, and I_W, and $\{f_n\}$ is a monotone increasing sequence converging to a real-valued f. Also suppose that $\sup\{I_W f_n : n \in N\} < \infty$. Then f has the Fubini property with respect to I_U, I_V, and I_W.

Proof: Exercise.   □

We next establish results concerning the standardizations $(\hat{L}_U, \hat{I}_U)$, $(\hat{L}_V, \hat{I}_V)$, and $(\hat{L}_W, \hat{I}_W)$ of internal integration structures (L_U, I_U), (L_V, I_V), and (L_W, I_W)

on the internal sets U, V, and $W = U \times V$, respectively, in an $\aleph_1$-saturated enlargement. These will be used to establish results on Euclidean spaces via the results of §IV.3. We assume that the function 1 (i.e., the function which is identically 1) is in L_W and that $°I_W 1 < \infty$. This will allow us to apply Theorem 1.16 when $\phi \in L_W$ by taking $\psi = 1$. We also assume that each function in L_W has the internal strong Fubini property (as in the case, for example, with Riemann integration of continuous functions). In particular, 1 is in L_U and L_V and $°I_U 1 < \infty$ and $°I_V 1 < \infty$.

5.3 Lemma Suppose that ϕ is a finite-valued function in L_W. Then $°\phi$ has the strong Fubini property with respect to $\hat{I}_U$, $\hat{I}_V$, and $\hat{I}_W$.

Proof: Since, by assumption, $\phi(u, \cdot) \in L_V$ for each $u \in U$, we see that $°\phi(u, \cdot) \in \hat{L}_V$ by Theorem 1.16. Similarly, using Theorem 1.16 where necessary, we have $\hat{I}_V(°\phi) = °I_V\phi$ in L_U, $\hat{I}_U(°\phi) = °I_U\phi$ in L_V, and $\hat{I}_W(°\phi) = °I_W(\phi) = °I_U I_V(\phi) = \hat{I}_U °I_V(\phi) = \hat{I}_U \hat{I}_V(°\phi)$. The same argument with U and V reversed yields the result. $\square$

For the next lemma we use the fact (Exercise IV.1.10) that if h is real-valued and nonnegative, then h is a null function (Definition 1.7) with respect to an integration structure (L, I) iff $h \in \hat{L}$ and $\hat{I}(h) = 0$.

5.4 Lemma Suppose that h is a bounded real-valued null function on W. Then h has the Fubini property with respect to $\hat{I}_U$, $\hat{I}_V$, and $\hat{I}_W$.

Proof: We may assume that $h \geq 0$ by considering $h = h^+ - h^-$ and using the fact that the Fubini property is preserved under sums (exercise). Then we have $0 \leq h \leq K$ for some standard integer K. Since h is null there is a decreasing sequence $\langle \phi_n : n \in N \rangle$ of functions $\phi_n \in L_W$ with $h \leq \phi_n \leq K$ for all n, and $\lim °I_W(\phi_n) \, (n \in N) = 0$. Since h is real-valued there is a real-valued $H \in \hat{L}_W$ to which the sequence $\langle °\phi_n \rangle$ monotonically decreases, and $0 \leq h \leq H$. Now H also has the strong Fubini property by Lemmas 5.3 and 5.2 (appropriately modified), and $\hat{I}_W(H) = 0$. It follows from Theorem 4.2 that for almost all $u \in U$ (in the measure induced by $\hat{L}_U$, $\hat{I}_U$), $\hat{I}_V H(u, \cdot) = 0$, whence $h(u, \cdot)$ is null on V. Therefore $\hat{I}_U \hat{I}_V h = 0$. The same argument works with U and V reversed, and we conclude that the Fubini property holds for h. $\square$

Our main theorem generalizes a result of H. J. Keisler [25, p. 7]

5.5 Nonstandard Fubini Theorem Let (L_U, I_U), (L_V, I_V), and (L_W, I_W) be internal integration structures on the internal sets U, V, and $W = U \times V$, respectively, with 1 in L_W and $°I_W 1 < \infty$. Assume that every finite-valued function ϕ in L_W has the internal strong Fubini property with respect to I_U, I_V, and I_W. Then any $f \in \hat{M}_W$ for which $\hat{J}_W|f| < \infty$ has the Fubini property with respect to $\hat{I}_U$, $\hat{I}_V$, and $\hat{I}_W$.

Proof: Using the fact that the Fubini property is preserved under sums and writing $f = f^+ - f^-$, we may assume that f is positive. Also, we may assume that f is bounded by first proving the result for $f \wedge n$ and using Lemma 5.2 to pass to the limit.

Suppose then that $f \in \hat{L}_W$ is a bounded nonnegative function. Then f has a decomposition $f = \phi + h$ with $\phi \in L_W$ bounded and h a bounded null function (check). Now $f = {}^\circ\phi + (\phi - {}^\circ\phi) + h$, and since the null function $(\phi - {}^\circ\phi) + h$ is real-valued, the theorem follows from 5.3 and 5.4. $\square$

We will now apply Theorem 5.5 to prove a Fubini theorem for integration structures in Euclidean spaces. In the following, X and Y will denote closed and bounded (and thus compact) subsets of R^n and R^m, respectively, and $Z = X \times Y$. Notice that 1 belongs to $C(X)$, $C(Y)$, and $C(Z)$. Given positive linear functionals I_X, I_Y, and I_Z on $C(X)$, $C(Y)$, and $C(Z)$, we obtain integration structures $(C(X), I_X)$, $(C(Y), I_Y)$, and $(C(Z), I_Z)$. These structures have $*$-transforms on $*X$, $*Y$, and $*Z$, namely, $(*C(X), *I_X)$, $(*C(Y), *I_Y)$, and $(*C(Z), *I_Z)$, respectively. For example, $*C(X)$ is the set of all $*$-continuous functions on $*X$. Using the techniques of §§IV.1 and IV.3, we find that these internal structures induce integration structures $(\hat{L}_X, \hat{I}_X)$, $(\hat{L}_Y, \hat{I}_Y)$, and $(\hat{L}_Z, \hat{I}_Z)$ on $*X$, $*Y$, and $*Z$, which in turn induce integration structures (L_X, J_X), (L_Y, J_Y), and (L_Z, J_Z) on X, Y, and Z, respectively. The latter structures extend $(C(X), I_X)$, $(C(Y), I_Y)$, and $(C(Z), I_Z)$. The reader should recall (Remark 2.9.2) that every real-valued function in $(\hat{L}_Z)_1$ is in $\hat{L}_Z$. We remark that for $f \in C(Z)$ the equality of the iterated integrals always nolds [34, 16B, p. 44]. If that common value is I_Z then the strong Fubini property holds for f.

5.6 Standard Fubini Theorem Assume that X and Y are compact. Suppose that each $f \in C(Z)$ has the strong Fubini property with respect to I_X, I_Y, and I_Z. Then each $f \in M_Z$ such that $J_Z|f| < \infty$ has the Fubini property with respect to J_X, J_Y, and J_Z.

Proof: It suffices to prove the result for f bounded and hence in L_Z. The assumptions of Theorem 5.5 are satisfied with $*X = U$, $*Y = V$, and $*Z = W$, since the strong Fubini property for each $f \in C(Z)$ transfers to the internal strong Fubini property for each $\phi \in *C(Z)$. Let $f \in L_Z$. Then $\tilde{f} \in \hat{L}_Z$ has the Fubini property with respect to $\hat{I}_X$, $\hat{I}_Y$, and $\hat{I}_Z$. If ${}^\circ x_1 = {}^\circ x_2$ then $\tilde{f}(x_1, y) = \tilde{f}(x_2, y)$ for all $y \in *Y$. Thus there is a standard set $A \subset X$ such that $\tilde{f}(x, \cdot) \in \hat{L}_Y$ for all $x \in *X - \tilde{A}$. Also $\tilde{A}$ is null in $*X$ so A is null in X. If $x \in X - A$ then $\widetilde{f(x, \cdot)} = \tilde{f}(x, \cdot)$ on $^*Y$ so $J_Y f(x, \cdot) = \hat{J}_Y \tilde{f}(x, \cdot)$. Set $J_Y f(x, \cdot) = 0$ for $x \in A$. Since $\widehat{J_Y f(x, \cdot)} = \hat{J}_Y \tilde{f}(x, \cdot)$ for $x \in {}^*X - \tilde{A}$, we have $J_Y f(x, \cdot) \in L_X$ and $J_X J_Y f = \hat{J}_X \hat{J}_Y \tilde{f} = \hat{J}_Z \tilde{f} = J_Z f$. The same argument with the roles of X and Y reversed gives the result. $\square$

We have established the Fubini theorem for the case that X and Y are compact subsets of R^n and R^m, respectively. The extension of this result for the case that X and Y are both open or both closed in R^n and R^m is a standard exercise, which we leave to the reader (Exercise 3).

Exercises IV.5

1. Prove Lemma 5.2.
2. Show that the Fubini property is preserved under sums.
3. Use Theorem 5.6, Exercise IV.3.8, and the obvious extension of Lemma 5.2 (for the case of $\bar{R}$-valued functions) to establish Fubini's theorem for integrable f on $X \times Y$, when X and Y are both open or both closed in R^n and R^m, respectively.
4. (Nonstandard version of Tonelli's theorem)   In the notation of this section, assume that $1 \in L_W$ with $^\circ I_W 1 < \infty$, and the other assumptions of Theorem 5.5 hold. Show that if $f \in \hat{M}_W^+$, then

   (a) $f(u, \cdot) \in \hat{M}_V$ for a.e. $u \in U$, and $f(\cdot, v) \in \hat{M}_U$ for a.e. $v \in V$,
   (b) $\hat{J}_V f(u, \cdot) \in \hat{M}_U$, and $\hat{J}_U f(\cdot, v) \in \hat{M}_V$,
   (c) $\hat{J}_W f = \hat{J}_U \hat{J}_V f = \hat{J}_V \hat{J}_U f$.

5. In the context of Exercise 4, show that if $f \in \hat{M}_W$ and either of the repeated integrals $\hat{J}_U \hat{J}_V |f|$ or $\hat{J}_U \hat{J}_V |f|$ is finite, then $|\hat{J}_W f| < \infty$ and $\hat{J}_W f = \hat{J}_U \hat{J}_V f = \hat{J}_V \hat{J}_U f$.
6. State and prove a standard version of Tonelli's theorem extending Theorem 5.6 for the case that $f \geq 0$.
7. (Standard)   (a) Let f be the function on $[0, 1] \times [0, 1]$ defined by

$$f(x, y) = \begin{cases} \dfrac{x^2 - y^2}{(x^2 + y^2)^2}, & (x, y) \neq (0, 0), \\[2mm] 0, & (x, y) = (0, 0). \end{cases}$$

Use trigonometric substitutions to show that (with Lebesgue integration)

$$\int_0^1 \left[\int_0^1 f(x, y)\, dy \right] dx = \frac{\pi}{4},$$

$$\int_0^1 \left[\int_0^1 f(x, y)\, dx \right] dy = -\frac{\pi}{4}.$$

Conclude that f is not Lebesgue integrable on $[0, 1] \times [0, 1]$.

(b) Let f be the function on $S = [-1,1] \times [-1,1]$ defined by

$$f(x) = \begin{cases} \dfrac{xy}{(x^2 + y^2)^2}, & (x, y) \neq (0,0) \\ 0, & (x, y) = (0,0) \end{cases}$$

Show that the iterated integrals of f over S are equal, but f is not integrable.

*IV.6   Applications to Stochastic Processes

In this section we present a few examples which show how the theory of integration structures and the associated measure theory as developed in §§IV.1 and IV.2 can be applied to problems in probability theory, and in particular to stochastic processes. The essential idea is to extend the concepts of elementary probability theory on finite sample spaces to situations in which the sample space is a *-finite set in some enlargement. By transfer, this allows us to use the techniques of calculation and also the conceptual simplicity of the finite cases to deal with probabilistic situations in which the sample spaces are intrinsically infinite. The standard treatment of the problems we present below, and especially Brownian motion, can be a little complicated. Following the nonstandard treatment of coin tossing and Poission processes in [27], a nonstandard approach to the theory of Brownian motion was developed by Robert Anderson in [2]. This work has since led to a sequence of papers on nonstandard probability theory (see, for instance, the survey article [39] and other related papers [19]) and, in particular, has resulted in the solution of some difficult questions in the theory of stochastic processes by Keisler [25] and Perkins [33].

We begin with a very quick survey of probability theory. This theory was developed in order to provide a mathematical foundation for the study of problems in which the outcomes of certain experiments or measurements cannot be determined with certainty. To illustrate, we consider two typical examples from elementary probability theory:

1. A die is tossed at random and the upturned face is recorded.
2. A marksman is shooting at a target, and the resulting hole in the target is noted. Each shot is subject to unpredictable effects of wind.

In example 1 the words "at random" are meant to convey that no device (for example, weighting the die, or influencing it with magnets) is in operation. We are interested in the likelihood of a particular number or set of numbers occurring on any given toss. It is almost evident that one is more likely to toss an element from the set $\{2, 4, 6\}$ than that the number 3 will turn up. If the die is tossed n times and even numbers turn up n_1 times, then n_1/n "should" turn out to be quite close to $\frac{1}{2}$ (and "should" approach $\frac{1}{2}$ as $n \to \infty$. Thus, the ratio $\frac{1}{2}$ is a measure of the likelihood of an even number turning up and is called the *probability* of that event. To attack the problem mathematically and, in particular, to attach a meaning to the word "should" used above, we consider the set $\{1, 2, 3, 4, 5, 6\}$ consisting of all the possible outcomes of the experiment. Because of the randomness we argue that each face is equally likely to turn up, and so the probability of any outcome is $\frac{1}{6}$. Using the idea that the probability of an even number turning up is the sum of the probabilities of a $2, 4$, or 6 turning up, we see that the probability of an even number turning up is $\frac{1}{2}$. Similarly, we can attach a number $P(A)$ between 0 and 1 to any subset A of $\{1, 2, \ldots, 6\}$ which will be a measure of the likelihood of a number in A turning up and will be called the probability of the event A. Given the random nature of the experiment, $P(A)$ will be $|A|/6$, where $|A|$ is the number of elements in A. For multiple tosses of the die, we must consider a product space and corresponding probabilities.

In example 2 the analogue of the set $\{1, 2, \ldots, 6\}$ in example 1 is the set of points in the target. Each point has zero probability of being hit, but sets with positive Lebesgue measure are assigned a positive probability of being hit. A typical event is the event A of hitting a particular set in the target. The probability of the event A should again be a number between 0 and 1 which could be approximately determined by performing the experiment many times.

In general, an abstract model for problems in probability is constructed as follows:

(a) We construct a space Ω, called the *sample space*, whose points consist of all of the outcomes of the experiment. In example 1, Ω consists of the possible six faces (or, equivalent, the numbers from 1 to 6). In example 2, Ω consists of all of the points in the target.

(b) The events to which we wish to assign a probability are subsets of Ω. Given events A and B, A' is the event that A does not occur, and $A \cup B$ is the event that A or B occurs. More generally we require the set of events to be closed under complementation and countable unions and thus be a σ-algebra $\mathscr{E}$.

(c) The probability of an event A is a number $P(A)$ satisfying $0 \leq P(A) \leq 1$. Since $\mathscr{E}$ is a σ-algebra we can and do require P to be a measure on $\mathscr{E}$. In particular, $P(\bigcup_1^\infty A_n) = \sum_{i=1}^\infty P(A_n)$ for disjoint events A_n. This generalizes the procedure for constructing P used in example 1.

Now we have the following definition.

6.1 Definition A *probability space* is a measure space $(\Omega, \mathscr{E}, P)$, where P is a measure on $(\Omega, \mathscr{E})$ satisfying $P(\Omega) = 1$. The σ-algebra $\mathscr{E}$ is called the collection of *events* and P the *probability measure*.

When the space Ω is finite and $\mathscr{E}$ is the set of all subsets of Ω, then P is completely determined by its values on the points in Ω. A particularly important situation, as represented above, occurs when all the points of Ω are assigned equal probabilities (the *equiprobability model*), in which case $P(A) = |A|/|\Omega|$. In the examples we will consider below, the nonstandard models will be hyperfinite analogues of the equiprobability model. Following a standard convention for probability theory, we will use ω to denote elements of Ω. Thus ω will no longer be used for elements of $*N_\infty$.

In applications, we are usually concerned with functions defined on the sample space Ω. For instance, in example 2 the target might be divided into three concentric regions A_1 (the central circle), A_2, and A_3, and the marksman could score 10, 5, or 1 depending on whether he hit A_1, A_2, or A_3. The expected average score of the marksman if the shooting is performed many times would be $10P(A_1) + 5P(A_2) + P(A_3)$. This leads us to the following definition.

6.2 Definition A *random variable* is a real-valued measurable function X on the probability space $(\Omega, \mathscr{E}, P)$. The *expected value* $E(X)$ of X is $\int X \, dP$ (when the integral is defined).

In many situations we are more interested in a particular random variable than in the underlying probability space $(\Omega, \mathscr{E}, P)$ on which it is defined. The probabilistic information involving a random variable X is contained in its *distribution* P_X, which is a probability measure defined on the collection $\mathscr{M}$ of Borel-measurable subsets of the real line R (i.e., the smallest σ-algebra containing the open sets in R) by the formula $P_X(A) = P(\{\omega \in \Omega : X(\omega) \in A\})$, $A \in \mathscr{M}$. It turns out that P_X is completely determined by its value on all intervals in R. Thus in many applications the properties of a random variable are defined in terms of the function $F_X(x) = P_X((-\infty, x])$, which is called the *distribution function* of X. When X takes on only finitely many values $\{a_1, \ldots, a_n\}$ then P_X is completely determined by the values $P_X(a_i) = P(\{\omega \in \Omega : X(\omega) = a_i\})$, $i = 1, \ldots, n$.

We are now ready for the definition of a stochastic process.

6.3 Definition A *stochastic process* is a family $\{X_t : t \in I\}$ of random variables all defined on a common probability space $(\Omega, \mathscr{E}, P)$. I is called the *parameter set*.

In the following examples I will be either the positive integers (for infinite coin tossing) or a subset of the real line (for the Poisson and Brownian motion processes). In the case of the coin-tossing and Poisson processes, the random variables will take values in the integers.

A fundamental notion in probability theory, and especially important for stochastic processes, is the notion of independence.

6.4 Definition A collection $X_1, \ldots, X_n$ of random variables is *independent* if for any $x_1, \ldots, x_n$

$$P(\{\omega \in \Omega : X_1(\omega) \le x_1, \ldots, X_n(\omega) \le x_n\}) = \prod_{i=1}^{n} P(\{\omega \in \Omega : X_i(\omega) \le x_i\}).$$

If the X_i are integer-valued we can replace the inequalities $\le$ by equality.

Suppose, for example, a coin is tossed n times and X_k, $1 \le k \le n$, is the random variable which records a 1 or -1 if the outcome of the kth toss is a head or a tail, respectively. Here Ω is the set of sequences of 1's or -1's of length n and contains 2^n elements. It is clear that, for any k, $P(\{\omega \in \Omega : X_k(\omega) = 0\})$ is $2^{n-1}/2^n = 1/2$. More generally, if x_i is fixed as 1 or -1 for $1 \le i \le k$, then $P(\{\omega \in \Omega : X_{n_1} = x_1, \ldots, X_{n_k} = x_k\}) = 2^{n-k}/2^n = 1/2^k$, so the X_k are independent.

A common practice is to define a stochastic process $\{X_t\}$ by properties involving the distribution functions of certain combinations of the X_t, for example the *increments* $X_t - X_s$. The Poisson and Brownian motion processes are ones in which the increments over a finite number of disjoint intervals are independent.

One last notion, which is central to probability, is that of conditional probability. Suppose, for example, that we want to compute the probability of a 5 turning up in example 1 given the extra information that an odd number will turn up. The answer is clearly $\frac{1}{3}$. In general the probability of A given B is denoted by $P(A|B)$ and is computed as follows.

6.5 Definition The *conditional probability* of the event A given the event B is given by $P(A|B) = P(A \cap B)/P(B)$ if $P(B) \ne 0$.

In the standard approach to the problems to follow it is sometimes difficult to define a suitable space $(\Omega, \mathscr{E}, P)$ on which the process is defined. One advantage of the nonstandard approach is that this step is relatively easy.

6.6 Example (Infinite Coin Tossing)

In the elementary theory of probability (for finite sample spaces), one encounters the experiment of tossing a fair (i.e., unbiased) coin a finite number of times. If the coin is tossed n times, then, as just remarked, a sample space for the experiment can be taken to be the set Ω_n of all sequences $\langle e_1, e_2, \ldots, e_n \rangle$, where e_i is either $+1$ or -1 depending on whether a head or a tail is obtained on the ith toss; thus Ω consists of 2^n points (sequences). Specifying any event, for example the event of obtaining exactly two heads in n tosses, is the same as specifying a subset A of Ω. Since the coin is fair, it is argued that each sequence is equally likely, and so the probability $P_n(A)$ of an event A is measured by $P_n(A) = (1/2^n)|A|$, where $|A|$ is the cardinality of A.

Suppose now that a coin is tossed an infinite number of times. We may define an associated stochastic process $\{X_n : n \in N\}$ by putting $X_n(\omega) = +1$ or -1 depending on whether a head or a tail occurs on the nth toss. Thus $\{X_n\}$ is a discrete parameter stochastic process in which the X_n take on the values 1 and -1. We would now like to define a probability space $(\Omega, \mathcal{E}, P)$ on which this process is defined. In the standard theory one takes Ω to be the (infinite) set of all infinite sequences $\langle e_1, e_2, \ldots \rangle$ of $+1$'s and -1's. Now, however, the specification of the set of events and the probability of each event is not so clear. It is required that the set of events form a σ-algebra $\mathcal{E}$ of subsets of Ω, and that the probability is a countably additive measure P on $\mathcal{E}$ with $0 \leq P(A) \leq P(\Omega) = 1$ for each $A \in \mathcal{E}$. Also, $\mathcal{E}$ should contain any event A, which depends on only a finite number of tosses, for example the event of getting two heads in the first 10 tosses, and P should assign to this event A the probability obtained by using only the finite theory. In the standard theory, the existence of an $\mathcal{E}$ and P satisfying these conditions is a consequence of a general theorem of Kolmogoroff.

We now show that the nonstandard theory provides an appropriate $\mathcal{E}$ and P, and that these have conceptual as well as calculational advantages. Our sample space Ω is the internal set of all internal sequences $\langle e_1, e_2, \ldots, e_\eta \rangle$ of $+1$'s and -1's of length η, where $\eta = \zeta!$ and ζ is an infinite integer. The lattice L is the set of all hyperreal-valued internal functions on Ω; we define $I(\phi) = (1/2^\eta) \sum \phi(e_i)$ if $\phi \in L$. As noted in 1.6, (L, I) is an internal integration structure. We denote by $\mathcal{E}$ the collection of internal subsets of Ω and put $P(A) = I(\chi_A)$ for $A \in \mathcal{E}$. The associated collection $\hat{\mathcal{E}}$ of measurable sets will be the collection of events, and the measure $\hat{P}$ on $\hat{\mathcal{E}}$ coming from $(\hat{L}, \hat{I})$ as in §IV.2 will define the probability. The reader should check that $0 \leq P(A) \leq 1$ for each $A \in \mathcal{E}$, and that any internal set A is in $\hat{\mathcal{E}}$ with $\hat{P}(A) = \mathrm{st}(|A|/2^\eta)$, where $|\cdot|$ is the internal cardinality, i.e., the $*$-transfer of the standard cardinality function. It is not hard to show that if A is an internal set in $\hat{\mathcal{E}}$ which depends on only the first n tosses then $P(A)$ equals the probability obtained using the finite theory on Ω_n. Thus $(\Omega, \hat{\mathcal{E}}, \hat{P})$ is an alternative to the standard space mentioned above.

We can use $(\Omega, \hat{\mathcal{E}}, \hat{P})$ to compute the probabilities of events depending on an infinite number of tosses. As an example, let A_n be the event "The first

$n - 1$ tosses are tails, then the nth toss is a head" in Ω. Then $\hat{P}(A_{2n}) = 1/2^{2n}$. The event $A = \bigcup_{n=1}^{\infty} A_{2n}$ corresponds to the standard event of getting at least one head in an infinite number of tosses, the first one occurring at an even-numbered toss, and $\hat{P}(A) = \sum_{1}^{\infty} (1/2^{2n}) = \frac{1}{3}$. Note that the internal set $B = \bigcup_{n=1}^{\eta/2} A_{2n}$ is the event of getting at least one head in η tosses, the first one occurring at an even-numbered toss, and we also have $\hat{P}(B) = \text{st}(\sum_{n=1}^{\eta/2} (1/2^{2n}) = \text{st}(\frac{1}{3} - 1/(3 \cdot 2^{\eta})) = \frac{1}{3}$.

We may now consider the original stochastic process $\{X_n : n \in N\}$ as a process on $(\Omega, \hat{\mathscr{E}}, \hat{P})$ by putting $X(\omega) = \{_1^{-1}$ according as $e_n = \{_1^{-1}$ in $\omega = \langle e_1, \ldots, e_\eta \rangle$. For each n, $P(\{\omega \in \Omega : X_n(\omega) = e_n\}) = \text{st}(2^{\eta-1}/2^{\eta}) = \frac{1}{2}$ for $e_n = -1$ or 1. Similarly any finite set of the X_n's is independent.

6.7 Example   (The Poisson Process)

The Poisson process is a stochastic process which is intended to model situations in which isolated events occur randomly in time. Imagine, for instance, an experiment in which we record the time $t \geq 0$ of arrival of each telephone call to an office. We can define a set $\{\tilde{N}(\omega, t) : t \in [0, \infty)\}$ of random variables by specifying that, for any particular ω in the as yet unspecified probability space $(\tilde{\Omega}, \tilde{\mathscr{E}}, \tilde{P})$ (ω should represent a particular selection from the set of ways the calls come in), $\tilde{N}(\omega, t)$ equals the number of incoming calls in the time interval $[0, t]$. Then, for $s < t$, $\tilde{N}(\omega, t) - \tilde{N}(\omega, s)$ equals the number of calls in the interval $(s, t]$. In many situations it is found that $\tilde{N}(\omega, t)$ has the following properties, which define a Poisson process, and in particular, force the measurability of the process.

(6.1) for each ω, $\tilde{N}(\omega, t) \geq 0$, $\tilde{N}(\omega, 0) = 0$, and $\tilde{N}(\omega, t)$ is integer-valued,

(6.2) if $s < t$ and $\omega \in \tilde{\Omega}$, then $\tilde{N}(\omega, s) \leq \tilde{N}(\omega, t)$ and $\tilde{N}(\omega, t)$
        is right continuous (see Exercise 3(a) for definition),

(6.3) for each $t_1 < t_2 < \cdots < t_n \in R$ the random variables
        $\tilde{N}(\cdot, t_2) - \tilde{N}(\cdot, t_1), \ldots, \tilde{N}(\cdot, t_n) - \tilde{N}(\cdot, t_{n-1})$ are independent
        [i.e., the $\tilde{N}(\cdot, t)$ have independent increments)],

(6.4) $\tilde{P}(\{\omega : \tilde{N}(\omega, s + t) - \tilde{N}(\omega, s) = k\}) = e^{-\lambda t}((\lambda t)^k / k!)$.

The assumption (6.3) says that what happens in one time interval is not affected by what happens in a disjoint time interval. The assumption (6.4) says in particular that the probability of n calls occurring in the interval $(s, s + t]$ is independent of s and depends only on the length t of the interval and the parameter λ in the manner indicated. We call λ the *rate* of the process.

We now present a nonstandard model for discussing the Poisson process. In doing so we will specify an appropriate probability space $(\tilde{\Omega}, \tilde{\mathscr{E}}, \tilde{P})$ on which the process can take place.

As in Example 6.6, let $\eta = \zeta!$ be an infinite factorial in $*N$. For simplicity we choose a standard positive rational number λ as the rate for our process, and we let γ be the infinite integer $\lambda\eta$. Divide the interval $[0, \eta)$ into η^2 intervals $[0, 1/\eta), [1/\eta, 2/\eta), \ldots, [(\eta - 1)/\eta, \eta)$, and let Ω be the internal set of all internal ways that γ distinguishable points can be put into the η^2 intervals $[k/\eta, (k + 1)/\eta)$. That is, Ω consists of internal sequences $\omega = \langle \omega_i : 1 \leq i \leq \gamma \rangle$ with $1 \leq \omega_i \leq \eta^2$ for each i. By transfer, the internal cardinality $|\Omega|$ of Ω is $\eta^{2\gamma}$. Again we use the counting measure to induce an internal integration structure on Ω. Let L denote the internal lattice of all internal $*R$-valued functions on Ω, and for $f \in L$ define $If = (1/\eta^{2\gamma}) \sum_{i=1}^{\eta^{2\gamma}} f(\omega_i)$. Then (L, I) is an internal integration structure on Ω. We let the set of internal subsets of Ω (internal events) be denoted by $\mathscr{E}$. If $A \in \mathscr{E}$ then $\chi_A \in L$ and we define the internal probability of A by $P(A) = I\chi_A$. The standardization $(\hat{L}, \hat{I})$ of (L, I) leads to the measure space $(\Omega, \hat{\mathscr{E}}, \hat{P})$, where $\hat{\mathscr{E}}$ is the collection of measurable sets and $\hat{P}$ is the measure on $\hat{\mathscr{E}}$ obtained from $(\hat{L}, \hat{I})$ by the methods of §IV.2. Since $P(\Omega) = 1$, we see that $0 \leq \hat{P}(A) \leq 1$ for all $A \in \hat{\mathscr{E}}$, and $(\Omega, \hat{\mathscr{E}}, \hat{P})$ is a probability space. Also $\hat{P}(A) = {}^\circ P(A)$ for all $A \in \mathscr{E}$.

We now define an internal stochastic process $\{N_t : t \in I\}$ on Ω which is an internal analogue of the Poisson process. Here I is the internal set $\{k/\eta : 1 \leq k \leq \eta^2\}$. For any $\omega \in \Omega$ and $t \in I$, we define $N_t(\omega)$ to be the number of points which the outcome ω places in the interval $[0, t)$. For the event $\omega = \langle \omega_i \rangle$, a point "lies" in $[k/\eta, (k + 1)/\eta)$ if $\omega_j = k + 1$ for some j, $1 \leq j \leq \gamma$. Note that $N_t(\omega)$ can be an infinite integer for some ω even for finite t. Also note that since η is an infinite factorial, any positive, standard rational number is of the form $k/\eta \in I$.

We want first to compute the P and $\hat{P}$ probabilities of the internal set $A = \{\omega : N_{t_2}(\omega) - N_{t_1}(\omega) = k\}$, i.e., the probabilities that k points fall in $[t_1, t_2)$, where $t_2 - t_1$ is finite and k is an ordinary natural number. Let $s = t_2 - t_1$. For simplicity we assume that t_1 and t_2 are rational numbers. Then there are exactly $s\eta$ of the η^2 intervals inside $[t_1, t_2)$, and the P-probability of any one of the γ points being put in $[t_1, t_2)$ is $s\eta/\eta^2 = s/\eta = \lambda s/\gamma$. Now by (transfers of) elementary counting and independence,

$$
\begin{aligned}
P(A) &= \frac{\gamma!}{(\gamma - k)!k!} \left(\frac{\lambda s}{\gamma}\right)^k \left(1 - \frac{\lambda s}{\gamma}\right)^{\gamma - k} \\
&= \frac{(\lambda s)^k}{k!} \frac{\gamma!}{\gamma^k(\gamma - k)!} \left(1 - \frac{\lambda s}{\gamma}\right)^\gamma \left(1 - \frac{\lambda s}{\gamma}\right)^{-k} \\
&\simeq \frac{(\gamma s)^k}{k!} \left(1 - \frac{\lambda s}{\gamma}\right)^\gamma \simeq \frac{(\lambda s)^k}{k!} e^{-\lambda s} = \hat{P}(A).
\end{aligned}
$$

This establishes the analogue of (6.4) for N_t.

The analogue of (6.3) can be established in the same way. Let $t_{i+1} - t_i = s_i$ and $s = s_1 + \cdots + s_n$, where $t_1 < t_2 < \cdots < t_n$, n is an ordinary natural number, and $t_i \in I$. Assume s is finite. If $A_i = \{\omega : N_{t_{i+1}}(\omega) - N_{t_i}(\omega) = k_i\}$, where the k_i are ordinary natural numbers, with $k = k_1 + \cdots + k_{n-1}$ and $i = 1, \ldots, n-1$, then

$$P(A_1 \cap \cdots \cap A_{n-1})$$

$$= \frac{\gamma!}{k_1! k_2! \cdots k_{n-1}! (\gamma - k)!} \left(\frac{\lambda s_1}{\gamma}\right)^{k_1} \left(\frac{\lambda s_2}{\gamma}\right)^{k_2} \cdots \left(\frac{\lambda s_{n-1}}{\gamma}\right)^{k_{n-1}} \left(1 - \frac{\lambda s}{\gamma}\right)^{\gamma - k}$$

$$= \frac{\gamma!}{(\gamma - k)! \gamma^k} \frac{(\gamma s_1)^{k_1}}{k_1!} \frac{(\lambda s_1)^{k_2}}{k_2!} \cdots \frac{(\lambda s_{n-1})^{k_{n-1}}}{k_{n-1}!} \left(1 - \frac{\lambda s}{\gamma}\right)^{\gamma} \left(1 - \frac{\lambda s}{\gamma}\right)^{-k}$$

$$\simeq \frac{(\gamma s_1)^{k_1}}{k_1!} \cdots \frac{(\lambda s_{n-1})^{k_{n-1}}}{k_{n-1}!} e^{-\lambda s} \simeq \hat{P}(A_1) \cdots \hat{P}(A_{n-1}).$$

We want to use $\{N_t(\omega) : t \in I\}$ to define a standard Poisson process. Unfortunately, for some $\omega \in \Omega$, $N_t(\omega)$ will take infinite values even for a finite $t \in I$. Another possibility is that for some $\omega \in \Omega$ there can be many points falling in an infinitesimal interval. We will show in the next paragraph that these abnormalities happen only on a set of $\hat{P}$ measure zero in Ω; we can define a standard Poisson process on the remainder.

Given $\omega \in \Omega$, we order the distinguishable points b_i by the order in which they fall in the line $*R$. Thus $b_i \leq b_{i+1}$ and $b_i = b_{i+1}$ if and only if b_i and b_{i+1} are in the same interval $[k/\eta, (k+1)/\eta)$. Again fix $j > 0$ and $k \geq 0$ in N. Given $t_0 \in I$ and $t > 0$ with t finite and $t_0 + t \in I$, let C_{t_0} be the event "$b_j \in [t_0, t_0 + 1/\eta)$", and let D_{t_0} be the event "If $j + 1 \leq i \leq j + k$, $b_i \in [t_0, t_0 + t)$, and $b_{j+k+1} \notin [t_0, t_0 + t)$." Let $\gamma' = \gamma - j$. Given C_{t_0}, the conditional probability of getting a given point of the remaining γ' points in $[t_0, t_0 + t)$ is

$$\frac{t\eta}{\eta^2 - t_0\eta} = \frac{t}{\eta - t_0} = \frac{\lambda t}{\gamma - t_0\lambda} = \frac{\lambda t}{\gamma' + j - t_0\lambda}.$$

Therefore, for all finite t_0, and hence for all $t_0 < \tau$ for some infinite τ, the conditional probability

$$P(D_{t_0} | C_{t_0}) = \frac{\gamma'}{(\gamma' - k)! k!} \left(\frac{\lambda t}{\gamma' + j - t_0\lambda}\right)^k \left(1 - \frac{\lambda t}{\gamma' + j - t_0\lambda}\right)^{\gamma - k}$$

$$\simeq \frac{(\lambda t)^k}{k!} e^{-\lambda t}.$$

On the other hand, $\sum_{t_0 < \tau} P(C_{t_0}) \simeq 1$, and so $\sum_{t_0 < \tau} P(D_{t_0} | C_{t_0}) P(C_{t_0}) \simeq (\lambda t)^k e^{-\lambda t}/k!$. That is, the $\hat{P}$-probability of having exactly k more distinguishable points in the interval of length t after the jth point is $(\lambda t)^k e^{-\lambda t}/k!$. Since

$$\sum_{k=0,\,k\in N}^{\infty} \frac{(\lambda t)^k}{k!} e^{-\lambda t} = e^{-\lambda t} \cdot e^{\lambda t} = 1,$$

the $\hat{P}$-probability of having only a finite number of distinguishable points in any finite interval $[0, t]$ is 1. Moreover, since $\lim_{t \to 0} e^{-\lambda t} = 1$, the $\hat{P}$-probability of having point b_{j+1} infinitely close to b_j is 0. Since this is true for each $j \geq 1$ in N, it follows that the $\hat{P}$-probability of having two distinguishable points in the same monad is 0.

We now let $E \subset \Omega$ denote that set of measure zero consisting of those ω for which $N_t(\omega)$ is infinite for some finite t or for which two or more distinguishable points fall in the same monad. Since $E \in \hat{\mathscr{E}}$, we define a new probability space $(\tilde{\Omega}, \tilde{\mathscr{E}}, \tilde{P})$ by putting $\tilde{\Omega} = \Omega - E$, $\tilde{\mathscr{E}} = \{A : A \subseteq \tilde{\Omega}, A \in \hat{\mathscr{E}}\}$, and $\tilde{P}(A) = \hat{P}(A)$ for $A \in \tilde{\mathscr{E}}$.

We now use $N_t(\omega)$ to define a process $\{\tilde{N}_t : t \in R\}$ on $(\tilde{\Omega}, \tilde{\mathscr{E}}, \tilde{P})$. For $\omega \in \tilde{\Omega}$ and $t \in R^+$ we put $\tilde{N}_t(\omega) = \sup N_s(\omega)$ $(s \simeq t, s \in I)$. By the above remarks, $\tilde{N}_t(\omega)$ is finite and integer-valued for any $\omega \in \tilde{\Omega}$ and $t \in R$, and $\tilde{N}_t(\omega) = N_s(\omega)$ for some $s \in I$, $s \simeq t$. We leave it to the reader to show that $\tilde{N}_t(\omega)$ is right continuous (Exercise 3) and that (6.3) and (6.4) are satisfied. Thus $\{\tilde{N}_t\}$ is a Poisson process on $(\tilde{\Omega}, \tilde{\mathscr{E}}, \tilde{P})$.

6.8 Example   (Anderson's Construction of Brownian Motion)

Brownian motion is a stochastic process which is intended to model the behavior of a particle (for example, a small particle suspended in water). The particle is subject to random disturbances (for example, collisions with the water molecules) which cause its position to change with time. For simplicity, we consider the one-dimensional case, and denote the random position of the particle on the real line at time $t \geq 0$ by $X(t)$. Again for simplicity we follow the particle only for a unit time interval. Then $\{X_t : t \in [0, 1]\}$ is to be a stochastic process on an as yet unspecified probability space $(\Omega, \mathscr{E}, P)$. A (standard) Brownian motion $\{X_t : t \in [0, 1]\}$ must satisfy the following conditions:

(6.5) $X_0 = 0$,

(6.6)   if $s_1 < t_1 \leq s_2 < t_2 \leq \cdots \leq s_n < t_n$ are points in $[0, 1]$ then
the random variables $X(t_1) - X(s_1), X(t_2) - X(s_2), \ldots, X(t_n) - X(s_n)$
are independent random variables, which we denote by $X_{t_1} - X_{s_1}$, etc.,

(6.7)   if $t > s$ are points in $[0, 1]$ then $P(\{\omega \in \Omega : X_t(\omega) - X_s(\omega) \leq \alpha\}) =$
$\psi(\alpha/\sqrt{t - s})$, where $\psi(x) = (1/\sqrt{2\pi}) \int_{-\infty}^{x} e^{-u^2/2} \, du$.

Condition (6.5) locates the particle at the origin at $t = 0$. Condition (6.6) says that the probability of a change in position of the particle in any time interval $(s_i, t_i]$ is unaffected by the changes in position in other disjoint intervals. Condition (6.7) indicates how closely the position of the particle at time t can be determined if its position at time s is known. The probability distribution function $\psi(x)$ is known as the *normal distribution* with mean 0 and variance 1. One should note that $\psi(x/\sigma) = (1/\sigma\sqrt{2\pi}) \int_{-\infty}^{x} e^{-u^2/2\sigma^2} du$, which is the normal distribution with mean 0 and variance σ^2.

In [2], Robert M. Anderson used the measure space construction of §IV.2 to obtain, among other things, a nonstandard representation of Brownian motion. We give here a brief account of some of his results, which is necessarily incomplete since we refer to his nonstandard version of the central limit theorem (Theorem 6.11), which is crucial to the development. The central limit theorem is one of the deeper results in probability theory and to prove it here woule lead us too far from the main theme of these examples.

A Brownian motion can now be defined as follows. Fix $\eta = \zeta!$, an infinite factorial in $*N$; and let $(\Omega, \mathscr{E}, P)$ be the internal space for infinite coin tossing of Example 6.6 (with Ω being all sequences $\omega = \langle \omega_1, \ldots, \omega_\eta \rangle$, and $\omega_i = +1$ or -1) constructed from the internal integration structure (L, I). Let $(\Omega, \hat{\mathscr{E}}, \hat{P})$ be the corresponding standardization of $(\Omega, \mathscr{E}, P)$ constructed from $(\hat{L}, \hat{I})$ as in Example 6.6. Let $\chi(t, \cdot)$ denote the internal random variable (function in L) defined by setting

$$\chi(t, \omega) = \frac{1}{\sqrt{\eta}} \sum_{i=1}^{[\eta t]} X_i(\omega), \qquad t \in *[0, 1],$$

where $X_i(\omega) = \omega_i$. Here $[\eta t]$ denotes the largest element of $*N$ less than or equal to ηt. Thus, for any $\omega = \langle \omega_1, \omega_2, \ldots, \omega_\eta \rangle$, the particle located by $\chi(t, \omega)$ starts at the origin at $t = 0$ [i.e., $\chi(0, \omega) = 0$], and at each time $t_i = i/\eta$ $(i = 1, 2, 3, \ldots, \eta)$ the particle moves to the right or left a distance $1/\sqrt{\eta}$, depending on whether ω_i is $+1$ or -1; at times lying between the t_i the particle remains fixed. The resulting motion is an internal analogue of a standard "symmetric random walk."

We now define $\beta(t, \omega) = {}^{\circ}\chi(t, \omega)$ for $t \in [0, 1]$ and $\omega \in \Omega$. We will show that $\beta(t, \cdot)$ is a Brownian motion on $(\Omega, \hat{\mathscr{E}}, \hat{P})$. To do so we need the following results.

6.9 Definition An *internal random variable* on $(\Omega, \mathscr{E}, P)$ is a function $X \in L$. A collection $\{X_i : i \in I\}$ of internal random variables is $*$-*independent* if for every $*$-finite internal subcollection $\{X_1, \ldots, X_m\}$ $(m \in *N)$ and every internal

m-tuple$\langle \alpha_1, \ldots, \alpha_m \rangle \in {}^*R^m$ we have

(6.8)
$$P(\{\omega \in \Omega : X_1(\omega) < \alpha_1, \ldots, X_m(\omega) < \alpha_m\})$$

$$= \prod_{k=1}^{m} P(\{\omega \in \Omega : X_k(\omega) < \alpha_k\}).$$

$\{X_i : i \in I\}$ is *S-independent* if, for every finite subcollection $\{X_1, \ldots, X_m\}$ ($m \in N$) and every m-tuple $\langle \alpha_1, \ldots, \alpha_m \rangle \in R^m$, (6.8) holds with $=$ replaced by $\simeq$.

6.10 Lemma Suppose $\{X_i : i \in I\}$ is S-independent. Then $\{{}^\circ X_i : i \in I\}$ is an independent collection of random variables on $(\Omega, \hat{\mathscr{E}}, \hat{P})$.

Proof: Suppose $m \in N$, $\langle \alpha_1, \ldots, \alpha_m \rangle \in R^m$. Then

$$\hat{P}(\{\omega : {}^\circ X_{i_1}(\omega) < \alpha_1, \ldots, {}^\circ X_{i_m}(\omega) < \alpha_m\})$$

$$= \lim_{n \to \infty} {}^\circ P\left(\left\{\omega : X_{i_1}(\omega) < \alpha_1 - \frac{1}{n}, \ldots, X_{i_m} < \alpha_m - \frac{1}{n}\right\}\right)$$

$$= \lim_{n \to \infty} {}^\circ\left(\prod_{j=1}^{m} P\left(\left\{\omega : X_{i_j}(\omega) < \alpha_j - \frac{1}{n}\right\}\right)\right)$$

$$= \prod_{j=1}^{m} \lim_{n \to \infty} {}^\circ P\left(\left\{\omega : X_{i_j}(\omega) < \alpha_j - \frac{1}{n}\right\}\right)$$

$$= \prod_{j=1}^{m} \hat{P}(\{\omega : {}^\circ X_{i_j}(\omega) < \alpha_j\}). \quad \square$$

6.11 Theorem Let $\{X_n : n \in N\}$ be an internal sequence of $*$-independent random variables on $(\Omega, \mathscr{E}, P)$. Assume that there is a standard distribution function F such that ${}^*F$ is the distribution of X_n, $E(X_n) = 0$, and $E(X_n^2) = 1$ for each $n \in {}^*N$. Let ψ denote the standard normal distribution. Then for any $m \in {}^*N - N$ and any $\alpha \in {}^*R$

$$P\left(\left\{\omega \in \Omega : \frac{1}{\sqrt{m}} \sum_{n=1}^{m} X_n(\omega) \le \alpha\right\}\right) \simeq {}^*\psi(\alpha).$$

Proof: See Theorem 21 in [2]. $\square$

6.12 Theorem If $\eta \in {}^*N - N$, then $\beta(t, \cdot)$ is a Brownian motion on $(\Omega, \hat{\mathscr{E}}, \hat{P})$.

Proof: (i) Given $t \in [0, 1]$, $\chi(t, \cdot)$ is $\mathscr{E}$-measurable, and so $\beta(t, \cdot)$ is $\hat{\mathscr{E}}$-measurable by Proposition 2.31.

(ii) By transfer from the case of finite coin tossing we see that the X_i have identical distributions. Also if $S_k = \sum_{i=1}^{k} X_i$ for any $k \in {}^*N$ then the S_k have independent increments by the transfer of Exercise 1. Thus if $s_1 < t_1 \leq s_2 < t_2 \leq \cdots \leq s_n < t_n$ are points in $[0, 1]$, then $\{\chi(t_1, \cdot) - \chi(s_1, \cdot), \ldots, \chi(t_n, \cdot) - \chi(s_n, \cdot)\}$ are *-independent and so S-independent, and condition (6.6) follows by Lemma 6.10.

(iii) Given $s < t$ in $[0, 1]$, $\lambda = [\eta t] - [\eta s]$, and $\alpha \in R$,

$$\hat{P}(\{\omega \in \Omega : \beta(t, \omega) - \beta(s, \omega) \leq \alpha\})$$

$$= \hat{P}(\{\omega : {}^\circ\chi(t, \omega) - {}^\circ\chi(s, \omega) \leq \alpha\})$$

$$= \hat{P}\left(\left\{\omega : {}^\circ\left(\sum_{k=[\eta s]}^{[\eta t]} \frac{\omega_k}{\sqrt{\eta}}\right) \leq \alpha\right\}\right)$$

$$= \lim_{n \to \infty} \hat{P}\left(\left\{\omega : \frac{1}{\sqrt{\lambda}} \sum_{k=[\eta s]}^{[\eta t]} \omega_k \leq \sqrt{\frac{\eta}{\lambda}}\left(\alpha + \frac{1}{n}\right)\right\}\right)$$

$$= \lim_{n \to \infty} {}^\circ({}^*\psi)\left(\sqrt{\frac{\eta}{\lambda}}\left(\alpha + \frac{1}{n}\right)\right) \qquad \text{(by Theorem 6.11)}$$

$$= \lim_{n \to \infty} \psi\left({}^\circ\left(\sqrt{\frac{\eta}{\lambda}}\left(\alpha + \frac{1}{n}\right)\right)\right)$$

$$= \lim_{n \to \infty} \psi\left[\frac{\alpha + 1/n}{\sqrt{t-s}}\right] = \psi\left[\frac{\alpha}{\sqrt{t-s}}\right].$$

This establishes condition (6.7) $\square$

In general, by a *path* of a stochastic process $\{X_t : t \in I\}$, we mean a function $f(t) = X_t(\omega)$, for some particular $\omega \in \Omega$. The last result of this section shows that almost all of the paths of Brownian motion are continuous.

6.13 Theorem There is a set $\Omega' \in \hat{\mathscr{E}}$ with $\hat{P}(\Omega') = 1$ such that $\beta(\cdot, \omega)$ is a continuous and finite function on $[0, 1]$ for all $\omega \in \Omega'$.

Proof: For each $m, n \in N$, let Ω_{mn} be the internal set given, using the internal extensions of sup and inf, by

$$\Omega_{mn} = \left\{\omega \in \Omega : \sup_{t \in [i/n, (i+1)/n]} \chi(t, \omega) - \inf_{t \in [i/n, (i+1)/n]} \chi(t, \omega) > \frac{1}{m}\right.$$

$$\left. \text{for some } i < n\right\}.$$

Then for $\lambda = \eta/n$,

$$P(\Omega_{mn}) \leq nP\left(\left\{\omega: \sup_{t \in [0,1/n]} \chi(t,\omega) - \inf_{t \in [0,1/n]} \chi(t,\omega) > \frac{1}{m}\right\}\right)$$

$$\leq nP\left(\left\{\omega: \max_{1 \leq k \leq \lambda} \left|\sum_1^k \omega_i\right| > \frac{\sqrt{\eta}}{2m}\right\}\right)$$

$$\leq nP\left(\left\{\omega: \max_{1 \leq k \leq \lambda} \sum_1^k \omega_i > \frac{\sqrt{\eta}}{2m}\right\}\right) + nP\left(\left\{\omega: \min_{1 \leq k \leq \lambda} \sum_1^k \omega_i < -\frac{\sqrt{\eta}}{2m}\right\}\right)$$

$$\leq 2nP\left(\left\{\omega: \sum_1^\lambda \omega_i > \frac{\sqrt{\eta}}{2m}\right\}\right) + 2nP\left(\left\{\omega: \sum_1^\lambda \omega_i < -\frac{\sqrt{\eta}}{2m}\right\}\right)$$

$$= 4nP\left(\left\{\omega: \frac{1}{\sqrt{\lambda}}\sum_1^\lambda \omega_i > \frac{\sqrt{\eta/\lambda}}{2m}\right\}\right)$$

$$\simeq 4n\left(1 - {}^*\psi\left(\frac{\sqrt{\eta/\lambda}}{2m}\right)\right) \simeq 4n\left(1 - \psi\left(\frac{\sqrt{n}}{2m}\right)\right)$$

$$= \frac{4n}{\sqrt{2\pi}} \int_{\sqrt{n}/2m}^\infty e^{-t^2/2}\, dt.$$

For $\sqrt{n}/2m > 1$

$$\hat{P}(\Omega_{mn}) \leq 2n \int_{\sqrt{n}/2m}^\infty e^{-t/2}\, dt = 4ne^{-\sqrt{n}/4m}.$$

Let $\Omega' = \Omega - \bigcup_{m=1}^\infty \bigcap_{n=1}^\infty \Omega_{mn}$. Then

$$\hat{P}(\Omega') \geq 1 - \sup_m \inf_n \hat{P}(\Omega_{mn}) \geq 1 - \sup_m \inf_n 4ne^{-\sqrt{n}/4m} = 1.$$

Fix $\omega \in \Omega$. If for some $t \in {}^*[0,1]$ we have ${}^\circ\chi(t,\omega) = +\infty$ or ${}^\circ\chi(t,\omega) = -\infty$, then $\omega \in \Omega_{m,n}$ for all standard m and $n \in N$, whence $\omega \notin \Omega'$. If for some s and $t \in {}^*[0,1]$ with $s \simeq t$ we have ${}^\circ|\chi(s,\omega) - \chi(t,\omega)| = a > 0$, then for $m > 2/a$ we have $\omega \in \Omega_{mn}$ for all $n \in N$ (exercise), whence $\omega \notin \Omega'$.

Now suppose $\omega \in \Omega'$. By the preceding paragraph, $\beta(t,\omega)$ is finite for all $t \in [0,1]$. Fix $\varepsilon > 0$ in R. Then the set $\{n \in {}^*N : |t - s| < 1/n \Rightarrow |\chi(t,\omega) - \chi(s,\omega)| < \varepsilon/2\}$ is internal and contains all infinite n. Hence it contains a finite n by II.7.2(ii). Thus if $|t - s| < 1/n$, $|\chi(t,\omega) - \chi(s,\omega)| < \varepsilon/2$ and hence $|\beta(t,\omega) - \beta(s,\omega)| < \varepsilon$. It follows that $\beta(\cdot,\omega)$ is continuous on $[0,1]$. $\square$

Exercise IV.6

1. (Standard) Let X_i be defined on the space Ω_n of Example 6.6 by $X_i(\omega) = e_i$ if $\omega = \langle e_1, e_2, \ldots, e_n \rangle$. Show that the random variables $S_k = \sum_{i=1}^k X_i$,

$1 \leq k \leq n$, have independent increments, i.e., if $1 \leq k_1 < k_2 < k_3 < k_4 < \ldots < k_l \leq n$ then $S_{k_2} - S_{k_1}, S_{k_3} - S_{k_2}, \ldots, S_{k_l} - S_{k_{l-1}}$ are independent.

2. In Example 6.7, check that

$$P(D_{t_0}|C_{t_0}) \simeq \frac{\lambda t^k}{k!} e^{-\lambda t} \quad \text{and} \quad \sum_{t_0 < \tau} P(C_{t_0}) \simeq 1.$$

3. (a) Show that the process $\tilde{N}_t(\omega)$ defined on $(\tilde{\Omega}, \tilde{\mathscr{E}}, \tilde{P})$ in Example 6.7 is right continuous. That is, show that, for each fixed $\omega \in \tilde{\Omega}$, the function $f: R^+ \to Z$ defined by $f(t) = \tilde{N}_t(\omega)$ satisfies $\lim_{s \to t, s > t} f(s) = f(t)$.
   (b) Show that the process $\tilde{N}_t(\omega)$ satisfies Properties (6.3) and (6.4).

4. (Inter-arrival times)   Define the process $\{\tilde{T}_n : n \in N\}$ on the space $(\tilde{\Omega}, \tilde{\mathscr{E}}, \tilde{P})$ of Example 6.7 as follows. For $\omega \in \tilde{\Omega}$, $\tilde{T}_n(\omega)$ is the time between the $(n-1)$st and the nth jump of $\tilde{N}_t(\omega)$.

   (a) Define the internal analogues $\{T_n : n \in {}^*N\}$ of $\{\tilde{T}_n : n \in N\}$ on $(\Omega, \mathscr{E}, P)$.
   (b) Show that $\tilde{P}\{\tilde{T}_1 > t_1\} = e^{-\lambda t_1}$ and $\tilde{P}\{\tilde{T}_n > t_n | \tilde{T}_1 = t_1, \ldots, \tilde{T}_{n-1} = t_{n-1}\} = e^{-\lambda t}$.
   (c) Use (b) to show that $\tilde{P}\{\tilde{T}_1 > t_1, \tilde{T}_2 > t_2, \ldots, \tilde{T}_n > t_n\} = e^{-\lambda t_1} e^{-\lambda t_2} \cdots e^{-\lambda t_n}$, showing that the $\tilde{T}_n$ are independent, identically distributed random variables.

5. Prove the result tagged as an exercise in the proof of Theorem 6.13. (Note that we may have $s < i/n < t$ for same values of $n \in N$).

APPENDIX

Ultrafilters

In this appendix we present the essential facts concerning ultrafilters which are needed in the text.

In the following, I will be an arbitrary set.

A.1 Definition A nonempty collection $\mathscr{F}$ of subsets of I is a *filter* on I if

(i) $\varnothing \notin \mathscr{F}$,
(ii) $A, B \in \mathscr{F}$ implies $A \cap B \in \mathscr{F}$,
(iii) $A \in \mathscr{F}$ and $B \supseteq A$ implies $B \in \mathscr{F}$.

A filter $\mathscr{F}$ on I is an *ultrafilter* if it is maximal; i.e., whenever $\mathscr{G}$ is a filter on I and $\mathscr{F} \subseteq \mathscr{G}$ then $\mathscr{F} = \mathscr{G}$.

The following result shows that this definition of ultrafilter is equivalent to that of Definition 1.1 in Chapter I.

A.2 Proposition A filter $\mathscr{F}$ on I is an ultrafilter iff, for every subset A of I, either $A \in \mathscr{F}$ or $A' = I - A \in \mathscr{F}$.

Proof: Suppose that $\mathscr{F}$ is a filter such that for every $A \subset I$ either $A \in \mathscr{F}$ or $A' \in \mathscr{F}$. Let $\mathscr{G}$ be a filter with $\mathscr{G} \supseteq \mathscr{F}$ and suppose that $B \in \mathscr{G}$ and $B \notin \mathscr{F}$. But then $B' \in \mathscr{F} \subseteq \mathscr{G}$, and so $\varnothing = B \cap B' \in \mathscr{G}$, contradicting A.1(i) for a filter. Thus there is no filter $\mathscr{G}$ properly containing $\mathscr{F}$, and so $\mathscr{F}$ is an ultrafilter.

Conversely, suppose that $\mathscr{F}$ is an ultrafilter and $A \notin \mathscr{F}$. Let $\mathscr{G}$ be the set $\{X \subseteq I : A \cap F \subseteq X \text{ for some } F \in \mathscr{F}\}$. Then $\mathscr{F} \subseteq \mathscr{G}$ and $\mathscr{F} \neq \mathscr{G}$ (since, for example, $A \in \mathscr{G}$), and so $\mathscr{G}$ is not a filter since $\mathscr{F}$ is maximal. But $\mathscr{G}$ is not empty, and if $B, C \in \mathscr{G}$ and $D \supseteq B$ then $B \cap C \in \mathscr{G}$ and $D \in \mathscr{G}$. Thus $\mathscr{G}$ can fail to be a filter only if $\varnothing \in \mathscr{G}$. That is, we have $A \cap F = \varnothing$ for some $F \in \mathscr{F}$ for which we then must have $F \subseteq A'$. It follows that $A' \in \mathscr{F}$ by A.1(iii). $\quad\square$

We now want to prove the ultrafilter axiom, 1.2 of Chapter I. To do so we need Zorn's lemma, which is a variant of the axiom of choice. The statement of Zorn's lemma involves the idea of a partially ordered set and related concepts.

A.3 Definition A *partially ordered* set is a pair $(X, \leq)$, where X is a nonempty set and $\leq$ is a binary relation on X which is

(i) reflexive, i.e., $x \leq x$ for all $x \in X$,
(ii) antisymmetric, i.e., if $x \leq y$ and $y \leq x$ then $x = y$,
(iii) transitive, i.e., if $x \leq y$ and $y \leq z$ then $x \leq z$.

A subset C of X is a *chain* if for all $x, y \in C$ either $x \leq y$ or $y \leq x$. The element x is an *upper bound* for a subset $B \subseteq X$ if $b \leq x$ for all $b \in B$. An element $m \in X$ is *maximal* if, for any $x \in X$, $m \leq x$ implies $x = m$.

A.4 Zorn's Lemma Let $(X, \leq)$ be a partially ordered set. If each chain in X has an upper bound then X has at least one maximal element.

Zorn's lemma is equivalent to the axiom of choice.

A.5 Axiom of Choice For any set A of nonempty sets, there is a function f with domain A such that $f(x) \in x$ for each $x \in A$. The function f is called a *choice function* for A.

We now use Zorn's lemma to prove the ultrafilter axiom.

A.6 Ultrafilter Axiom If $\mathscr{F}$ is a filter on I then there is an ultrafilter $\mathscr{U}$ on I containing $\mathscr{F}$.

Proof: Let $\hat{\mathscr{F}}$ be the set of all filters which contain $\mathscr{F}$. $\hat{\mathscr{F}}$ is nonempty since $\mathscr{F} \in \hat{\mathscr{F}}$. We partially order $\hat{\mathscr{F}}$ by inclusion; i.e., if $\mathscr{A}, \mathscr{B} \in \hat{\mathscr{F}}$ then we say that $\mathscr{A} \leq \mathscr{B}$ if $A \in \mathscr{A}$ implies $A \in \mathscr{B}$. It is easy to check that $\leq$ is a partial ordering on $\hat{\mathscr{F}}$.

Now let $\tilde{\mathscr{C}}$ be a chain in $\hat{\mathscr{F}}$. To show that $\tilde{\mathscr{C}}$ has an upper bound consider $\tilde{\mathscr{F}} = \bigcup \mathscr{C}\ (\mathscr{C} \in \tilde{\mathscr{C}})$. Then $\mathscr{C} \leq \tilde{\mathscr{F}}$ for all $\mathscr{C} \in \tilde{\mathscr{C}}$. Also $\tilde{\mathscr{F}}$ is a filter. For if $A, B \in \tilde{\mathscr{F}}$ then $A \in \mathscr{C}_1$ and $B \in \mathscr{C}_2$ for some $\mathscr{C}_1$ and $\mathscr{C}_2$ in $\tilde{\mathscr{C}}$. Since $\tilde{\mathscr{C}}$ is a chain, we may assume without loss of generality that $\mathscr{C}_1 \leq \mathscr{C}_2$, and so $A, B \in \mathscr{C}_2$ and $A \cap B \in \mathscr{C}_2 \subseteq \tilde{\mathscr{F}}$. Similarly we check conditions (i) and (iii) of A.1.

We deduce from Zorn's lemma that $\hat{\mathscr{F}}$ contains a maximal element which is then an ultrafilter containing $\mathscr{F}$. $\square$

There are some ultrafilters on I which, for our purposes, are quite trivial. Consider, for example, the collection $\mathscr{U}_a = \{A \subseteq I : a \in A\}$ for some $a \in I$. It is easy to see that $\mathscr{U}_a$ is an ultrafilter.

A.7 Definition An ultrafilter $\mathscr{U}$ is *principal* or *fixed* if there is some $a \in I$ so that $\mathscr{U} = \{A \subseteq I : a \in A\}$. If the ultrafilter $\mathscr{U}$ is not principal it is called *free*.

A.8 Theorem Free ultrafilters exist on any infinite set I.

Proof: The collection $\mathscr{F}_I = \{F \subseteq I : I - F \text{ is finite}\}$ is a filter (check) called the cofinite or Fréchet filter. Let $\mathscr{U}$ be an ultrafilter containing $\mathscr{F}_I$. Then $\mathscr{U}$ cannot be principal. For if $\mathscr{U} = \{A \subseteq I : a \in A\}$ and $\mathscr{U} \supset \mathscr{F}_I$, then the set $F = \{a\}' \in \mathscr{U}$ (contradiction). $\quad\square$

References

1 Abian, A. Solvability of infinite systems of polynomial equations (to appear).

2 Anderson, R. M. A non-standard representation for Brownian motion and Itô integration. *Israel J. Math.* **25** (1976), 15–46.

3 Anderson, R. M. Star-finite representations of measure spaces. *Trans. A.M.S.* **271** (1982), 667–687.

4 Anderson, R. M., and Rashid, S. A nonstandard characterization of weak convergence. *Proc. A.M.S.* **69** (1978), 327–332.

5 Behrens, M. A local inverse function theorem. *In* "Victoria Symposium on Nonstandard Analysis," (Hurd A. E., and Loeb, P. A. eds.), Lecture Notes in Mathematics, Vol. 369. Springer, Berlin, 1974.

6 Benoit, E., Callot, J. L., Diener, F., and Diener, M. Chasse au canard. *Collectanea Mathematica*, **32** (1981), 37–74.

7 Bernstein A. R., and Robinson, A. Solution of an invariant subspace problem of K. T. Smith and P. R. Halmos, *Pacific J. Math.* **16** (1966), 421–431.

8 Coddington, E. A., and Levison, N. "Theory of Ordinary Differential Equations." McGraw-Hill, New York, 1955.

9 Constantinescu, C., and Cornea, A. "Ideale Ränder Riemannscher Flächen," Ergebnisse der Math., Vol. 32. Springer, Berlin, 1963.

10 Dacunha-Castelle, D., and Krivine, J.-L. Applications des ultraproduits a l'étude des espaces et des algèbres de Banach. *Studia Math.* **41** (1972), 315–334.

11 Daniel, P. A general form of the integral. *Ann. Math.* **19** (1917–18), 279–294.

12 Davis, M. "Applied Nonstandard Analysis." Wiley, New York, 1977.

13 De Bruijn, N. G., and Erdös, P. A color problem for infinite graphs and a problem in the theory of relations. *Proc. Kon. Nederl. Akad. v. Wetensch.*, Ser. A, **54** (1951), 371–373.

14 Dunford, N., and Schwartz, J. T. "Linear Operators," Vol. 1. Interscience, New York, 1958.

15 Gonshor, H. Enlargements contain various kinds of completions. *In* "Victoria Symposium on Nonstandard Analysis," (A. E. Hurd and Loeb, P. A., eds.), Lecture Notes in Mathematics, Vol. 369. Springer, Berlin, 1974.

16 Henson, C. W., and Moore, L. Nonstandard analysis and the theory of Banach spaces. *In* "Nonstandard Analysis—Recent Developments," (Hurd A. E., ed.), Lecture Notes in Mathematics, Vol. 983. Springer, Berlin, 1983.

17 Hewitt, E. Rings of real-valued continuous functions, Vol. I. *Trans. Amer. Math. Soc.* **64** (1948), 45–99.

18 Hirschfeld, J. "A Non-Standard Smorgasbord" (unpublished notes). Tel-Aviv University.

19 Hurd, A. E. (ed.) "Nonstandard Analysis—Recent Developments," Lecture Notes in Mathematics, Vol. 983. Springer, Berlin, 1983.

20 Kelley, J. L. "General Topology." Van Nostrand, New York, 1955.

21 Keisler, H. Jerome. Good ideals in fields of sets. *Ann. Math.* **79** (1964), 338–359.

22 Keisler, H. Jerome. Ultraproducts and saturated models, *Proc. Kon. Nederl. Akad. v. Wetensch.*, Ser. A, **67** (1964), 178–186.

23 Keisler, H. Jerome. "Elementary Calculus." Prindle, Weber and Schmidt, Boston, 1976.

24 Keisler, H. Jerome. "Foundations of Infinitesimal Calculus." Prindle, Weber and Schmidt, Boston, 1976.

25 Keisler, H. Jerome. "An infinitesimal approach to stochastic analysis," Memoirs Amer. Math. Soc., No. 297, American Mathematical society, Providence, Rhode Island, Vol. 48, 1984.

26 Lebesgue, H. Sur une généralization de l'intégrale définie, *Comptes Rendus Acad. Sci. Paris,* **132** (1901), 1025–28.

27 Loeb, P. A. Conversion from nonstandard to standard measure spaces and applications in probability theory. *Trans. Amer. Math. Soc.* **211** (1975), 113–122.

28 Loeb, P. A. "An introduction to nonstandard analysis and hyperfinite probability theory. *In* "Probabilistic Analysis and Related Topics, Vol. 2, (edited by A. T. Bharucha-Reid), Academic Press, New York, 1979.

29 Loeb, P. A. Weak limits of measures and the standard part map. *Proc. Amer. Math. Soc.* **77** (1979), 128–135.

30 Loeb, P. A. Review of [23] and [24]. *J. Symbolic Logic,* **46** (1981), 673–676.

31 Loeb, P. A. Measure spaces in nonstandard models underlying standard stochastic processes. "Proceedings of the International Congress of Mathematicians." Warsaw, 1983.

32 Loeb, P. A. A functional approach to nonstandard measure theory. *Contemporary Math.,* **26** (1984), 251–261.

33 Loeb, P. A. A nonstandard functional approach to Fubini's theorem. *Proc. Amer. Math. Soc.* **93** (1985), 343–346.

34 Loomis, L. H. "An Introduction to Abstract Harmonic Analysis." Van Nostrand, New York, 1953.

35 Luxemburg, W. A. J. A remark on a paper by N. G. De Bruijn and P. Erdös, *Proc. Kon. Kederl. Akad. v. Wetensch.*, Ser. A, **65** (1962), 343–345.

36 Luxemburg, W. A. J. A general theory of monads, *In* "Applications of Model Theory to Algebra, Analysis, and Probability, (W. A. J. Luxemburg, ed.). Holt, Rinehart, and Winston, New York, 1969.

37 Machover, M. and Hirschfeld, J. "Lectures on Non-standard Analysis," Lecture Notes in Mathematics, Vol. 94. Springer, Berlin, 1969.

38 Perkins, E. A. A global intrinsic characterization of Brownian local time. *Ann. Probability* **9** (1981), 800–817.

39 Perkins, E. A. Stochastic processes and nonstandard analysis. *In* "Nonstandard Analysis—Recent Developments," (Hurd A. E., ed.), Lecture Notes in Mathematics, Vol. 983. Springer, Berlin, 1983.

40 Robinson, A. "Introduction to Model Theory and to the Metamathematics of Algebra." North-Holland, Amsterdam, 1963.

41 Robinson, A. On generalized limits and linear functionals. *Pacific J. Math.* **14** (1964), 269–283.

42 Robinson, A. "Non-standard Analysis." North-Holland, Amsterdam, 1966.

43 Robinson, A. Compactifications of groups and rings and nonstandard analysis. *J. Symbolic Logic* **34** (1969), 576–588.

44 Robinson, A. and Lightstone, A. H. "Nonarchimedean Fields and Asymptotic Expansions." North-Holland, Amsterdam, 1975.

45 Robinson, A. and Zakon, E. A set-theoretical characterization of enlargements. *In* "Applications of Model Theory to Algebra, Analysis, and Probability," (W. A. J. Luxemburg, ed.), Holt, Rinehard and Winston, New York, 1969.

46 Stroyan, K. and Luxemburg, W. A. J. "Introduction to the Theory of Infinitesimals." Academic Press, New York, 1976.

47 Tacon, D. G. Weak compactness in normed linear spaces. *J. Australian Math. Soc.* **14** (1972), 9–12.

48 Zakon, E. A new variant of nonstandard analysis. *In* "Victoria Symposium on Nonstandard Analysis," (Hurd A. E. and Loeb, P. A. eds.), Lecture Notes in Mathematics, Vol. 369. Springer, Berlin, 1974.

List of Symbols

225

Index

Pure and Applied Mathematics

A Series of Monographs and Textbooks

Editors **Samuel Eilenberg and Hyman Bass**

Columbia University, New York

RECENT TITLES

CARL L. DEVITO. Functional Analysis

MICHIEL HAZEWINKEL. Formal Groups and Applications

SIGURDUR HELGASON. Differential Geometry, Lie Groups, and Symmetric Spaces

ROBERT B. BURCKEL. An Introduction to Classical Complex Analysis: Volume 1

JOSEPH J. ROTMAN. An Introduction to Homological Algebra

C. TRUESDELL and R. G. MUNCASTER. Fundamentals of Maxwell's Kinetic Theory of a Simple Monatomic Gas: Treated as a Branch of Rational Mechanics

BARRY SIMON. Functional Integration and Quantum Physics

GRZEGORZ ROZENBERG and ARTO SALOMAA. The Mathematical Theory of L Systems

DAVID KINDERLEHRER and GUIDO STAMPACCHIA. An Introduction to Variational Inequalities and Their Applications

H. SEIFERT and W. THRELFALL. A Textbook of Topology; H. SEIFERT. Topology of 3-Dimensional Fibered Spaces

LOUIS HALLE ROWEN. Polynominal Identities in Ring Theory

DONALD W. KAHN. Introduction to Global Analysis

DRAGOS M. CVETKOVIC, MICHAEL DOOB, and HORST SACHS. Spectra of Graphs

ROBERT M. YOUNG. An Introduction to Nonharmonic Fourier Series

MICHAEL C. IRWIN. Smooth Dynamical Systems

JOHN B. GARNETT. Bounded Analytic Functions

EDUARD PROGOVÉCKI. Quantum Mechanics in Hilbert Space, Second Edition

M. SCOTT OSBORNE and GARTH WARNER. The Theory of Eisenstein Systems

K. A. ZHEVLAKOV, A. M. SLIN'KO, I. P. SHESTAKOV, and A. I. SHIRSHOV. Translated by HARRY SMITH. Rings That Are Nearly Associative

JEAN DIEUDONNÉ. A Panorama of Pure Mathematics; Translated by I. MACDONALD

JOSEPH G. ROSENSTEIN. Linear Orderings

AVRAHAM FEINTUCH and RICHARD SAEKS. System Theory: A Hilbert Space Approach

ULF GRENANDER. Mathematical Experiments on the Computer

HOWARD OSBORN. Vector Bundles: Volume 1, Foundations and Stiefel-Whitney Classes

K. P. S. BHASKARA RAO and M. BHASKARA RAO. Theory of Charges

RICHARD V. KADISON and JOHN R. RINGROSE. Fundamentals of the Theory of Operator Algebras, Volume I

EDWARD B. MANOUKIAN. Renormalization

BARRETT O'NEILL. Semi-Riemannian Geometry: With Applications to Relativity